Studies in Systems, Decision and Control

Volume 269

Series Editor

Janusz Kacprzyk, Systems Research Institute, Polish Academy of Sciences, Warsaw, Poland

The series "Studies in Systems, Decision and Control" (SSDC) covers both new developments and advances, as well as the state of the art, in the various areas of broadly perceived systems, decision making and control–quickly, up to date and with a high quality. The intent is to cover the theory, applications, and perspectives on the state of the art and future developments relevant to systems, decision making, control, complex processes and related areas, as embedded in the fields of engineering, computer science, physics, economics, social and life sciences, as well as the paradigms and methodologies behind them. The series contains monographs, textbooks, lecture notes and edited volumes in systems, decision making and control spanning the areas of Cyber-Physical Systems, Autonomous Systems, Sensor Networks, Control Systems, Energy Systems, Automotive Systems, Biological Systems, Vehicular Networking and Connected Vehicles, Aerospace Systems, Automation, Manufacturing, Smart Grids, Nonlinear Systems, Power Systems, Robotics, Social Systems, Economic Systems and other. Of particular value to both the contributors and the readership are the short publication timeframe and the world-wide distribution and exposure which enable both a wide and rapid dissemination of research output.

** Indexing: The books of this series are submitted to ISI, SCOPUS, DBLP, Ulrichs, MathSciNet, Current Mathematical Publications, Mathematical Reviews, Zentralblatt Math: MetaPress and Springerlink.

More information about this series at http://www.springer.com/series/13304

Nabil Derbel · Jawhar Ghommam ·
Quanmin Zhu
Editors

Diagnosis, Fault Detection & Tolerant Control

Springer

Editors
Nabil Derbel
Sfax Engineering School
University of Sfax
Sfax, Tunisia

Jawhar Ghommam
Department of Electrical
and Computer Engineering
Sultan Qaboos University
Al Khoudh, Oman

Quanmin Zhu
Department of Engineering
Design and Mathematics
University of the West of England
Bristol, UK

ISSN 2198-4182 ISSN 2198-4190 (electronic)
Studies in Systems, Decision and Control
ISBN 978-981-15-1745-7 ISBN 978-981-15-1746-4 (eBook)
https://doi.org/10.1007/978-981-15-1746-4

This Springer imprint is published by the registered company Springer Nature Singapore Pte Ltd.
The registered company address is: 152 Beach Road, #21-01/04 Gateway East, Singapore 189721, Singapore

Preface

This book entitled *Diagnosis, Fault Detection & Tolerant Control* is dedicated to unhealthy cyber-physical systems. In fact, when a system becomes faulty, it becomes dangerous and can cause its damage. As a consequence, the detection and the handling of faults play an increasing role in modern technology. A main concern on the question of how to accommodate and overcome the performance deterioration caused by faulty sensors or actuator faults has been the focus of many researchers in recent years. The practical emphasis of the book is to give deep insight into how the fault-tolerant methods institute rigorous program to minimize the occurrence of fault situations in a myriad of system ranging from control to power systems.

This book is made of three distinct parts, and it unfolds into 14 chapters organized as follows.

The first part deals with diagnosis and fault detection. Chapters 1–3 are dedicated to the use of sensors for the system diagnosis. Chapters 4 and 5 use is concerned by the development of observers for fault detection. Chapters 6 and 7 consider two applications of fault detection. This part comprises seven chapters.

Chapter 1, "A Nondestructive Technique to Identify and Localize Microscopic Defects on a Microstrip Line", presents a nondestructive technique that can identify and localize geometrical microscopic defects in a silver microstrip line. The proposed technique is based on measuring the reflection coefficients in the frequency domain. Then, the results will be processed by the inverse Fourier transform to finally obtain the reflection coefficient in the time domain.

Chapter 2, "Sensor Fault Detection and Isolation Based on Variable Moving Window KPCA", introduces an online Kernel principal component analysis (KPCA) methods for fault detection. The moving window KPCA (MWKPCA) and the variable moving window KPCA (VMWKPCA) methods are used to update the KPCA model according to the process status. To locate the faulty sensor, a fault isolation algorithm is enabled once a fault is detected. Thus, a partial VMWKPCA is proposed to achieve the fault isolation task.

Chapter 3, "Sensor Fault Detection and Estimation Based on UIO for LPV Time Delay Systems Using Descriptor Approach", investigates fault detection of linear parameter varying (LPV) systems with varying time delay. The proposed approach is based on the use of an unknown input observer to estimate faults and to generate residuals by elaborating an augmented system which considers faults as axillary states. The convergence has been ensured using Lyapunov quadratic stability by generating linear matrix inequalities.

Chapter 4, "Fault Diagnosis of Linear Switched Systems Based on Hybrid Observer", deals with fault detection and localization method for switched linear systems which are a class of hybrid dynamical systems involving interaction between continuous dynamics and discrete dynamics. Two hybrid observer modules are proposed for the diagnosis. The first module is used to identify the current operating mode, and the second one is a bank of Luenberger observer which aims to detect and localize the fault.

Chapter 5, "Neutral Time-Delay System: Diagnosis and Prognosis Using UIO Observer", develops a new prognosis method for a class of neutral time-delay system. An unknown input observer (UIO) is designed to achieve actuator fault detection. Another UIO with a new form is used to generate an algebraic method of system prognosis. The two observers are applied to the transmission line system as an example of neutral time-delay system.

Chapter 6, "Intrinsic Mode Function Selection and Statistical Information Analysis for Bearing Ball Fault Detection", presents a methodology for incipient ball faults detection based on Empirical Mode Decomposition (EMD) that splits the original signal in different components called Intrinsic Mode Functions (IMFs). At first, the number of IMFs is reduced by keeping only the most energized. In a second stage, using the Kullback–Leibler divergence (KLD), the less robust to load variations is eliminated. Finally, the most sensitive IMFs are then selected and used for ball fault detection. To extract the fault features, the first four statistical moments of the retained components are computed and analyzed.

Chapter 7, "Fault Detection and Localization of Centrifugal Gas Compressor System Using Fuzzy Logic and Hybrid Kernel-SVM Methods", investigates a method that realizes a fault detection and localization strategy applied to a centrifugal gas compressor system. The proposed strategy is based on the method of hybrid Kernel-SVM method, where this method is used in many industrial sectors for detecting and diagnosing faults system.

The second part is dedicated to the fault detection and the diagnosis of power systems. It includes three chapters.

Chapter 8, "Performance Investigation of an Improved Diagnostic Method for Open IGBT Faults in VSI-Fed IM Drives", investigates the performance of an improved diagnosis method for open IGBT faults in VSI-fed IM controlled by the direct rotor flux oriented control (RFOC) in attempt to bypass the drawbacks of the existing diagnostic methods. The proposed diagnostic method is based on the information of the normalized currents combined with a new diagnosis variable which can be provided using the information of the slope of the current vector in (α, β) frame.

Chapter 9, "Study of an HVDC Link in Dynamic State Following AC Faults and Commutation Failures Based on Modeling and Real-Time Simulation", introduces an SIL approach to examine the dynamic performance of the HVDC inverter and to demonstrate relationship between the DC control, the stability, and the recovery from both the instability and the commutation failures.

Chapter 10, "Influence of Design Variables on the Performance of Permanent Magnet Synchronous Motor with Demagnetization Fault", emphasizes the effects of design variables on the permanent magnet (PM) demagnetization of a permanent magnet synchronous motor (PMSM) using a two-dimensional (2D) finite element analysis (FEA) software package. The aim is to investigate the shape design of a permanent magnet considering the effect of the pole embrace and the magnet thickness on demagnetization by using finite element method (FEM). As a result, it is shown that pole embrace and magnet thickness are the most important geometrical dimensions of the PMSM in terms of demagnetization fault. The effects of demagnetization fault on the magnetic field distribution, the back electromotive force (EMF) induced, and the electromagnetic torque of PMSM are examined. Using the simulated model, a technical method, based on fast Fourier transform (FFT) analysis of induced voltage and electromagnetic torque, is exported to detect the demagnetization fault.

The third and last part is concerned about recent developments on the fault-tolerant control of systems. This part is made of four chapters.

Chapter 11, "Extended Kalman Filtering for Remaining Useful Lifetime Prediction of a Pipeline in a Two-Tank System", introduces the Wiener processes to model faults in two-tank system, and then the extended Kalman filter (EKF) algorithm is used to generate the residual for detection and diagnosis. As the deterioration states are hidden, the maximum likelihood estimation (MLE) is used to estimate the unknown model parameters.

Chapter 12, "Fault-Tolerant Control of Two-Time-Scale Systems", presents fault-tolerant control of two-time-scale systems in both linear and nonlinear cases. An adaptive approach for fault-tolerant control of singularly perturbed systems is used in linear case, where both actuator and sensor faults are examined in the presence of external disturbances. For sensor faults, an adaptive controller is designed based on an output-feedback control scheme. The feedback controller gain is determined in order to stabilize the closed-loop system in the fault-free case and vanishing disturbance, while the additive gain is updated using an adaptive law to compensate for the sensor faults and the external disturbances. To correct the actuator faults, a state-feedback control method based on adaptive mechanism is considered.

Chapter 13, "Reliable Control of Power Systems", proposes different designs in the frequency domain and time domain for reliable control of power systems. The designs consider two cases: system without uncertainty and with uncertainty due to load changes. The solution employs Kharitonov's theorem, linear matrix inequalities (LMI). It also tackles a more difficult case when one controller is fast, whereas the other is slow due to its inherent time delay. Lyapunov–Krasovski method is used to handle this case.

Chapter 14, "Active FTC of LPV System by Adding Virtual Components", develops an FTC control using real modeling or close to the reality of the process. A control method uses virtual components to tolerate a defect that appears on the blade control system of a wind turbine. This command is based on an LPV model of a wind turbine benchmark.

These chapters were thoroughly reviewed. Based on comments of the reviewers, the chapters were modified by the authors and the modified chapters are examined for incorporation in a compendium of interesting contributions to fault diagnosis, fault detection, and tolerant control.

The editorial team members would like to extend gratitude and sincere thanks to all contributed authors, reviewers, and panelist, for paying attention to the quality of the publication. Hoping this book will be a good material on the state-of-the-art research.

Sfax, Tunisia — Nabil Derbel
Al Khoudh, Oman — Jawhar Ghommam
Bristol, UK — Quanmin Zhu
October 2019

Contents

Chapter 1
A Nondestructive Technique to Identify and Localize Microscopic Defects on a Microstrip Line

Leila Troudi, Khaled Jelassi and Mohamed Sellaouti

Abstract In this chapter, we present an industrial nondestructive technique to can identify and localize geometrical microscopic defects in a silver microstrip line. We worked on two types of defects: the narrow transverse slits and the overflows. It is true that there are several techniques to detect defects on the microstrip line, but these techniques can be destructive and do not allow us to localize and characterize the defect like the measure of the impedance or to identify the defect when the type of defect is unknown like the reflectometry. The proposed technique is based on the measure of the scatting parameters, the calculation of the characteristic impedance of each model of the microstrip line (undamaged and defective), and the comparison between the different results obtained from the practical and the theoretical tests.

Keywords *S* parameters · Impedance · Overflows · Narrow transverse slits · Microstrip line

1.1 Introduction

In the flexible printed circuit industry with the screen printing method, the most common and easiest method to diagnose defects is taking measures of the loop resistance of the conductors. This technique can detect global impedance variation, but it cannot allow the localization of the defects or the characterization of their depth or width.

Also, sometimes, it is dangerous for insulators or electrical varnish that protects the tracks that can slam electrically or in the worst case are weakened without the operator noticing.

L. Troudi · K. Jelassi (✉) · M. Sellaouti
Laboratoire des Systèmes Electriques, Ecole Nationale d'Ingénieurs de Tunis, University Tunis El Manar, Tunis, Tunisia
e-mail: khaled.jelassi@enit.utm.tn

L. Troudi
e-mail: leila.troudii@gmail.com

N. Derbel et al. (eds.), *Diagnosis, Fault Detection & Tolerant Control*, Studies in Systems, Decision and Control 269,
https://doi.org/10.1007/978-981-15-1746-4_1

When the defects are microscopic, this technique becomes inoperative because these types of defects are characterized by a small impedance variation which makes the reflection coefficient very low and this reflection can be masked by the noise. Moreover, the reflectometry is one of the most popular techniques to detect and localize defects of all the electrical circuits (Kwon et al. 2017). This technique works by injecting a signal into the system to be diagnosed. This signal propagates according to the laws of the propagation. When it encounters discontinuous impedance, a part of its energy is reflected to the injection point.

The analysis of reflected signals helps us to get information about the tested circuits (Auzanneau et al. 2016). The reflectometry can be applied in two domains: the time-domain reflectometry (TDR) and the frequency-domain reflectometry (FDR). Both are based on the method described before. The difference between the TDR and FDR is in the injection procedure and the signal processing.

For the time-domain reflectometry (TDR), the analysis of the reflected wave (shape, duration, magnitude) provides information about the impedance variation. However, the frequency-domain reflectometry is the analysis of the standing wave that provides information (Farhat et al. 2016). This technique of testing has many advantages. First, it is very fast to detect defaults when scanning the circuit and it allows the localization of the defects with high precision (Tze Mei et al. 2017). But this technique has some limits, for example, it cannot characterize the type of the defect when the model of the circuit is unknown.

In this work, we propose an industrial nondestructive technique that can be a solution to detect, identify, and localize the defects even when they are microscopic. This technique is based on measuring the reflection coefficients $S_{11}(f)$ in the frequency domain. Then, the results will be processed by the inverse Fourier transform to finally obtain the reflection coefficient $S_{11}(t)$ in the time domain. Afterward, we will calculate the characteristic impedance and we will compare the results obtained from the practical values and the theoretical values.

1.2 The Transmission Lines

1.2.1 Definition

A transmission line is a set of two conductors or more. Its main role is to transfer the information without disturbance. It means convey the electrical signal from the source (transmitter) to the load (receiver) with the least attenuation and deformation, but it can be used to make filters, impedance transformers, and couplers (Raju 2009).

The transmission lines that allow the channeling of the electromagnetic waves represent the base of study and analysis of circuit at high frequencies. They are characterized by the attenuation constant and the propagation velocity, which depends on the used dielectric and the characteristic impedance (Chandra et al. 2008).

1.2.2 Types of Transmission Lines

1.2.2.1 Coaxial Cables

They are the most used transmission lines. They are characterized by a very important frequency band (Natarajan 2013). Coaxial cables are used in several fields such as the computer field, video cabling, low-frequency electronics, and the microwave domain when the frequency can reach several tens of gigahertz. The losses in this type of lines depend on the used dielectrics. In the microwave, we use a special dielectric with a very large diameters (Natarajan 2013) (Fig. 1.1).

1.2.2.2 Parallel Plates

They are not used a lot in practice, but they can help us simplify the analysis of very complex waveguide (Keqian Zhang 2008) (Fig. 1.2).

1.2.2.3 Twin-Lead

They are the least used. They allow us mainly to connect an antenna to a television (Fig. 1.3) (Bagad 2008).

1.2.2.4 Microstrip Lines

They are especially used in high frequency in integrated circuits to connect the components and the chips. The microstrip lines are composed of a substrate of Si,

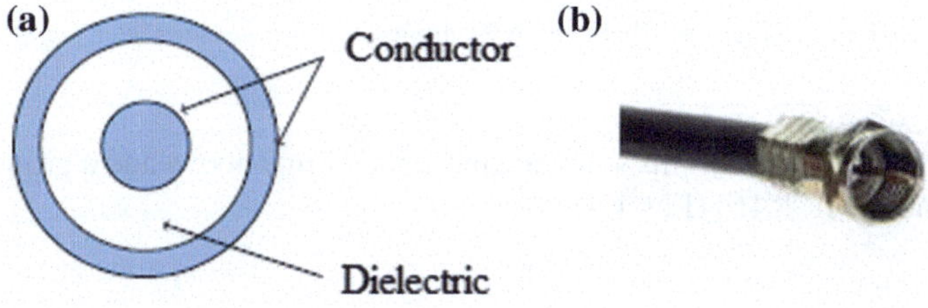

Fig. 1.1 Coaxial cables: **a** Schematic, **b** Example

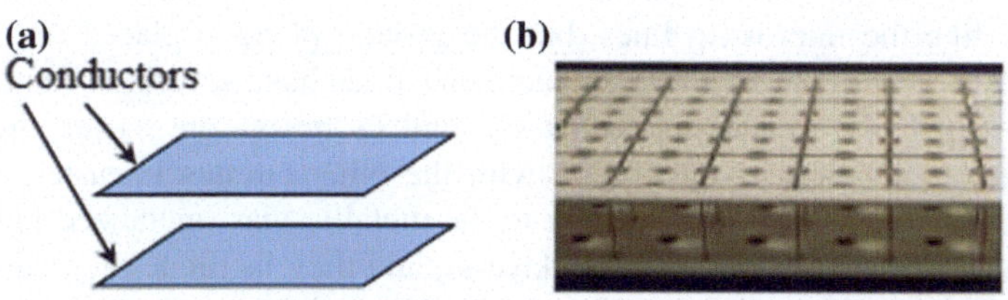

Fig. 1.2 Parallel plates: **a** Schematic, **b** Example

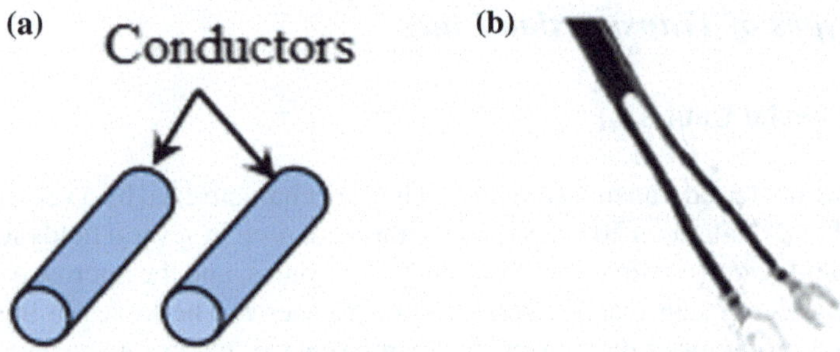

Fig. 1.3 Twin-lead: **a** Schematic, **b** Example

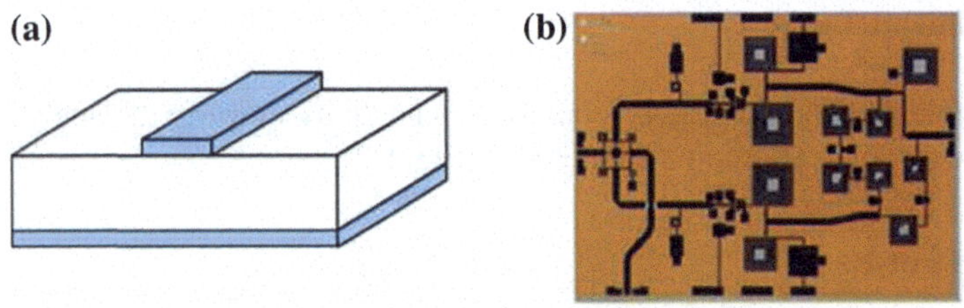

Fig. 1.4 Microstrip lines: **a** Schematic, **b** Example

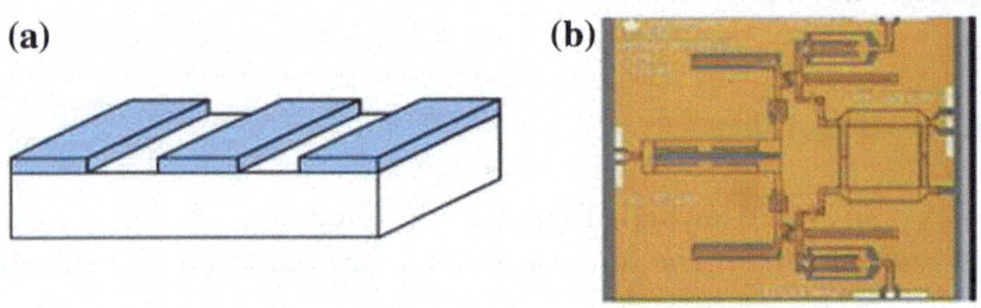

Fig. 1.5 Coplanar waveguides: **a** Schematic, **b** Example

GaAs, or InP, for example. This substrate has a metal line above and a ground below (Edwards and Steer 2016) (Fig. 1.4).

1.2.2.5 Coplanar Waveguides

They look like the microstrip lines, but the ground plane is placed on each side of the metallic line as in Fig. 1.5. The coplanar lines have several advantages over the microstrip lines because its realization and its design are easier, the cost of manufacture is smaller, its integration with the other circuits is easier, thanks to its unipolar character without resorting to the metallic vias (metalized hole allows the electrical connection between two layers), and they facilitate the realization of complex circuits such as filters and insulators (Wen 1969).

1.3 Microstrip Line Model

1.3.1 Modelling of a Microstrip Line

In the microstrip line, the current and the voltage varied with the length of the line x. We composed the length x into small sections called "dx", and we modeled every section by ideal elements that present the characteristics of a microstrip line like in Fig. 1.6 (Robertson et al. 2016).

These elements are as follows:

- R is the linear resistance which presents the losses in the conductors by the Joule heating (Ω/m).
- L is the linear inductance which presents the magnetic effects due to the passage of the current in the conductor (H/m).
- G is the linear conductance which presents the dielectric losses (Siemens/m).
- C is the linear capacitance between conductors (F/m).

Generally, parameters R, C, L, and G depend on the frequency (Chaturvedi et al. 2016).

- G presents the dielectric losses strongly increases with frequency.
- R, also, increases with frequency due to the Joule heating.
- L and C which present the reactive elements do not depend a lot on the frequency.

According to the Kirchhoff's circuit laws (voltage laws, current laws), we have

$$-V(x,t) = L.dx\frac{dI(x,t)}{dt} + R.dxI(x,t) + V(x+dx,t) \tag{1.1}$$

$$\frac{V(x+dx,t) - V(x,t)}{dx} = L\frac{dI(x,t)}{dt} + RI(x,t) \tag{1.2}$$

$$\frac{-dV}{dx} = \frac{L(dI(x,t))}{dt} + RI(x,t) \tag{1.3}$$

$$I(x,t) = \frac{C.dxdV(x,t)}{dt} + C.dxV(x,t) + I(x+dx,t) \tag{1.4}$$

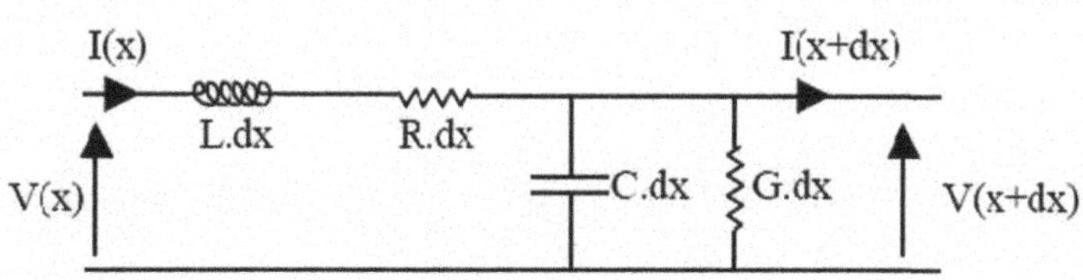

Fig. 1.6 The modeling of the microstrip line

$$\frac{-(I(xdx,t)-I(x,t))}{dx}=\frac{CdV(x,t)}{dt}+GI(x,t) \tag{1.5}$$

$$\frac{-dI}{dx}=\frac{CdV(x,t)}{dt}+GI(x,t) \tag{1.6}$$

In harmonic regime, we have

$$V(x,t)=V(x)e^{j\omega t} \tag{1.7}$$

$$I(x,t)=I(x)e^{j\omega t} \tag{1.8}$$

Equations (1.3) and (1.6) become

$$\frac{dV(x)}{dx}=-(R+j\omega L)I(x,t) \tag{1.9}$$

$$\frac{dI(x)}{dx}=-(G+j\omega C)V(x,t) \tag{1.10}$$

We derive these two equations, and we obtain

$$\frac{d^2V(x)}{dx}=(R+j\omega L)(G+j\omega C)V(x,t) \tag{1.11}$$

$$\frac{d^2I(x)}{dx}=(R+j\omega L)(G+j\omega C)I(x,t) \tag{1.12}$$

$$\gamma=\sqrt{(R+j\omega L)(G+j\omega C)}=\alpha+j\beta \tag{1.13}$$

with α is the attenuation constant and β is the propagation constant.

$$\alpha=\sqrt{\frac{1}{2}\left(\sqrt{(R^2+L^2\omega^2)(G+C^2\omega^2)}+(RG-LC\omega^2)\right)} \tag{1.14}$$

$$\beta=\sqrt{\frac{1}{2}\left(\sqrt{(R^2+L^2\omega^2)(G+C^2\omega^2)}-(RG-LC\omega^2)\right)} \tag{1.15}$$

$$\frac{d^2V(x)}{dx}-\gamma^2V(x,t)=0 \tag{1.16}$$

$$\frac{d^2I(x)}{dx}-\gamma^2I(x,t)=0 \tag{1.17}$$

The solutions of Eqs. (1.16) and (1.17) are

$$V(x) = V_1 e^{-\gamma x} + V_2 e^{\gamma x} \tag{1.18}$$

$$I(x) = I_1 e^{-\gamma x} + I_2 e^{\gamma x} \tag{1.19}$$

$e^{-\gamma x}$ represents the propagation of the wave in the $x+$ direction and $e^{\gamma x}$ represents the propagation of the wave in $x-$ direction.

- Now, we inject the expression of $V(x)$ in Eq. (1.9) to obtain the relation between the current and the voltage:

$$I(x) = \frac{\gamma}{(R + j\omega L)[V_1 e^{-\gamma x} - V_2 e^{\gamma x}]} = \frac{1}{Z_0[V_1 e^{-\gamma x} - V_2 e^{\gamma x}]} \tag{1.20}$$

with Z_0 is the characteristic impedance of the microstrip line.

$$Z_0 = \frac{R + j\omega L}{\gamma} = \sqrt{\frac{R + j\omega L}{G + j\omega C}} \tag{1.21}$$

- We can say that the characteristic impedance Z_0 of the transmission line depends only on the electrical characteristics of the line. Also, it varies with the frequency (Robertson et al. 2016).
- In the general case, when we have losses, it is complex (Robertson et al. 2016).

1.3.2 Particular Case: Lossless Line

The analysis of the lossless line provides a precise approximation for the real transmission lines. This approximation simplifies the mathematical calculations, especially, in the modeling of transmission lines. A lossless line is a transmission line that has no line resistance and no dielectric loss. This means that the dielectric acts like a perfect dielectric and the conductor acts like a perfect conductor (Huang and Boyle 2008).

In this case, we have $R = G = 0$ and therefore

$$\gamma = j\omega\sqrt{LC} \tag{1.22}$$

$$\beta = \omega\sqrt{LC} \tag{1.23}$$

$$\alpha = 0 \tag{1.24}$$

The characteristic impedance becomes, in this case,

$$Z_0 = \sqrt{\frac{L}{C}} \tag{1.25}$$

In this case, Z_0 no longer depend on the frequency and it became purely resistive. The model of the lossless line can be useful for many practical cases, such as high-frequency transmission lines and low-loss transmission lines. In these two cases, R and G are, respectively, smaller than ωL and ωC (Huang and Boyle 2008).

1.3.3 *The Impedance and the Reflection Coefficient*

1.3.3.1 The Reflection Coefficient

The presence of a reflected wave on the line can be explained by the presence of a disturbing element on the transmission line like the presence of a load at the end of the line or a discontinuity in the characteristic of the line. We will model this reflection by the reflection coefficient, which is the ratio of the reflected wave and the incident wave:

$$\Gamma = \frac{V_{reflected}}{V_{incident}} \tag{1.26}$$

The reflection coefficient depends on the position on the line.

$$\Gamma = \frac{V_2 e^{\gamma x}}{V_1 e^{-\gamma x}} = \frac{V_2}{V_1} e^{2\gamma x} \tag{1.27}$$

1.3.3.2 The Impedance on the Line

At any point of the transmission line, we can define the impedance Z which is the ratio of the voltage and the current at a point x by

$$Z(x) = \frac{V(x)}{I(x)} \tag{1.28}$$

So

$$Z(x) = \frac{V_1 e^{-\gamma x} + V_2 e^{\gamma x}}{\frac{1}{Z_0}(V_1 e^{-\gamma x} - V_2 e^{\gamma x})} = Z_0 \frac{V_1 e^{-\gamma x} + V_2 e^{\gamma x}}{V_1 e^{-\gamma x} - V_2 e^{\gamma x}} \tag{1.29}$$

1.3.3.3 Relation Between Impedance and Reflection Coefficient

- **General case**

According to Eq. (1.29), we can write the impedance Z in the following form:

$$Z(x) = Z_0 \frac{1 + \dfrac{V_2 e^{\gamma x}}{V_1 e^{-\gamma x}}}{1 - \dfrac{V_2 e^{\gamma x}}{V_1 e^{-\gamma x}}} \tag{1.30}$$

we have seen that

$$\Gamma = \frac{V_2}{V_1}\, e^{2\gamma x} \tag{1.31}$$

So, the relation between the impedance Z and the reflection coefficient Γ is

$$Z(x) = Z_0 \frac{1 + \Gamma(x)}{1 - \Gamma(x)} \quad , \quad Z(x) = \frac{Z(x) - Z_0}{Z(x) + Z_0} \tag{1.32}$$

- **At the end of the line**

The relation between the transmission coefficient and the impedance becomes

$$\Gamma_e(x) = \frac{Z_e(x) - Z_0}{Z_e(x) + Z_0} \tag{1.33}$$

- **Change of variables**

The impedance and the reflection coefficient of the transmission line depend on the characteristics of the line, but especially the load placed at the end of the line. We found previously

$$\Gamma(x) = \frac{V_2}{V_1}\, e^{2\gamma x} \tag{1.34}$$

At the end of the line, when $x = l$

$$\Gamma(l) = \frac{V_2}{V_1} e^{2\gamma l} = \Gamma_e \tag{1.35}$$

We can calculate the value of

$$\frac{V_2}{V_1} = \Gamma_e e^{-2\gamma l} \tag{1.36}$$

So

$$\Gamma(x) = \Gamma_e e^{2\gamma (x-l)} \tag{1.37}$$

This relationship is not practical because it depends on the length l. So, we will change the scale. We will define a new scale S directed from the load to the generator.

In this case, we will take $s = l - x$ and the reflection coefficient becomes

$$\Gamma(s) = \Gamma_e e^{-2\gamma s} \tag{1.38}$$

In this new scale,

$$\begin{aligned} V(x) &= V_1 e^{-\gamma x} + V_2 e^{\gamma x} \\ &= V_1 e^{-\gamma(l-s)} + V_2 e^{\gamma(l-s)} \\ &= V_1 e^{-\gamma l} e^{\gamma s} + V_2 e^{\gamma l} e^{-\gamma s} \\ &= V_{11} e^{\gamma s} + V_{22} e^{-\gamma s} \end{aligned} \tag{1.39}$$

V_{11} and V_{22} are two complex constants that can be determined using the boundary conditions. For simplification, we take $V_{11} = V_1$ and $V_{22} = V_2$.

So, the voltage and the current are written in the following form:

$$V(s) = V_1 e^{\gamma s} + V_2 e^{-\gamma s} \tag{1.40}$$

$$I(x,t) = \frac{1}{Z_0}[V_1 e^{\gamma s} - V_2 e^{-\gamma s}] \tag{1.41}$$

The reflection coefficient:

$$\Gamma(s) = \frac{V_2}{V_1} e^{-2\gamma s} \tag{1.42}$$

The impedance Z:

$$Z(s) = Z_0 \frac{V_1 e^{\gamma s} + V_2 e^{-\gamma s}}{V_1 e^{\gamma s} - V_2 e^{-\gamma s}} \tag{1.43}$$

and the relation between the impedance Z and the reflection coefficient $\Gamma(s)$:

$$Z(s) = Z_0 \frac{1 + \dfrac{V_2 e^{-\gamma s}}{V_1 e^{\gamma s}}}{1 - \dfrac{V_2 e^{-\gamma s}}{V_1 e^{\gamma s}}} \tag{1.44}$$

$$Z(s) = Z_0 \frac{1 + \Gamma(s)}{1 - \Gamma(s)} \quad , \quad Z(s) = \frac{Z(s) - Z_0}{Zs) + Z_0} \tag{1.45}$$

- **particular value of Z_e**
  - The line ends with a short circuit.

$$\text{If } Z_e = 0\,,\ \Gamma_e = \frac{Z_e - Z_0}{Z_e + Z_0} = -1 \text{ or } \Gamma_e = \frac{V_2}{V_1}e^{-2\Gamma s}\bigg|_{s=0} = \frac{V_2}{V_1} \rightarrow V_2 = -V_1 \tag{1.46}$$

– The line ends with an open circuit.

$$\text{If } Z_e = \infty\,,\ \Gamma_e = \frac{Z_e - Z_0}{Z_e + Z_0} \rightarrow 1 \text{ or } \Gamma_e = \frac{V_2}{V_1}e^{-2\Gamma s}\bigg|_{s=0} = \frac{V_2}{V_1} \rightarrow V_2 = V_1 \tag{1.47}$$

→ The wave is totally reflected by the open circuit.

– The line ends with the characteristic impedance.

$$\text{If } Z_e = Z_0\,,\ \Gamma_e = \frac{Z_e - Z_0}{Z_e + Z_0} = 0 \text{ or } \Gamma_e = \frac{V_2}{V_1}e^{-2\Gamma s}\bigg|_{s=0} = \frac{V_2}{V_1} \rightarrow V_2 = 0 \tag{1.48}$$

→ There is no reflection in this case, and the wave is totally transmitted in the load.

1.4 The *S* Parameters

1.4.1 Definition

There are several well-known parameters to study and characterize the linear multi-port network such as *H*, *Z*, and *Y* parameters. To determine these parameters, the measurements must be made in short circuit or in open circuit. But when the frequency exceeds 100 MHz, the realization of tests in open circuit becomes very difficult and in short circuit may cause the oscillation of the electrical circuit, while the S parameters adapt with the input and the output of the two-port network avoiding all these difficulties (Hickman 2007).

The *S* parameters essentially allow us to describe the electrical behavior of transmission lines. The calculation of these parameters requires the measure of the incident and the reflected waves. Also, through these parameters, we can extract several electrical properties such as power, gain, reflection coefficient, impedance, the insertion, and the return loss (Hickman 2007).

The *S* parameters are applicable for all frequencies, but they are regularly used for very high frequencies. Among the measuring devices of *S* parameters in the high frequencies is the vector network analyzer (VNA).

1.4.2 The S Matrix

The S parameters are calculated through a vector network analyzer; we can put them in the form of matrix to finally get the S matrix (the distribution matrix). In the general case, the relation between the incident, reflected, and transmitted waves of a N port device are described by the following relation.

$$\begin{pmatrix} b_1 \\ \vdots \\ b_N \end{pmatrix} = \begin{pmatrix} S_{11} & \cdots & S_{1N} \\ \vdots & \ddots & \vdots \\ S_{N1} & \cdots & S_{NN} \end{pmatrix} \begin{pmatrix} a_1 \\ \vdots \\ a_N \end{pmatrix} \tag{1.49}$$

The element S_{ij} of S the matrix is written in the following form:

$$S_{ij} = \frac{b_i}{a_j} \tag{1.50}$$

The parameters a_i and b_i represent, respectively, the complex normalized waves incident and reflected.

1.4.3 The S Matrix for a Two-Port Network

To better understand the calculation of the S matrix, we will work on a two-port network (Fig. 1.7).

For a two-port network, the relations between the incident, reflected, and transmitted waves are described by the following relation:

$$\begin{pmatrix} b_1 \\ b_2 \end{pmatrix} = \begin{pmatrix} S_{11} & S_{12} \\ S_{21} & S_{22} \end{pmatrix} \begin{pmatrix} a_1 \\ a_2 \end{pmatrix} \tag{1.51}$$

and

$$b_1 = S_{11}a_1 + S_{12}a_2 \tag{1.52}$$

Fig. 1.7 A two-port network

$$b_2 = S_{21}a_1 + S_{22}a_2 \tag{1.53}$$

The waves a and b are measured using the incident voltages V_i and reflected voltages V_r of each port:

$$a_1 = \frac{V_{i1}}{\sqrt{Z_0}} = \frac{V_1 + Z_0 I_1}{2\sqrt{Z_0}} \qquad a_2 = \frac{V_{i2}}{\sqrt{Z_0}} = \frac{V_2 + Z_0 I_2}{2\sqrt{Z_0}} \tag{1.54}$$

$$b_1 = \frac{V_{r1}}{\sqrt{Z_0}} = \frac{V_1 - Z_0 I_1}{2\sqrt{Z_0}} \qquad b_2 = \frac{V_{r2}}{\sqrt{Z_0}} = \frac{V_2 + Z_0 I_2}{2\sqrt{Z_0}} \tag{1.55}$$

The impedance Z_0 is the characteristic impedance of the line.

– **Physical properties of the S parameters**

- $S_{11} = \left[\frac{b_1}{a_1}\right]_{a_2=0}$ is the reflection coefficient at port 1.
- $S_{12} = \left[\frac{b_1}{a_2}\right]_{a_1=0}$ is the transmission coefficient from port 2 to port 1.
- $S_{21} = \left[\frac{b_2}{a_1}\right]_{a_2=0}$ is the transmission coefficient from port 1 to port 2.
- $S_{22} = \left[\frac{b_2}{a_2}\right]_{a_1=0}$ is the reflection coefficient at port 2.

– **Relation between *S* parameters**

$$|S_{11}|^2 + |S_{21}|^2 = 1 \qquad |S_{22}|^2 + |S_{21}|^2 = 1 \qquad S_{11}^* S_{21} + S_{22}^* S_{21} = 1$$

Figure 1.8 shows a demonstrative diagram of the *S* parameters of a two-port network.

1.4.4 The S Matrix Properties

- **Reciprocal**

  The *S* matrix is reciprocal, when $S_{ij} = S_{ji}$ for all i and j. The passive components are reciprocal like the resistance, transformers, and capacitors except for the structures involving magnetized ferrites or plasmas. The active components such as amplifiers are generally non-reciprocal (Gustrau 2012).

- **Symmetrical**

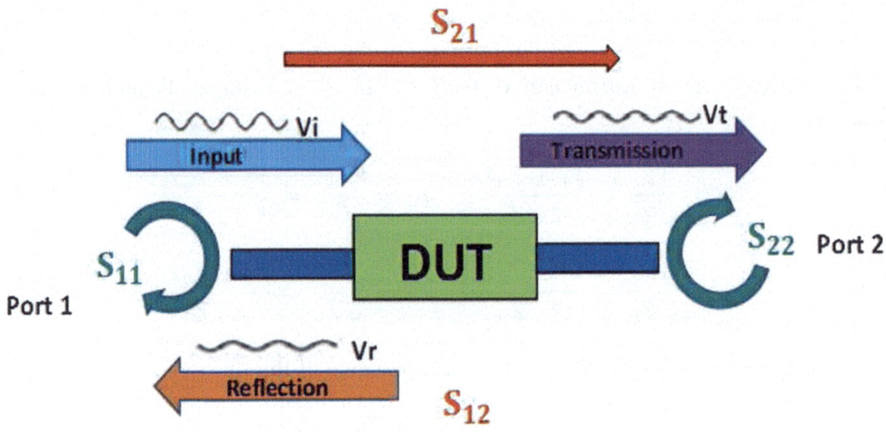

Fig. 1.8 The S parameters of a two-port network

The S matrix is symmetrical, when it is reciprocal and the output and input reflection coefficients are equal (Gustrau 2012).
In a two-port network,

$$S_{11} = S_{22}$$

- **Passive and lossless**

The S matrix is passive and lossless, when it is unitary: $(S^*)^T S = 1$.
$(S^*)^T$ is the conjugate transpose of S.
In a two-port network, we have

$$(S^*)^T = \begin{pmatrix} S_{11}^* & S_{21}^* \\ S_{12}^* & S_{22}^* \end{pmatrix} \begin{pmatrix} S_{11} & S_{12} \\ S_{21} & S_{22} \end{pmatrix} = \begin{pmatrix} 1 & 0 \\ 0 & 1 \end{pmatrix} \tag{1.56}$$

1.5 The Used Model of the Microstrip Line

For the simulation and the practical tests, we chose to work on the microstrip line presented in Fig. 1.9

The used microstrip line was modeled on a polyester substrate (the dielectric constant $(r) = 2.89$, the thickness $(h) = 500\,\mu\text{m}$), and the conducting strip was made with silver (length $(l) = 300\,\text{mm}$, the width $(w) = 1.285$, the thickness $(T) = 5\,\mu\text{m}$).

The various defects are placed at the half of the conducting strip length. The defect represents a discontinuity which has undesirable effects on the transmission line. This discontinuity can be presented by an equivalent circuit in a point on the microstrip.

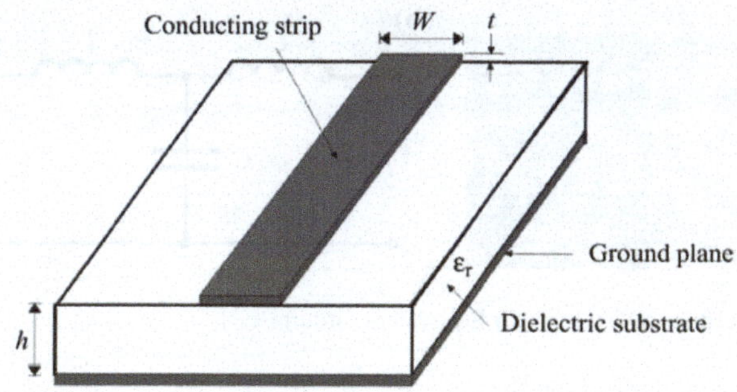

Fig. 1.9 The microstrip line model

The equivalent circuits can be a simple shunt or series components on the transmission line. The values of the component of the equivalent circuit depend on the parameters of the transmission line, the discontinuity, and the operating frequency.

1.6 Defects Description

1.6.1 The Narrow Transverse Slit

In our prototype, we fixed the width L of the narrow transverse slit to 500 μm and then to 250 μm. Then, we varied the depth D to 60 and 80% of the width of the conducting strip (w = 1.285 μm) of the microstrip line.

The narrow transverse slit is characterized by a variation of the characteristic impedance that is relatively bigger than the characteristic impedance of the undamaged conducting strip (inductive behavior) (Hoefer 1977). This type of defect can be presented by a circuit equivalent to the Pi model (Terry and Michael 2016) presented in Fig. 1.10.

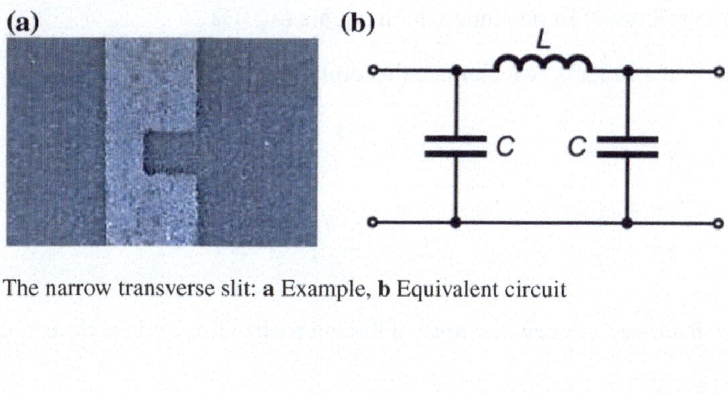

Fig. 1.10 The narrow transverse slit: **a** Example, **b** Equivalent circuit

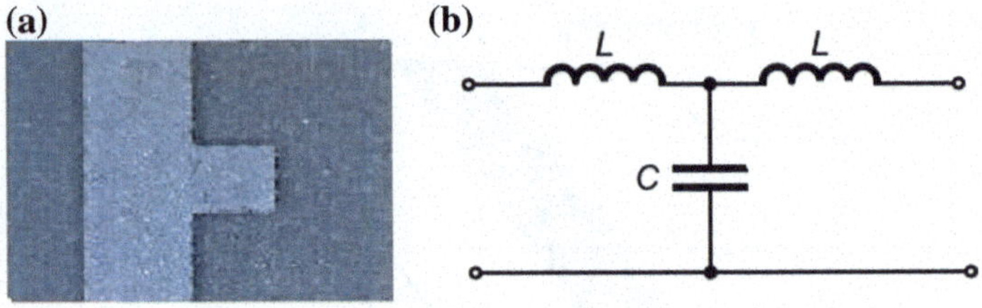

Fig. 1.11 The overflow: **a** Example, **b** Equivalent circuit

1.6.2 The Overflow

In our study, we fixed the length L of the overflow to 500 μm and then to 250 μm. Then, we varied the height H to 60 and 80% of the width of the conducting strip ($w = 1.285\,\mu$m).

The overflow is characterized by a variation of the characteristic impedance that is smaller than the characteristic impedance of the undamaged conducting strip (Capacitive behavior) (Hoefer 1977). This type of defect can be presented by an equivalent circuit to the Π model (Terry and Michael 2016) presented in Fig. 1.11.

1.7 The Solution Description

For the experimental work, we measured the S parameters of every model mentioned before in the frequency domain. Then, we calculated the reflection coefficient $S_{11}(t)$ in the time domain by using the inverse Fourier transform and the characteristic impedance defined by this equation:

$$Z_0 = Z_{ref}\,\frac{1+S_{11}(t)}{1-S_{11}(t)} \tag{1.57}$$

- Z_0 is the characteristic impedance.
- Z_{ref} is the reference impedance which equals to 50 Ω.

To localize the defects, we can use this equation:

$$T_d = \frac{T}{2} \tag{1.58}$$

$$L_d = \frac{C\,T_d}{\sqrt{\varepsilon_r}} \tag{1.59}$$

- L_d is the distance between the input of the microstrip line and the defect, expressed in (m).

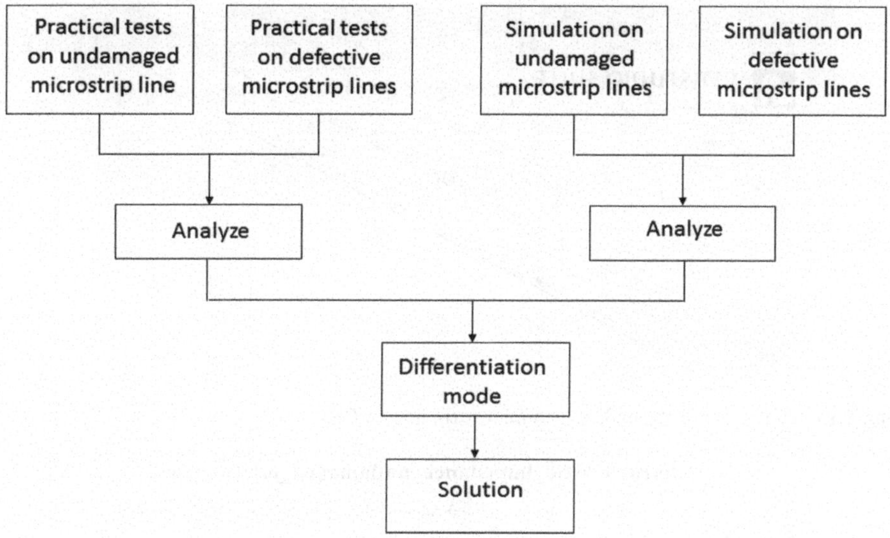

Fig. 1.12 Illustration of the principle of the solution

- C is the celerity, expressed in (m/s).
- T_d is the moment when the defects appear, expressed in (s).
- ε_r is the relative permittivity.

To do all that, we used AWR design environment software that can read and generate a touchstone file and allows us to calculate the impedance and trace it. For the simulation, we used CST MICROWAVE STUDIO and we repeated the same steps of the experimental work.

The base of our method of testing is to compare the results obtained from the practical tests and the simulation to detect and localize the defects (Fig. 1.12).

1.8 The Simulation Results with CST Software

1.8.1 The Simulation of Undamaged Microstrip Line

Figure 1.13 shows the model of the undamaged microstrip line on CST.

In Fig. 1.14, we observed that the characteristic impedance equals to 50 Ω approximately, but we can see a big perturbation between 2800 ps and 3300 ps that indicates that the end of the microstrip line was reached.

For example, at 3150 ps, we applied Eqs. (1.58) and (1.59):

- $T = 3150 \times 10^{-12}$ s.

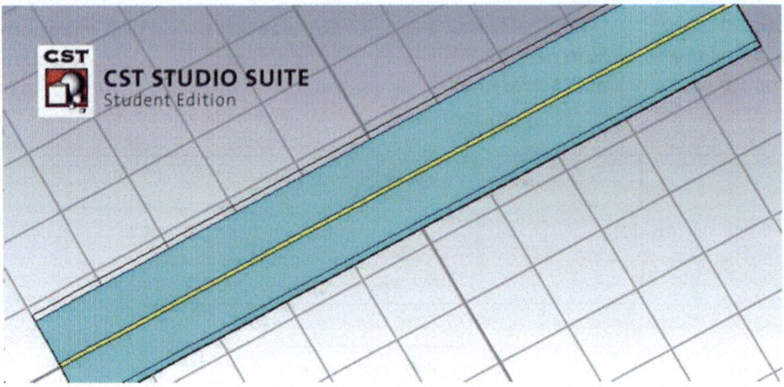

Fig. 1.13 The model of the undamaged microstrip line on CST

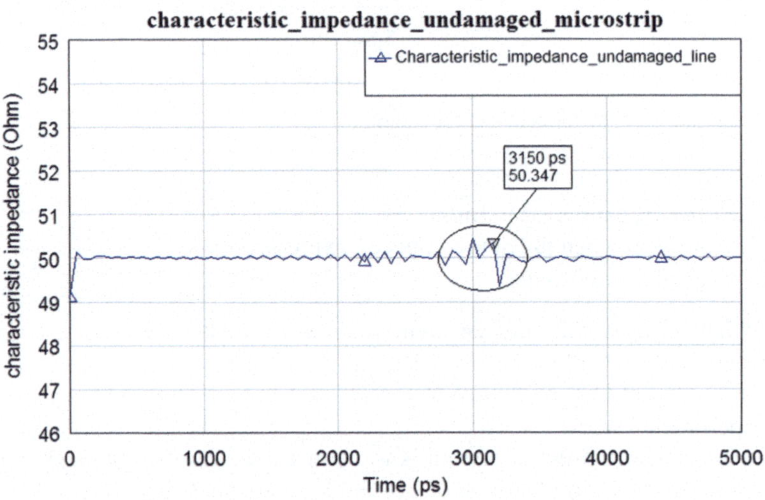

Fig. 1.14 The characteristic impedance of the undamaged microstrip line

- $T_d = \dfrac{3150}{2}$.
- L_d=0.2779 m $\simeq$ 0.3 m the length of the line.

1.8.2 The Simulation of a Microstrip Line with an Overflow

Figure 1.15 shows the model of the microstrip line with an overflow on CST. In Fig. 1.16, we compare the undamaged microstrip line and a microstrip line that have

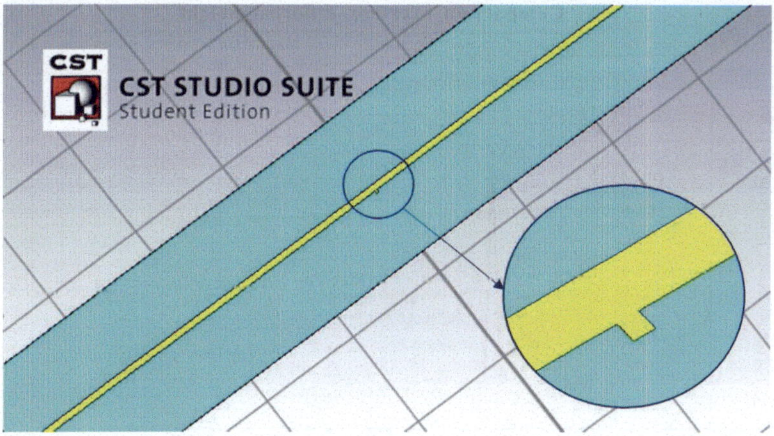

Fig. 1.15 The model of the microstrip line with an overflow on CST

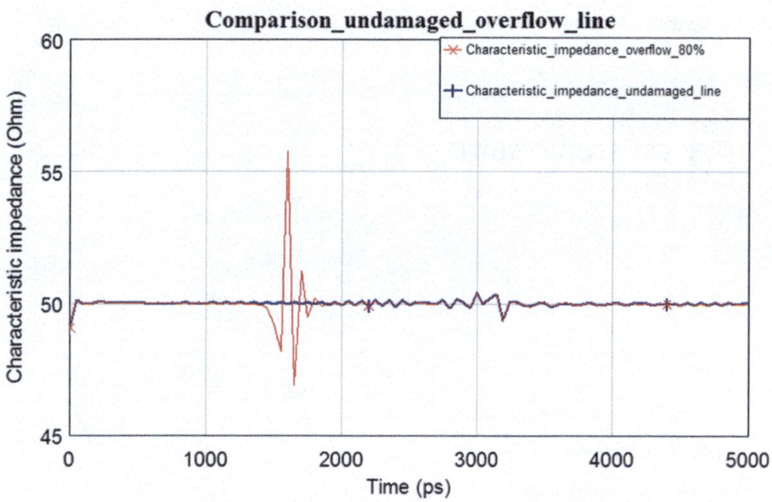

Fig. 1.16 Comparison between an undamaged microstrip line and a microstrip line with an overflow

an overflow. We observe that the characteristic impedance decreases; then it presents a big increase and after that it decreases again.

In Fig. 1.17, we compare two microstrip lines with different height of overflows. We saw that when the height of the overflow increases, the characteristic impedance increases. When the height of the overflow equals to 60% of the width of the conducting strip, the characteristic impedance equals to 54.46 Ω but when it equals to 80% of the width of the conducting strip, the characteristic impedance equals to 55.66 Ω.

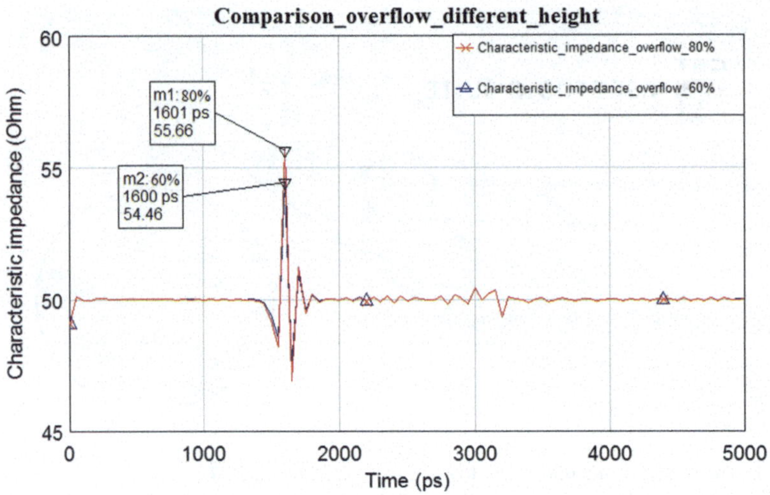

Fig. 1.17 Comparison between two microstrip lines with different heights of overflows

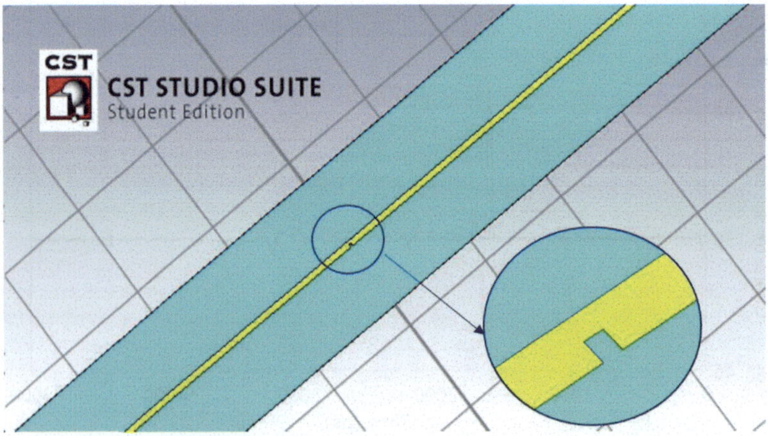

Fig. 1.18 The model of the microstrip line with a narrow transverse slit on CST

1.8.3 The Simulation of a Microstrip Line with a Narrow Transverse Slit

Figure 1.18 shows the model of the microstrip line with a narrow transverse slit on CST.

In Fig. 1.19, we do a comparison between the undamaged microstrip line and a microstrip line that have a narrow transverse slit.

We observe that the characteristic impedance increases, and then it presents a big decrease and after that it increases again.

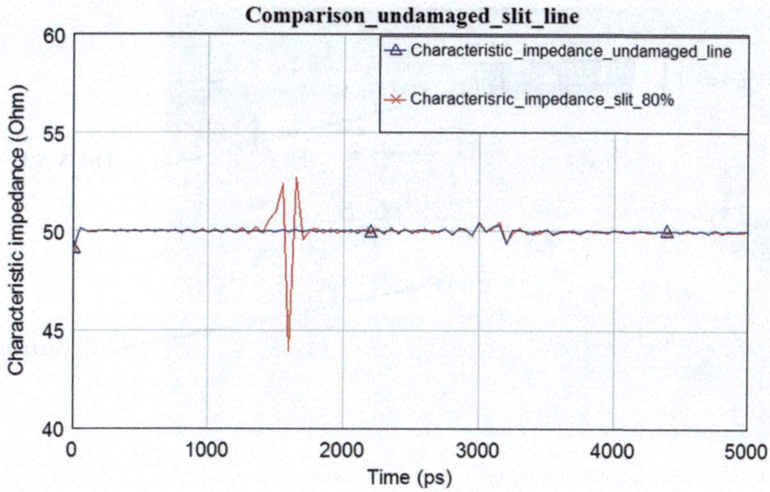

Fig. 1.19 Comparison between an undamaged microstrip line and a microstrip line with a narrow transverse slit

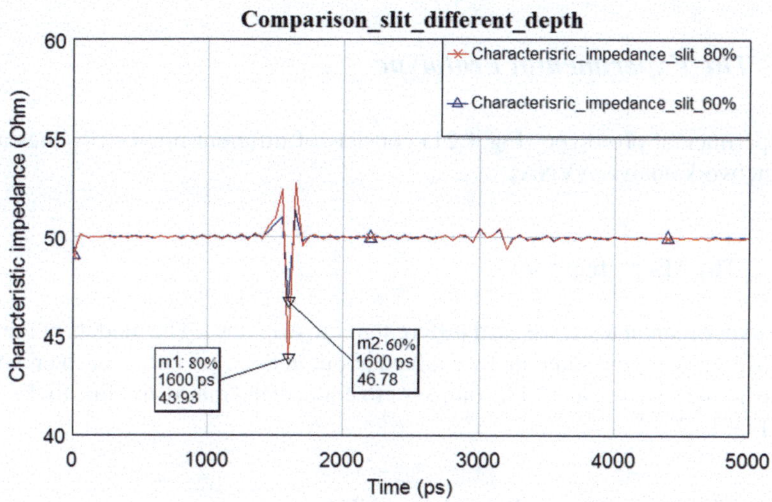

Fig. 1.20 Comparison between two microstrip lines with different depths of narrow transverse slits

In Fig. 1.20, we compare two microstrip lines with different depths of narrow transverse slit. We saw that when the depth of the narrow transverse slit increases, the characteristic impedance decreases.

When the depth of the narrow transverse slit equals to 60% of the width of the conducting strip, the characteristic impedance equals to 46.78 Ω but when it equals to 80% of the width of the conducting strip, the characteristic impedance equals to 43.93 Ω.

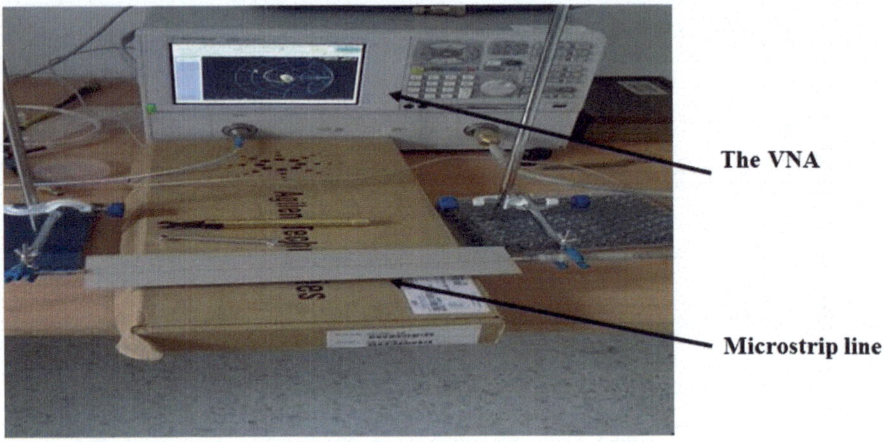

Fig. 1.21 The experimental prototype

1.9 The Experimental Results

1.9.1 The Experimental Prototype

The experimental prototype (Fig. 1.21) consists of different microstrip lines and a vector network analyzer(VNA).

1.9.1.1 The Microstrip Lines

For the experimental tests, we used different microstrip lines. The models of the used microstrip lines are the same that we talked about in Sect. 1.5 but, at each end of the microstrip lines, we put an SMA connector to connect the microstrip line to the VNA like in Fig. 1.22

1.9.1.2 The Vector Network Analyzer (VNA)

The vector network analyzer (VNA) is a measuring instrument that allows us to determine the parameters S (S11, S21) (Jamal Deen 2002).

The VNA is generally used for the two-port network characterization but can also be used for the multi-port network characterization (Giovanni Ghione 2018).

This measuring instrument is more used in the microwave and radio-frequency domains for the characterization of antennas, filters, and cables.

In our case, practical tests have been carried out in a frequency range, which varies between 10 MHz and 10 GHz. The operating principle of VNA is represented in Fig. 1.23.

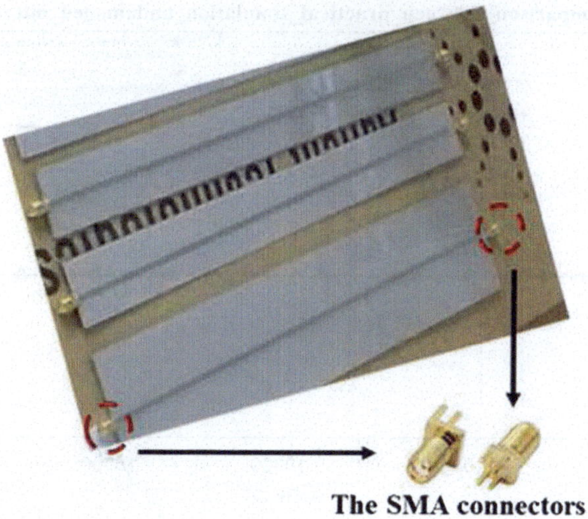

Fig. 1.22 The microstrip lines

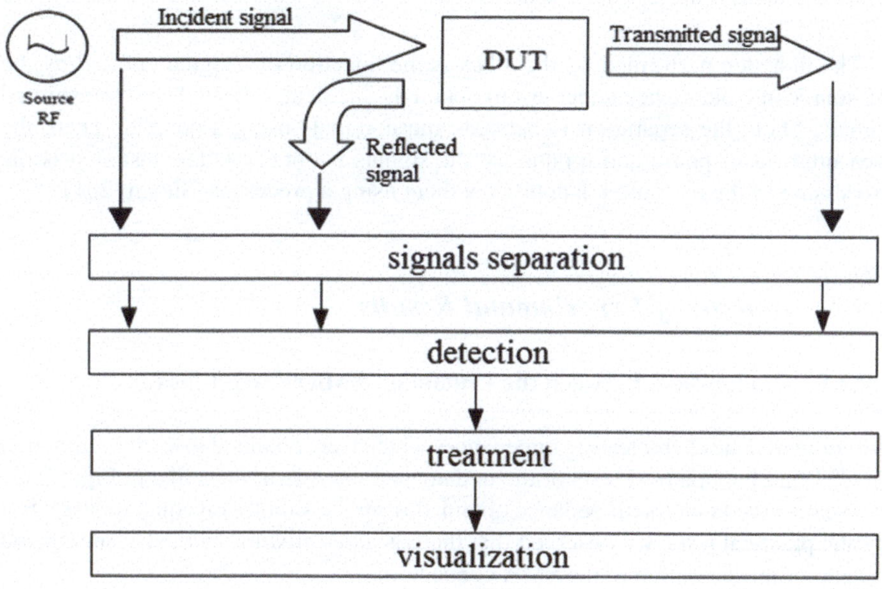

Fig. 1.23 The operating principle of VNA

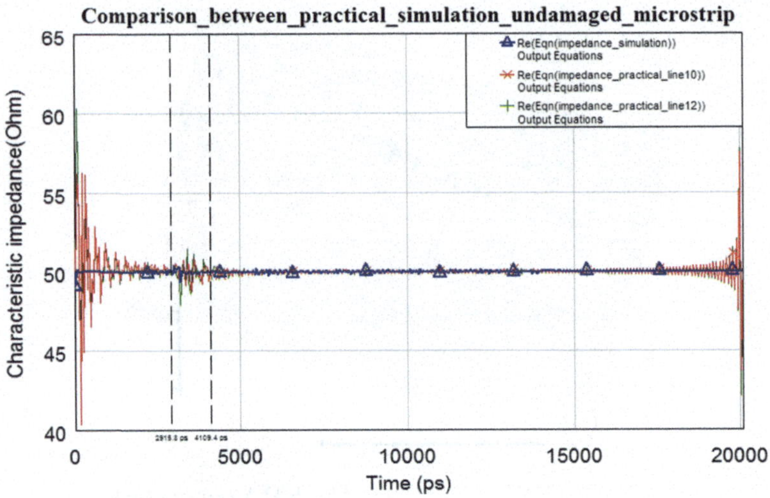

Fig. 1.24 Comparison between the undamaged microstrip lines (the blue curve: from simulation, the green and red curves: from the practical tests)

The first step performed by the VNA is the injection of a signal emitted by the RF source into our device under test to obtain the incident, reflected, and transmitted signals. Then, the separation is between these signals using a coupler. Then, the measurement of phase and module of the signals is obtained. The last step is the processing of the measured data to view them using a processor (Shoaib 2017).

1.9.2 Analysis of Experimental Results

1.9.2.1 Comparison Between the Undamaged Microstrip Lines

We compared the characteristic impedances which were obtained from the simulation on CST and the practical tests of the undamaged microstrip lines like in Fig. 1.24.

We observed that the impedance obtained from the simulation equal to 50 Ω. But for the practical tests, we observed that there is a perturbation in the beginning and the end of the curve due to the SMA connectors.

Also, between 2915.8 ps and 4109.4 ps, we saw a perturbation that indicates that the end of the microstrip line is reached in which its length is equal to 30 cm.

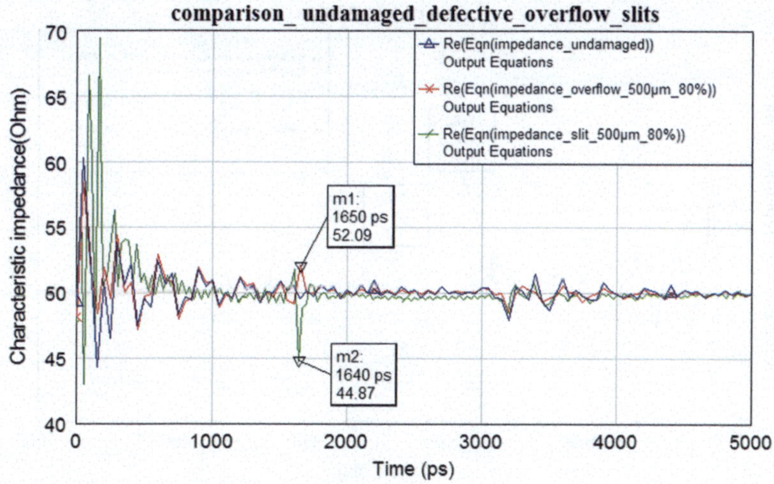

Fig. 1.25 Comparison between the undamaged and defective microstrip lines (the blue curve: the undamaged line, the red curve: the overflow, the green curve: the narrow transverse slits)

1.9.2.2 Comparison Between Undamaged and Defective Microstrip Lines

In models of the microstrip lines, we fixed the defects at the half of the microstrip line.

If we want to know the moment when the defects appear on the curves, we use Eqs. (1.58) and (1.59) that allow us to calculate the distance between the connectors SMA and the defects.

So, from the shape of curves and the theoretical results, we can know that the defects appear between 1640 and 1670 ps.

In Fig. 1.25, we compared the undamaged and the defective (overflow, narrow transverse slits) microstrip lines.

When we had an overflow, we have observed that the characteristic impedance increases (52.09 Ω). However, when we have a narrow transverse slit, the impedance decreases (44.87 Ω).

We used an impulse response low-pass filter because the filter type has a big influence on the results.

1.9.2.3 Comparison Between Microstrip Lines with Different Sizes of Overflows

Here, we compared microstrip lines that have overflowed with different sizes (two widths (500 μm, 250 μm) and two heights (80 and 60% of the width of the conducting strip)) like indicated in Figs. 1.26 and 1.27.

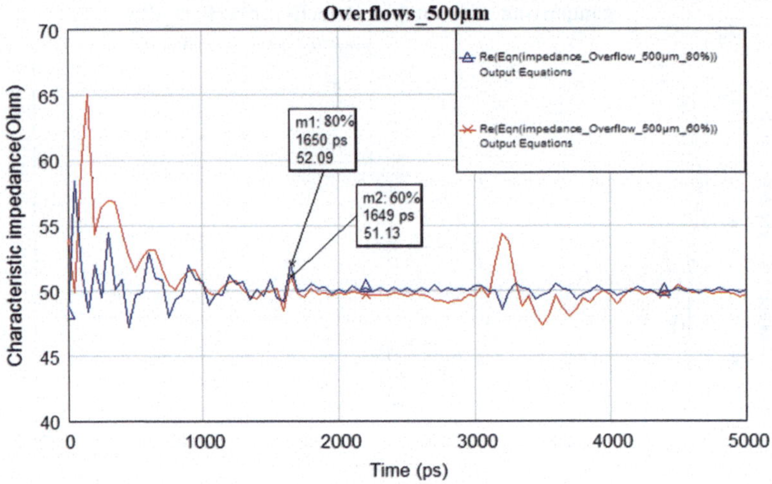

Fig. 1.26 Comparison between microstrip lines with different sizes of overflows (500 µm)

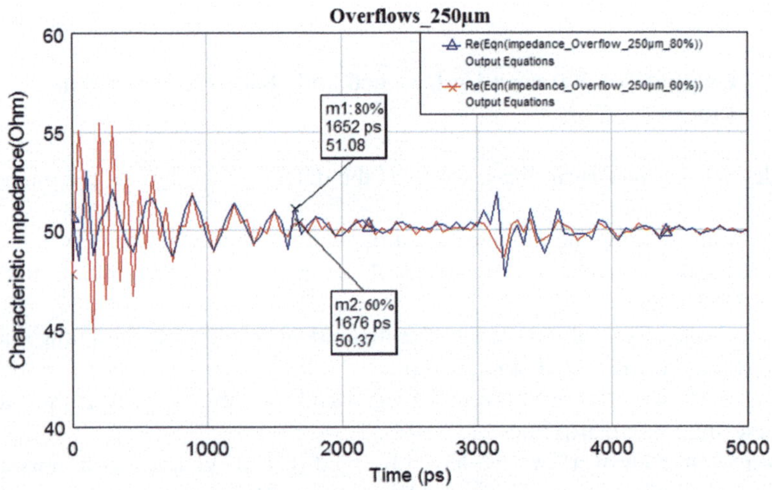

Fig. 1.27 Comparison between microstrip lines with different sizes of overflows (250 µm)

We have observed that the characteristic impedance of the microstrip line with the highest overflow is more important than the characteristic impedance of the microstrip line with the lowest overflow.

So, the more the height of the overflow increases, the more the characteristic impedance increases (Table 1.1).

Table 1.1 Characteristic impedance of microstrip lines with different sizes of overflows

Width (μm)	Characteristic impedance	
	The height is 60% of the width of the conducting strip (Ω)	The height is 80% of the width of the conducting strip (Ω)
250	50.37	51.08
500	51.13	52.09

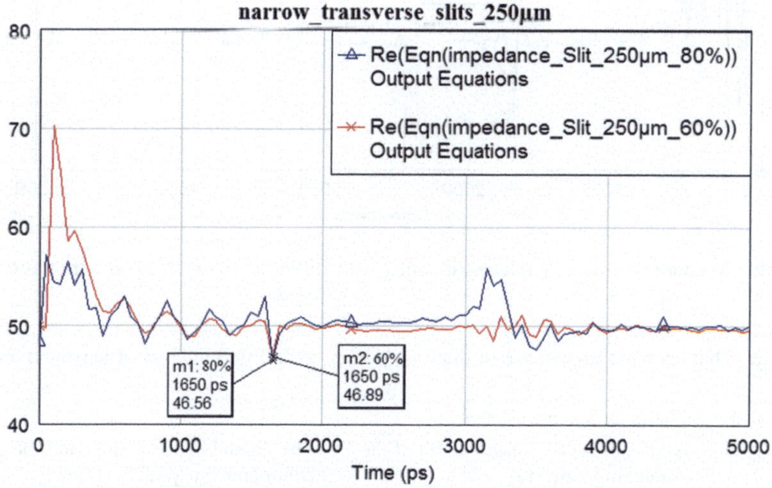

Fig. 1.28 Comparison between microstrip lines with different sizes of narrow transverse slits (250 μm)

1.9.2.4 Comparison Between Microstrip Lines with Different Sizes of Narrow Transverse Slits

Here, we compared microstrip lines that have narrow transverse slits with different sizes (two widths (500 μm, 250 μm) and two depths (80 and 60% of the width of the conducting strip)) as shown in Figs. 1.28 and 1.29.

We have observed that the characteristic impedance of the microstrip line that has the biggest narrow transverse slit is smaller than the characteristic impedance of the microstrip line that has the lowest narrow transverse slit.

So, the characteristic impedance decreases, when the depth of the narrow transverse slit increases (illustrated in Table 1.2).

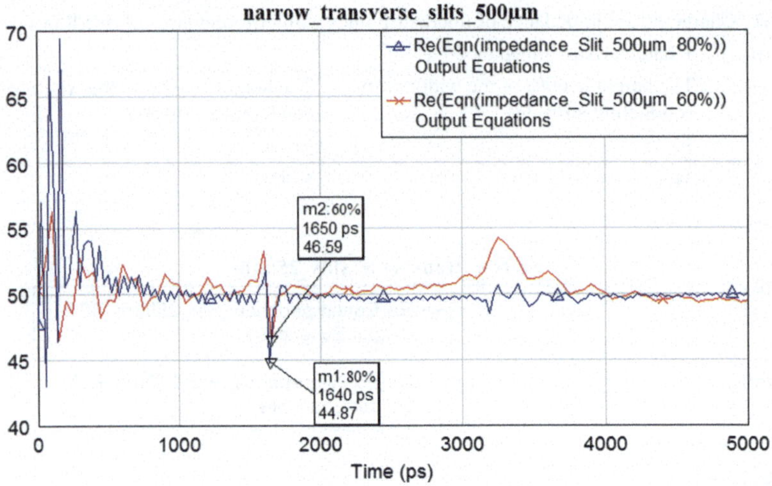

Fig. 1.29 Comparison between microstrip lines with different sizes of narrow transverse slits (500 μm)

Table 1.2 Characteristic impedance of microstrip lines with different sizes of narrow transverse slits

Width (μm)	Characteristic impedance	
	The depth is 60% of the width of the conducting strip (Ω)	The depth is 80% of the width of the conducting strip (Ω)
250	46.89	46.56
500	46.59	44.87

1.10 Conclusion

In this chapter, we presented an industrial nondestructive technique that can help us to identify and localize geometrical microscopic defects on microstrip line. There are many types of defects that can affect the microstrip line performance.

In our researches, we have worked on two types of defects: the narrow transverse slits and the overflows.

The presented solution is based on the measure of S parameters that will be used in the calculation of the characteristic impedance and the comparison between the simulation and the practical results. The analysis of the characteristic impedance curves allows us to detect and localize defects with enough precision. Moreover, unlike other techniques of test, we can identify the type of the defects.

In the case of overflows, we observe an increase of characteristic impedance and in the case of narrow transverse slits, there is a decrease of this impedance.

Acknowledgements This work is a part of a project proposed by the company APEM-SACELEC. A special thanks to Sys'Com, laboratory of ENIT, and ENSTA for their collaboration.

References

Alok Kumar, R., Munira, B., & Shanu, S. (2014). Design and simulation model for compensated and optimized t-junctions in microstrip line. *International Journal of Advanced Research in Computer Engineering and Technology, 3*, 4216–4220.

Auzanneau, F., Ravot, N., & Incarbone, L. (2016). Off-line fault localization technique on HVDC submarine cable via time-frequency domain reflectometry. *IEEE Sensors Journal, 16*, 8027–8034.

Bagad, V. S. (2008). *Microwaves and radar*. India: Technical Publications Pune.

Chandra, P., Dobkin, D., Bensky, D., Olexa, R., Lide, D., & Dowla, F. (2008). *Wireless networking: Know it all*. Amsterdam: Elsevier.

Chaturvedi, S., Bozanic, M., & Sinha, S. (2016). Transmission line parameters and effect of conductive substrates on their characteristics. *IRomanian Journal Of Information Science And Technology, 19*, 199–212.

Deen, M. J., & Fjeldly, T. A. (2002). *CMOS RF modeling characterization and applications*. USA: World Scientific.

Edwards, T. C., & Steer, M. B. (2016). *Foundations for microstrip circuit design*. United Kingdom: Wiley.

Farhat, R., Marcos, R., & Mario, P. (2016). *Electromagnetic time reversal: Application to EMC and power systems*. UK: Wiley.

Giovanni Ghione, M. P. (2018). *Microwave electronics*. UK: Cambridge University Press.

Gustrau, F. (2012). *RF and microwave engineering: Fundamentals of wireless communications*. UK: Wiley.

Hickman, I. (2007). *Practical RF handbook*. UK: Elsevier.

Hoefer, W. J. R. (1977). Equivalent series inductivity of a narrow transverse slit in microstrip. *MTT Transactions, 25*, 822–824.

Huang, Y., & Boyle, K. (2008). *Antennas: from theory to practice*. UK: Wiley.

Keqian Zhang, D. L. (2008). *Electromagnetic theory for microwaves and optoelectronic*. Germany: Springer.

Kwon, G.-Y., Lee, C.-K., Lee, G. S., Lee, Y. H., Chang, S. J., Jung, C.-K., et al. (2017). Chaos time domain reflectometry for online defect detection in noisy wired networks. *IEEE Transactions on Power Delivery, 32*(3), 1626–1635.

Natarajan, D. (2013). *A practical design of lumped, semi-lumped and microwave cavity filters*. Germany: Springer.

Raju, G. (2009). *Electromagnetic field theory and transmission lines*. India: Dorling Kindersley.

Robertson, I., Chongcheawchamnan, M., & Somjit, N. (2016). *Microwave and millimeter-wave design for wireless communications*. UK: Wiley.

Shoaib, N. (2017). *Vector network analyzer (VNA) measurements and uncertainty assessment*. Switzerland: Springer.

Terry, E., & Michael, S. (2016). *Foundations for microstrip circuit design*. UK: Wiley.

Tze Mei, K., Mohd, A., Ariffin, S. S., & Madelina, B. S. M. (2017). Improvement in cable defects assessment using time domain reflectometry technique. *Indian Journal of Science and Technology, 10*, 1–7.

Wen, C. (1969). Coplanar waveguide: A surface strip transmission line suitable for nonreciprocal gyromagnetic device applications. *IEEE Transactions on Microwave Theory and Techniques, 17*, 1087–1090.

Wolff, I., Kompa, G., & Mehran, R. (1972). Calculation method for microstrip discontinuities and t-junctions. *Electronics Letters, 8*, 177–179.

Chapter 2
Sensor Fault Detection and Isolation Based on Variable Moving Window KPCA

Radhia Fezai, Okba Taouali, Majdi Mansouri, Mohamed Faouzi Harkat and Nasreddine Bouguila

Abstract Kernel principal component analysis (KPCA) has been widely applied for fault detection. The time-varying property of industrial processes requires the adaptive ability of the KPCA model. This paper introduces online KPCA methods for fault detection. The moving window KPCA (MWKPCA) and variable moving window KPCA (VMWKPCA) methods update the KPCA model according to the process status. To locate the faulty sensor, a fault isolation algorithm should be carried out once a fault is detected. Thus, a partial VMWKPCA is proposed to achieve the fault isolation task. To demonstrate the performance of the proposed method, it is applied for fault diagnosis of air quality monitoring networks. The simulation results show that the proposed method effectively identifies the source of the fault.

Keywords Principal component analysis (PCA) · Kernel PCA (KPCA) · Fault detection · Moving window KPCA (MWKPCA) · Variable MWKPCA (VMWKPCA) · Fault isolation · Partial VMWKPCA

2.1 Introduction

Fault detection and isolation are important for the safe operation and control of a process. Multivariate statistical methods such as principal component analysis (PCA) (Sheriff et al. 2017; Dunia and Qin 1998), partial least squares (Joe Qin 2003;

R. Fezai (✉) · O. Taouali · N. Bouguila
Laboratory of Automatic Signal and Image Processing,
National School of Engineers of Monastir, University of Monastir, Monastir, Tunisia
e-mail: radhia.fazai@gmail.com

M. Mansouri
Electrical and Computer Engineering Program, Texas A&M University at Qatar, Doha, Qatar
e-mail: majdi.mansouri@qatar.tamu.edu

M. F. Harkat
Chemical Engineering Program, Texas A&M University at Qatar, Doha, Qatar
e-mail: mharkat@univ-annaba.org

N. Derbel et al. (eds.), *Diagnosis, Fault Detection & Tolerant Control*,
Studies in Systems, Decision and Control 269,
https://doi.org/10.1007/978-981-15-1746-4_2

Nomikos and MacGregor 1995), and independent component analysis (Lee et al. 2006; Widodo et al. 2007) have been developed and applied for this purpose. Among the statistical methods, a popular method used in industry is principal component analysis (Sheriff et al. 2017; Dunia and Qin 1998). PCA partitions the measurement space into a principal component subspace (PCS) and a residual subspace (RS). The PCS contains the normal or "common-cause" variability, while the RS contains noise that is normally uncorrelated. PCA transforms the original input variables into the independent variables in a new reduced space without losing generality of the system. It is one of the most popular modeling techniques applied to data-drivens of soft sensors.

However, PCA methods assume linear variable relationships in the process, which limit their application if these relationships are nonlinear. To cope with this problem, extended versions of PCA suitable for handling nonlinear systems have been developed (Kramer 1991; Dong and McAvoy 1996; Schölkopf et al. 1998; Nguyen and Golinval 2010; Lee et al. 2004b). One of these methods which has gained considerable interests in various research fields is kernel principal component analysis (KPCA) (Schölkopf et al. 1998; Nguyen and Golinval 2010; Lee et al. 2004b). The main idea of KPCA is first to map the input data into a feature space via a nonlinear function and then apply PCA in that feature space. By introducing a kernel function, the nonlinear mapping and the inner product computation can be avoided. Compared to other nonlinear methods, KPCA has the main advantage that no nonlinear optimization is involved. For KPCA, fault detection is performed with fault detection indices (Taouali et al. 2016; Fezai et al. 2018; Alcala and Qin 2010). Popular indices are the SPE index for monitoring the residual subspace and the Hotelling's T^2 statistic which monitors the principal component subspace. KPCA has been applied successfully for process monitoring and it demonstrated superior monitoring performance compared to linear PCA (Lee et al. 2004b). However, a major limitation of KPCA-based monitoring is that the KPCA model is time-invariant. Most real industrial processes are time-varying. The time-varying characteristics of industrial processes include changes in the mean, changes in the variance, changes in the correlation structure among variables and including changes in the number of significant principal components (PCs). To track the changing characteristics of the industrial process, attention should be paid more on the recent data. Moving window kernel principal component analysis (MWKPCA) is a popular method to adjust KPCA model with respect to the time-varying data (Liu et al. 2009; Jaffel et al. 2016). MWKPCA generates a new process model by including the newest kernel function and excluding the oldest one by moving a time window. However, the window has to cover a large number of samples in order to include sufficient process variation for modeling and monitoring purposes. On the contrary, a large window size can lead to significant reduction in computation time and when the system varies rapidly, a window containing too many outdated samples may result in the MWKPCA failing to follow exactly the dynamic process change. Furthermore, when a smaller window size is attempted to improve computational efficiency, data within the window then may not properly represent the underlying relationships between the process variables. An additional danger is that the resulting model may adapt to dynamic process changes so quickly that abnormal behavior remains undetected. Thus, choosing an appropriate moving window size depending on the variation of the dynamic process changes is an important issue. In this context, a variable moving window KPCA has been developed (Fazai et al. 2016). The VMWKPCA aims to update the KPCA model using a variable moving window size which is varied according to the dynamic changes of the process. After a fault is detected, it is necessary to isolate its provenance. Using KPCA technique, fault isolation is a much more difficult problem than in linear PCA and as a result, there has been little research into fault isolation (Alcala and Qin 2010; Cho et al. 2005). In this paper, a structured partial VMWKPCA for kernel PCA is proposed for online fault isolation. This method is based on the structured partial PCA approach to fault isolation proposed by Janos Gertler and Jin Cao for linear PCA (Gertler and Cao 2005).

The paper is structured as follows: Sect. 2.2 describes the Kernel principal component analysis. In Sect. 2.3, the online KPCA methods for fault detection are presented. Extension of the partial PCA method to the nonlinear case based on the VMWKPCA technique is addressed in Sect. 2.4. The performance of the proposed method is illustrated through an air quality monitoring network AIRLOR process in Sect. 2.5. The paper concludes with Sect. 2.6.

2.2 Kernel Principal Component Analysis

2.2.1 Identification of the KPCA Model

PCA is a simple linear transformation technique that compresses high-dimensional data with minimum loss of data information. PCA is performed in the original sample space, whereas kernel PCA is carried out in the extended feature space. To derive KPCA, we first map the input data $x(k) \in \mathbf{R}^m$ $k = 1, \ldots, N$ into a feature space F where N is the number of samples and m is the number of process variables using a nonlinear function ϕ. Linear PCA is then executed in F. An important property of the feature space is that the dot product of two vectors $\phi(x(i))$ and $\phi(x(j))$ can be determined as a function of the corresponding vectors $x(i)$ and $x(j)$ as follows:

$$\phi^T[x(i)]\phi[x(j)] = k[x(i), x(j)] \tag{2.1}$$

The function $k(.,.)$ is called the kernel function, and there exist several types of these functions. The popularly used kernel functions (Lee et al. 2004a) are as follows:

1. Linear kernel

$$k(x, y) = < x, y >$$

2. Polynomial kernel

$$k(x, y) = (< x, y > +1)^d$$

where d is a positive integer

3. Radial basis function (RBF)

$$k(x, y) = exp\left[-\frac{\|x - y\|^2}{\sigma^2}\right]$$

where σ is the parameter that controls the width of RBF.

4. Sigmoid kernel:

$$k(x, y) = tanh(\beta_0 < x, y > +\beta_1)$$

In this study, the RBF will be used.

In the feature space, assuming that the vectors are scaled to zero mean, the mapped data in F is arranged as follows:

$$\mathcal{X} = \left[\phi[x(1)]\ \phi[x(2)]\ \ldots\ \phi[x(N)]\right]^T \tag{2.2}$$

In kernel PCA, the covariance matrix Q of the mapped training data $\mathcal{X}$ is defined as follows:

$$Q = \frac{1}{N}\sum_{i=1}^{N} \phi[x(i)]\ \phi[x(i)]^T \tag{2.3}$$

To find the principal components, one has to solve the following eigenvector equation:

$$\begin{aligned}\mathcal{X}^T\mathcal{X}v &= \sum_{i=1}^{N} \phi[x(i)]\phi[x(i)]^T v \\ &= \lambda v\end{aligned} \tag{2.4}$$

where λ and v denote, respectively, the eigenvalue and the eigenvector of the covariance matrix Q.

In fact, ϕ is usually hard to obtain. To avoid this problem, a kernel matrix K is defined as follows:

$$K = \mathcal{X}\mathcal{X}^T = \begin{bmatrix} k[x(1), x(1)] & \cdots & k[x(1), x(N)] \\ \vdots & \ddots & \vdots \\ k[x(N), x(1)] & \cdots & k[x(N), x(N)] \end{bmatrix} \tag{2.5}$$

Premultiplying Eq. (2.4) by $\mathcal{X}$, the eigenvector problem can be expressed as follows:

$$\mathcal{X}\mathcal{X}^T\mathcal{X}v = \lambda\mathcal{X}v \tag{2.6}$$

Using the so-called kernel trick, Eq. (2.6) can be rewritten as follows:

$$K\mathcal{X}v = \lambda\mathcal{X}v \tag{2.7}$$

Let us define α as follows:

$$\alpha = \mathcal{X}v \tag{2.8}$$

By substituting Eq. (2.8) in Eq. (2.7), the eigenvector problem can be represented by the following simple form:

$$K\alpha = \lambda\alpha \tag{2.9}$$

where α and λ are an eigenvector and eigenvalue of K.

By multiplying with $\mathcal{X}$ from the left of both sides in Eq. (2.8), we obtain

$$\mathcal{X}^T\alpha = \mathcal{X}^T\mathcal{X}v = \lambda v \tag{2.10}$$

Therefore, the eigenvector v can be expressed as follows:

$$v = \lambda^{-1}\mathcal{X}^T\alpha \tag{2.11}$$

To determine the KPCA model (λ_i and v_i), we first perform eigendecomposition of Eq. (2.9) to find λ_i and v_i. Then use Eq. (2.11) to obtain v_i. In order to assure that $v_i^T v_i = 1$, Eqs. (2.8), (2.10), and (2.11) are used to derive the following:

$$\begin{aligned} v_i^T v_i &= \lambda_i^{-2}\alpha_i^T\mathcal{X}\mathcal{X}^T\alpha_i \\ &= \lambda_i^{-2}\alpha_i^T\mathcal{X}\mathcal{X}^T\mathcal{X}v_i \\ &= \lambda_i^{-2}\alpha_i^T\lambda_i\mathcal{X}v_i \\ &= \lambda_i\alpha_i^T\alpha_i \\ &= 1 \end{aligned} \tag{2.12}$$

Therefore, the eigenvector α_i needs to have a norm of $\sqrt{\lambda_i}$. Let α_i^0 be the unit norm eigenvector corresponding to λ_i, such that

$$\alpha_i = \sqrt{\lambda_i}\alpha_i^0 \tag{2.13}$$

In the feature space F, the ℓ first eigenvectors v_i form a matrix denoted as $\hat{P}_f = \begin{bmatrix} v_1 \; v_2 \; \ldots \; v_\ell \end{bmatrix}$. Using Eq. (2.11), the matrix P_f can be written as follows:

$$P_f = \left[\frac{1}{\lambda_1}\mathcal{X}^T\alpha_1 \; \frac{1}{\lambda_2}\mathcal{X}^T\alpha_2 \cdots \frac{1}{\lambda_\ell}\mathcal{X}^T\alpha_\ell \right] \tag{2.14}$$

Inserting Eq. (2.11) into Eq. (2.14), the matrix P_f can be rewritten as follows:

$$P_f = \left[\mathcal{X}^T\alpha_1^0\lambda_1^{-\frac{1}{2}} \; \mathcal{X}^T\alpha_2^0\lambda_2^{-\frac{1}{2}} \ldots \mathcal{X}^T\alpha_\ell^0\lambda_\ell^{-\frac{1}{2}} \right] \tag{2.15}$$

The matrix P_f can be represented by the following form:

$$P_f = \mathcal{X}^T P \Lambda^{-\frac{1}{2}} \tag{2.16}$$

where $P = \begin{bmatrix} \alpha_1^0 \ \alpha_2^0 \cdots \alpha_\ell^0 \end{bmatrix}$ and $\Lambda = diag(\lambda_1 \cdots \lambda_\ell)$ are the principal eigenvectors and eigenvalues of the matrix K.

The principal components t of a measurement x are then extracted by projecting $\phi(x)$ into principal space:

$$t = P_f^T \phi(x) \tag{2.17}$$

According to Eq. (2.17), the principal components t can be expressed as follows:

$$t = \Lambda^{-\frac{1}{2}} P^T \mathcal{X} \phi(x) \tag{2.18}$$

The vector $k(x)$ is defined by the following expression:

$$\begin{aligned} k(x) &= \mathcal{X}\phi(x) \\ &= \begin{bmatrix} \phi[x(1)] \ \phi[x(2)] \cdots \phi[x(N)] \end{bmatrix}^T \phi(x) \\ &= \begin{bmatrix} k[x(1), x] \ k[x(2), x] \cdots k[x(N), x] \end{bmatrix}^T \end{aligned} \tag{2.19}$$

Integrating Eq. (2.19) into Eq. (2.18), we obtain

$$t = \Lambda^{-\frac{1}{2}} P^T k(x) \tag{2.20}$$

2.2.2 *Scaling*

In the feature space F, it is preferred to have centered data. Thus, the vector $\phi(x)$ is scaled as follows:

$$\begin{aligned} \overline{\phi}(x) &= \phi(x) - \frac{1}{N}\sum_{i=1}^{N} \phi[x(i)] \\ &= \phi(x) - \begin{bmatrix} \phi[x(1)] \ \phi[x(2)] \ldots \phi[x(N)] \end{bmatrix} I_N \end{aligned} \tag{2.21}$$

where $I_N = \frac{1}{N} \begin{bmatrix} 1 \ldots 1 \end{bmatrix}^T \in \mathbf{R}^N$.

The scaled kernel function $\overline{k}[x(i), x(j)]$ of two scaled vectors $\overline{\phi}[x(i)]$ and $\overline{\phi}[x(j)]$ is given by

$$\begin{aligned} \overline{k}[x_i, x_j] &= \overline{\phi}[x(i)]^T \overline{\phi}[x(j)] \\ &= k[x(i), x(j) - k^T(x(i))] I_N - k^T[x(j)] I_N + I_N^T K I_N \end{aligned} \tag{2.22}$$

Using Eq. (2.22), the centered kernel vector $\overline{k}(x)$ is determined by

$$\begin{aligned}\overline{k}(x) &= \left[\overline{\phi}(x(1)) \ldots \overline{\phi}(x(N))\right]^T \overline{\phi}(x) \\ &= F[k(x) - KI_N]\end{aligned} \tag{2.23}$$

where $F = I - E$, I is the identity matrix and $E \in \mathbf{R}^{N\times N}$ is a matrix with elements $1/N$.

Finally, the scaled kernel matrix $\bar{K}$ is determined by

$$\begin{aligned}\overline{K} &= \left[\overline{\phi}(x_1) \ldots \overline{\phi}(x_N)\right]^T \left[\overline{\phi}(x_1) \ldots \overline{\phi}(x_N)\right] \\ &= FKF\end{aligned} \tag{2.24}$$

2.2.3 *Fault Detection Indices*

Fault detection with KPCA is performed using fault detection indices. In this case, a fault is detected when the fault detection statistic is beyond its control limit. The most used indices for fault detection are the squared prediction error (SPE), known as the Q statistic and Hotelling's T^2 statistic. The SPE and T^2 statistics are constructed and monitored based on the assumption that the training data have a multivariate normal distribution in the feature space.

2.2.3.1 The Squared Prediction Error (SPE)

The SPE index measures the projection of the sample vector on the residual subspace and it is expressed as follows:

$$SPE(x) = \widetilde{t}^T\widetilde{t} \tag{2.25}$$

where $\widetilde{t} = \widetilde{P}_f^T\overline{\phi}(x) = \left[\, v_{\ell+1} \; v_{\ell+2} \ldots \,\right]^T \overline{\phi}(x)$ is the residual components and $\widetilde{P}_f$ represent the eigenvectors defining the residual space (RS).

Substituting the expression of $\widetilde{t}$ into Eq. (2.25), we get

$$SPE(x) = \overline{\phi}^T(x)\widetilde{P}_f\widetilde{P}_f^T\overline{\phi}(x) \tag{2.26}$$

Since we do not know the dimension of the feature space F, it is not possible to know the number of residual components there. Therefore, we cannot determine explicitly the matrix $\widetilde{P}_f$. However, we can compute the product $\widetilde{P}_f\widetilde{P}_f^T$ as the projection orthogonal to the principal component space, which is given by

$$\widetilde{C}_f = \widetilde{P}_f\widetilde{P}_f^T = I - P_fP_f^T \tag{2.27}$$

According to Eq. (2.27), the SPE index can be rewritten as follows:

$$\begin{aligned} SPE(x) &= \overline{\phi}^T(x)(I - P_f P_f^T)\overline{\phi}(x) \\ &= \overline{\phi}^T(x)\overline{\phi}(x) - \overline{\phi}^T(x) P_f P_f^T \overline{\phi}(x) \end{aligned} \tag{2.28}$$

From Eqs. (2.1), (2.19), the SPE is calculated as follows:

$$SPE(x) = \overline{k}(x, x) - \overline{k}^T(x) P \Lambda^{-1} P^T \overline{k}(x) \tag{2.29}$$

The process is considered in normal operation if the following condition is satisfied:

$$SPE(x) < SPE_{lim} \tag{2.30}$$

The control limit of the SPE index is determined from the χ^2-distribution and it is defined as (Zhang et al. 2012) follows:

$$SPE_{lim} \sim g\chi_h^2 \tag{2.31}$$

where $g = \dfrac{b}{2a}$ and $h = \dfrac{2a^2}{b}$, where a and b are the mean and variance of the SPE statistic.

2.2.3.2 The Hotelling's T^2 Statistic

The Hotelling's T^2 statistic is used to measure the variation of a process in the principal component subspace. The T^2 statistic is the sum of normalized squared scores, and is defined as follows:

$$T^2(x) = t^T \Lambda^{-1} t \tag{2.32}$$

From Eq. (2.19), the T^2 statistic of an observation x is expressed as follows:

$$\begin{aligned} T^2(x) &= \overline{\phi}(x)^T \hat{P}_f \Lambda^{-1} \hat{P}_f^T \overline{\phi}(x) \\ &= \overline{k}(x)^T P \Lambda^{-2} P^T \overline{k}(x) \\ &= \overline{k}(x)^T D \overline{k}(x) \end{aligned} \tag{2.33}$$

where $D = P\Lambda^{-2}P^T$.

The process is considered normal if Eq. (2.34) is satisfied:

$$T^2(x) < T_\alpha^2 \tag{2.34}$$

where T_α^2 is the confidence limit for T^2.

The confidence limit for T^2 can be approximated by

$$T_{\alpha}^{2} = \frac{1}{N-1}\chi_{\ell,\alpha}^{2} \tag{2.35}$$

where $\chi_{\ell,\alpha}^{2}$ is a distribution of χ^2 with ℓ degrees of freedom and a confidence level of $(1-\alpha)\times 100\%$.

2.3 Online KPCA Methods for Fault Detection

2.3.1 Moving Window Kernel PCA (MWKPCA)

When slow and natural process changes occur in the processes, it is preferred to update the KPCA model by a moving window because the old kernel function cannot represent the current status of the process. That is, the newest kernel function is added to the transformed data matrix and the oldest one is discarded, keeping a fixed number of kernel functions in the transformed data matrix.

Let the data matrix with window size L, at instant k, be $X(k)$:

$$X(k) = \left[\, x(k)\ x(k+1)\ \ldots\ x(k+L-1) \,\right]^{T} \in \mathbb{R}^{L\times m} \tag{2.36}$$

The transformed data matrix $\Phi(X(k))$, at instant k, is defined by

$$\Phi[X(k)] = \left[\ \phi[x(k)]\ \phi[x(k+1)]\ \ldots\ \phi[x(k+L-1)]\ \right]^{T} \in \mathbb{R}^{L\times h} \tag{2.37}$$

The mean vector M_k of the transformed data matrix $\Phi(X(k))$ is given by

$$M_k = \frac{1}{L}\sum_{i=k}^{k+L-1}\phi[x(i)] \tag{2.38}$$

Therefore, the centered matrix $\overline{\Phi}(X(k))$ is determined as follows:

$$\overline{\Phi}(X(k)) = \Phi[X(k)] - I_L M_k \tag{2.39}$$

where $I_L = \left[\, 1\ 1\ \ldots\ 1 \,\right]^{T} \in \mathbb{R}^{L}$.

The scaled kernel matrix $\overline{K}_k$ is calculated as follows:

$$\overline{K}_k = \begin{bmatrix} \overline{k}[x(k), x(k)] & \cdots & \overline{k}[x(k), x(k+L-1)]] \\ \vdots & \ddots & \vdots \\ \overline{k}[x(k+L-1), x(k)] & \cdots & \overline{k}[x(k+L-1), x(k+L-1)] \end{bmatrix} \tag{2.40}$$

The procedure of updating a moving window takes place in two steps: the first step is to discard the oldest kernel function and the second step is characterized by adding newest sample to the transformed data matrix.

Case 1: $\phi[X(k)] \rightarrow \phi[X(k, k+1)]$

The result of removing the oldest kernel function $\phi(x(k))$ from the transformed data matrix $Phi(X(k))$ is given by

$$\phi[X(k, k+1)] = \begin{bmatrix} \phi[x(k+1)] \dots \phi[x(k+L-1)] \end{bmatrix}^T \in \mathbb{R}^{(L-1)\times h} \tag{2.41}$$

where $X(k, k+1) = \begin{bmatrix} x(k+1) \dots x(k+L-1) \end{bmatrix}^T \in \mathbb{R}^{(L-1)\times m}$ is the intermediate data matrix in the input space.

The mean vector $M_{k,k+1}$ of the intermediate transformed data matrix can be expressed as a function of the mean vector M_k by eliminating the impact of the oldest kernel function $\phi(x(k))$:

$$M_{k,k+1} = \frac{1}{L-1}\big(LM_k - \phi[x(k)]\big) \tag{2.42}$$

Then, the centered transformed data matrix $\overline{\phi}[X(k, k+1)]$ is defined as follows:

$$\overline{\Phi}[X(k, k+1)] = \Phi[X(k, k+1)] - I_{L-1}M_{k,k+1} \tag{2.43}$$

where I_{L-1} is a vector of size $L-1$ whose elements are equal to 1.

In this case, the centered gram matrix of the intermediate transformed data matrix is determined by eliminating the first row and the first column:

$$\overline{K}_{k,k+1} = \begin{bmatrix} \overline{k}[x(k+1), x(k+1)] & \cdots & \overline{k}[x(k+1), x(k+L-1)] \\ \vdots & \ddots & \vdots \\ \overline{k}[x(k+L-1), x(k+1)] & \cdots & \overline{k}[x(k+L-1), x(k+L-1)] \end{bmatrix} \tag{2.44}$$

Case 2: $\phi[X(k, k+1)] \rightarrow \phi[X(k+1)]$

By adding the new kernel function $\phi[x(k+L)]$ to the transformed data matrix $\phi[X(k, k+1)]$, the resulting transformed data matrix is given by

$$\phi[X(k+1)] = \begin{bmatrix} \phi[x(k+1)] \dots \phi[x(k+L)] \end{bmatrix}^T \in \mathbb{R}^{(L)\times h} \tag{2.45}$$

where $X(k+1) = \begin{bmatrix} x(k+1) \dots x(k+L) \end{bmatrix}^T \in \mathbb{R}^{L\times m}$.

The mean vector of the new transformed data matrix is defined as follows:

$$M_{k+1} = \frac{1}{L}\Big((L-1)M_{k,k-1} + \phi[x(k+L)]\Big) \tag{2.46}$$

According to Eq. (2.42), Eq. (2.46) is written as follows:

$$M_{k+1} = M_k + \frac{1}{L}\Big(\phi[x(k+L)] - \phi[x(k)]\Big) \tag{2.47}$$

The centered matrix $\overline{\phi}[x(k+L)]$ is determined by:

$$\overline{\phi}[X(k+1)] = \phi[X(k+1)] - I_L M_{k+1} \tag{2.48}$$

where I_L is the identity matrix whose dimension equal to L. Therefore, the resulting centered gram matrix $\overline{K}_{k+1}$ is determined by adding a row and a column to the matrix $\overline{K}_{k,k+1}$:

$$\overline{K}_{k+1} = \begin{bmatrix} K_{k,k+1} & a \\ a^T & b \end{bmatrix} \in \mathbb{R}^{L\times L} \tag{2.49}$$

where $a^T = \Big[\overline{k}[x(k+1), x(k+L)] \ldots \overline{k}[x(k+L-1), x(k+L)]\Big]^T \in \mathbb{R}^{1\times L-1}$ end $b = \overline{k}[x(k+L), x(k+L)]$.

2.3.2 *Variable Moving Window Kernel PCA (VMWKPCA)*

The variable moving window kernel PCA (VMWKPCA) updates the KPCA model by using a variable moving window size. The variation of the window size depends on the dynamic changes of the process. Thus, when the dynamic changes of the process are rapid, the window size should be small. Conversely, when the dynamic variation becomes slow, the window size should be large. To choose an optimal size of the moving window size, it has been developed to update it as follows:

$$L_{k+1} = L_{min} + (L_{max} - L_{min})\,\exp\left(-\alpha\frac{\|\Delta M_k\|}{\|\Delta M_0\|} + \beta\frac{\|\Delta \overline{K}_k\|}{\|\Delta \overline{K}_0\|}^{\gamma}\right) \tag{2.50}$$

where L_{max} and L_{min} are, respectively, the maximum and minimum values of the moving window,

α, β, and γ are three parameters determined previously by the user,

$\|\Delta M_k\| = \|M_k - M_{k-1}\|$ is an Euclidean vector norm of the difference between the mean vectors M_{k+1} and M_k,

$\|\Delta \overline{K}_k\| = \|\overline{K}_k - \overline{K}_{k-1}\|$ is the Euclidean matrix norm of the difference between the kernel matrices $\overline{K}_{k+1}$ and$\overline{K}_k$,

$\|\Delta M_0\|$ and $\|\Delta \overline{K}_0\|$ are the averages of $\|\Delta M\|$ and $\|\Delta \overline{K}\|$ determined by the training data set.

At time instant k, let $\phi(X(k)) = \left[\phi(x(k)) \ldots \phi(x(k+L_k-1))\right]^T \in \mathbb{R}^{(L_k)\times h}$, where $L_k$is the moving window size, be the transformed data matrix under normal operation.

The centered data matrix $\overline{\Phi}[X(k)]$ in the feature space is given by

$$\overline{\phi}(X(k)) = \phi[X(k)] - I_{L_k} M_k \tag{2.51}$$

where $I_{L_k} = \left[1 \ldots 1\right]^T \in \mathbb{R}^{L_k}$ and $M_k = \frac{1}{L_k} \sum_{i=k}^{k+L_k-1} \phi[x(i)]$ is the mean vector of the matrix $\Phi[X(k)]$ at instant k.

The scaled kernel matrix $\overline{K}_k$ is calculated as follows:

$$\overline{K}_k = \begin{bmatrix} \overline{k}[x(k), x(k)] & \cdots & \overline{k}[x(k), x(k+L_k-1)] \\ \vdots & \ddots & \vdots \\ \overline{k}[x(k+L_k-1), x(k)] & \cdots & \overline{k}[x(k+L_k-1), x(k+L_k-1)] \end{bmatrix} \tag{2.52}$$

The task for the VMWKPCA is to update the KPCA model depending on the moving window size. Therefore, according to the value of L_{k+1}, two cases may arise.

Case 1: If $L_k \geq L_{k+1}$

In this case, the VMWKPCA method consists first to discard the oldest data set $L_k - L_{k+1}$ from the matrix $\Phi(X(k))$, and then to add the new data $\phi(x(k+L_k))$ to the intermediate matrix $\Phi(X_{k,k+1})$. The resulting matrix after discarding the oldest kernel functions from $\Phi(X(k))$ is named as $\Phi(X_{k,k+1})$ and it is given by

$$\phi(X_{k,k+1}) = \left[\phi[x(k+L_k-L_{k+1})] \ldots \phi[x(k+L_k-1)]\right]^T \in \mathbb{R}^{(L_{k+1})\times h} \tag{2.53}$$

The mean vector of the matrix $\Phi(X_{k,k+1})$ can be expressed as follows:

$$M_{k,k+1} = \frac{1}{L_{k+1}} \left(L_k M_k - \sum_{i=k}^{k+L_k-L_{k+1}} \phi[x(i)] \right) \tag{2.54}$$

Then, the centered data matrix $\overline{\phi}[X(k, k+1)]$ in the feature space is given by

$$\overline{\Phi}[X(k, k+1)] = \phi[X(k, k+1)] - I_{L_{k+1}} M_{k,k+1} \tag{2.55}$$

where $I_{L_{k+1}} = \left[1 \ldots 1\right]^T \in \mathbb{R}^{L_{k+1}}$.

Therefore, the centered kernel matrix $\overline{K}_{k,k+1}$ can be represented by the elimination of the first $L_k - L_{k+1}$ rows and columns of the matrix $\overline{K}_k$:

$$\overline{K}_{k,k+1} = \begin{bmatrix} \overline{k}[x(k+L_k-L_{k+1}), x(k+L_k-L_{k+1})] & \cdots & \overline{k}[x(k+L_k-L_{k+1}), x(k+L_k-1)] \\ \vdots & \ddots & \vdots \\ \overline{k}[x(k+L_k-1), x(k+L_k-L_{k+1})] & \cdots & \overline{k}[x(k+L_k-1), x(k+L_k-1)] \end{bmatrix} \tag{2.56}$$

The result of adding the new function $\phi([(k+L_k)]$ can be represented by

$$\Phi(X_{k+1}) = \left[\phi[x(k+L_k-L_{k+1})] \ldots \phi[x(k+L_k)]\right]^T \in \mathbb{R}^{(L_{k+1}+1)\times h} \tag{2.57}$$

From Eq. (2.54), the mean vector M_{k+1} can be updated recursively as follows:

$$M_{k+1} = \frac{1}{L_{k+1}+1}\left(L_k M_k - \sum_{i=k}^{k+L_k-L_{k+1}} \phi[x(i)] + \phi[x(k+L_k)]\right) \tag{2.58}$$

According to Eq. (2.58), the centered data matrix in $\mathbb{F}$ is determined by

$$\overline{\Phi}[X(k+1)] = \Phi[X(k+1)] - I_{L_{k+1}+1}M_{k+1} \tag{2.59}$$

where $I_{L_{k+1}+1} = \left[1 \ldots 1\right]^T \in \mathbb{R}^{(L_{k+1}+1)}$.

The resulting gram matrix $\overline{K}_{k+1}$, at instant $k+1$ is given by

$$\overline{K}_{k+1} = \begin{bmatrix} \overline{K}_{k,k+1} & a \\ a^T & b \end{bmatrix} \in \mathbb{R}^{(L_{k+1}+1)\times(L_{k+1}+1)} \tag{2.60}$$

where

$$a^T = \left[\overline{k}[x(k+L_k-L_{k+1}), x(k+L_k)] \cdots \overline{k}[x(k+L_k-1), x(k+L_k)]\right]^T$$

where $a \in \mathbb{R}^{L_{k+1}\times 1}$ and $b = \overline{k}[x(k+L_k), x(k+L_k)]$.

Case 2: If $L_k < L_{k+1}$

In this case, only perform the updating procedure and do not carry out the downdating procedure. Thus, the matrix $\Phi[X(k+1)]$ is determined by adding the new kernel function $\phi[x(k+L_k)]$ to the data matrix $\Phi(X(k,k+1))$:

$$\Phi[X(k+1)] = \left[\phi[x(k)] \cdots \phi[x(k+L_k-1)] \; \phi[x(k+L_k)]\right]^T \in \mathbb{R}^{(L_k+1)\times h} \tag{2.61}$$

At instant $k+1$, the mean vector M_{k+1} is updated as follows:

$$M_{k+1} = \frac{1}{L_k+1}[L_k M_k + \phi[x(k+L_k)]] \tag{2.62}$$

According to Eq. (2.52), the gram matrix is updated using the following equation:

$$\overline{K}_{k+1} = \begin{bmatrix} \overline{K}_k & a \\ a^T & b \end{bmatrix} \in \mathbb{R}^{(L_k+1)\times(L_k+1)} \tag{2.63}$$

where

$$a^T = \left[\bar{k}[x(k), x(k+L_k)] \cdots \bar{k}[x(k+L_k-1), x(k+L_k)] \right]^T$$

with $a \in \mathbb{R}^{1\times(L_k+1)}$
and $b = \bar{k}[x(k+L_k), x(k+L_k)]$.

2.4 Fault Isolation

After a fault has been detected, it is important to isolate faulty sensor. In this study, we propose to use the partial VMPKPCA to isolate the faulty sensor. The aim of the proposed method is to perform a VMWKPCA on a reduced vector $x^r(k)$, where some variables in $x(k)$ are missing. Thus, the residual will only be sensitive to faults associated with the variables which are present in the reduced vector $x^r(k)$. Faults associated with variables discarded from the partial VMWKPCA will leave the SPE index within the nominal thresholds. The residual is generated from a strongly isolating matrix. The incidence matrix is presented in binary form whose lines indicate the structure of the residuals and the columns represent the signatures of the faults. The value "0" at the intersection means that the residual is insensitive to the faults. In addition, the value "1" indicates that the residual is sensitive to the fault.
To build an incidence matrix, it is necessary that

1. The number of zeros contained in each column is equal to that contained in each line.
2. The number of lines is equal to the number of faults. The incidence matrix is strongly isolating if no column can be obtained from another by replacing "1"s into "0"s.

The incidence matrix is strongly isolating if, no column can be obtained from another by replacing "1"s into "0"s.
The procedure of achieving a structured partial VMWKPCA set is as follows:

1. Perform a VMWKPCA;
2. Construct a strongly isolable incidence matrix;
3. Perform a set of partial VMWKPCAs with each one implementing a row of the incidence matrix;
4. Determine the thresholds beyond which abnormality is indicated.

This procedure is also illustrated in Fig. 2.1.

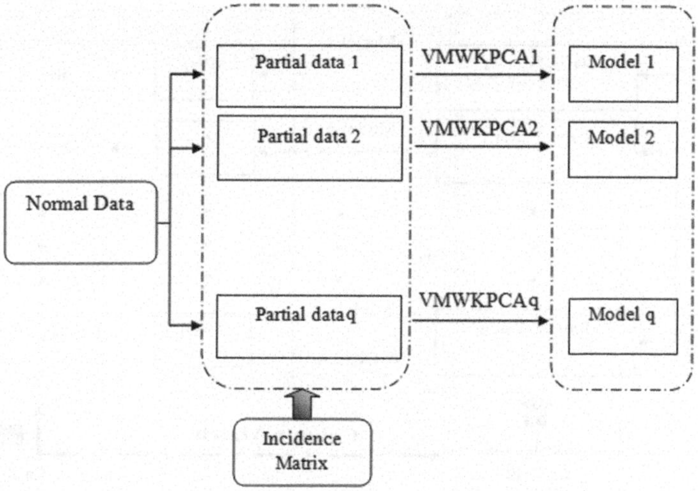

Fig. 2.1 The modeling procedure of structured partial VMWKPCA

After the structured partial VMWKPCA subspace set is obtained, it can be used for online fault isolation.
The fault isolation procedure can be done online, for each time:

1. Run the observed data against each partial VMWKPCA and compute the SPE indices;
2. Compare the SPE indices to appropriate thresholds and from the fault code SE_i according to

$$SE_i = \begin{cases} 0 & \texttt{if } SPE_i \le SPE_{lim,i} \\ 1 & \texttt{if } SPE_i > SPE_{lim,i} \end{cases} \tag{2.64}$$

3. Compare the experimental signature of the fault to the columns of the incidence matrix to determine the faulty variables.

The online fault isolation procedure by structured partial VMWKPCA is also illustrated in Fig. 2.2.

2.5 Application to an Air Quality Monitoring Network

In this section, the VMWKPCA technique is applied for fault detection and the partial VMWKPCA method is applied for fault isolation of air quality monitoring network (Harkat et al. 2006, 2009, 2018).

During the study, the air quality monitoring network AIRLOR working in Lorraine, (France), consists of 20 stations placed in rural, peri-urban, and urban sites.

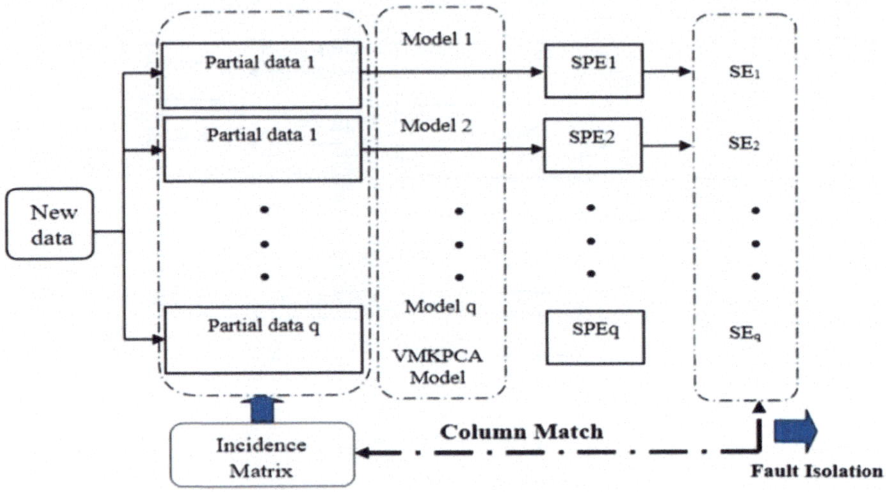

Fig. 2.2 The fault isolation procedure by structured partial VMWKPCA

Each monitoring station consists of a set of sensors for measuring concentrations of pollutants: carbon monoxide CO, oxides of nitrogen (NO and NO_2) measured by the same analyzer, the sulfur dioxide SO_2, and ozone O_3. Some stations also measure certain meteorological parameters (temperature °C, relative humidity (%), global solar radiation (W/m^2), wind direction (degree), and wind speed (m/s),...). The measures are averages calculated over 15 min. Ozone O_3 is a secondary pollutant produced by complex photochemical reactions between primary pollutants (i.e., NO, NO_2, and VOC) emitted into the atmosphere. The pollutants' concentrations depend mainly on the vertical and horizontal movements of the atmosphere that are related to the meteorological conditions.

In this work, only six stations are considered. The data matrix X contains 18 state variables, x_1 to x_{18}, which correspond, respectively, to ozone O_3 and nitrogen oxides (NO_2 and NO) gathered on these six stations. The number of observations generated from these six stations is equal to 1000. The first 100 observations are used as training data to construct the reference model and the last 900 observations are used as testing data. For the identification of the KPCA model, the cumulative percent variance (CPV) is used. Moreover, the kernel parameter of the RBF is set to 52. For the MWKPCA method, the moving window size is equal to 90 samples. The fault detection result obtained using the MWKPCA for the SPE index in case of normal operating conditions is given in Fig. 2.3. It is shown from this figure that the SPE index presents high false alarm rates. This is mainly due to the fixed moving window size which is not able to adapt to the dynamic changes of the AIRLOR system.

In order to reduce the false alarm rate of the SPE index under the normal operation of the process, the VMWKPCA method is applied for the diagnosis of the AIRLOR system. Thus, the parameters α, β, and γ are chosen such that the rate of false

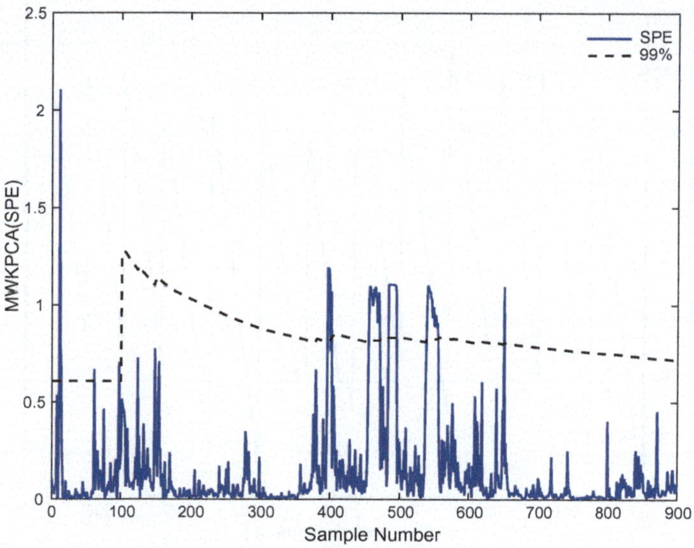

Fig. 2.3 Time evolution of the MWKPCA-based SPE statistic under normal operation

alarms of the process is minimal. Thus, these parameters are, respectively, set to 0.01, 0.01, and 0.3. The two parameters L_{max} and L_{min} determine the adjustment limits of the moving window size. When dynamic of the process has an extremely slow change, the change rates of the mean and the kernel matrix tend to be zero ($\Delta M_k \rightarrow 0$ $\Delta \overline{K}_k \rightarrow 0$), and therefore the window size tends to the maximum value ($L_{k+1} \rightarrow L_{max}$). When dynamic process changes are very fast, the change rates of the mean and the correlation matrix tend to be infinite ($\Delta M_k \rightarrow \infty$ $\Delta \overline{K}_k \rightarrow \infty$) and the window size tends to the minimum value ($L_{k+1} \rightarrow L_{min}$). In order to have a good estimate of the parameters L_{max} and L_{min}, these are chosen to represent, respectively, 180 and 8% of the set of training data. The variation of the moving window size is presented in Fig. 2.4. From this figure, we see that the window size varies between 40 and 115% with respect to the size of the initial data set.

Figure 2.5 shows the fault detection result of the VMWKPCA-based SPE technique. We can show from this figure that the evolution of the SPE index presents some false alarm rates.

The fault detection performance of the MWKPCA- and VMWKPCA-based SPE index will be evaluated in terms of average computation time (TC) and the false alarm rate (FAR). The false alarm rate is determined by calculating the percentage of incorrect faulty declarations in the non-faulty region (Jaffel et al. 2017). The performances of MWKPCA and VMWKPCA methods under normal operation are given in Table 2.1. From this table, it is clear that the VMWKPCA has less FAR and TC compared to the MWKPCA method.

Then, a sensor fault whose magnitude is approximately 25% of measurement for variable x_2, which corresponds to nitrogen oxide NO_2, is simulated between samples

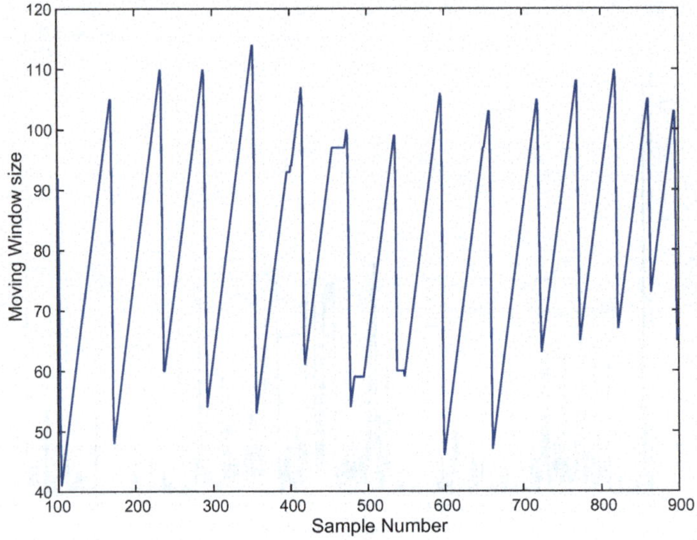

Fig. 2.4 Time evolution of the moving window size

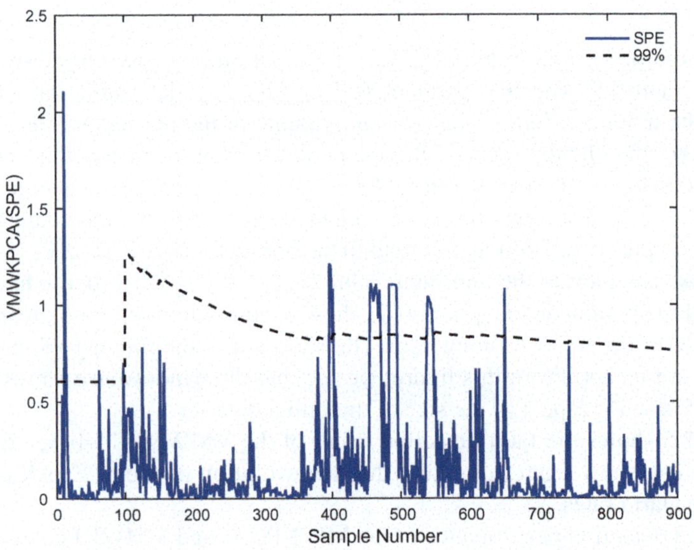

Fig. 2.5 Time evolution of the VMWKPCA-based SPE statistic under normal operation

300 and 600. Figures 2.6 and 2.7 show the fault detection results of the MWKPCA-based SPE and VMWKPCA-based SPE techniques. As we can see from these figures, the fault is clearly detected between samples 300 and 600.

Table 2.1 Performances of MWKPCA and VMWKPCA methods under normal operation

	MWKPCA	VMWKPCA
Average computation time (TC) (s)	0.0048	0.0019
False alarm rate (%)	6.2222	5.3333

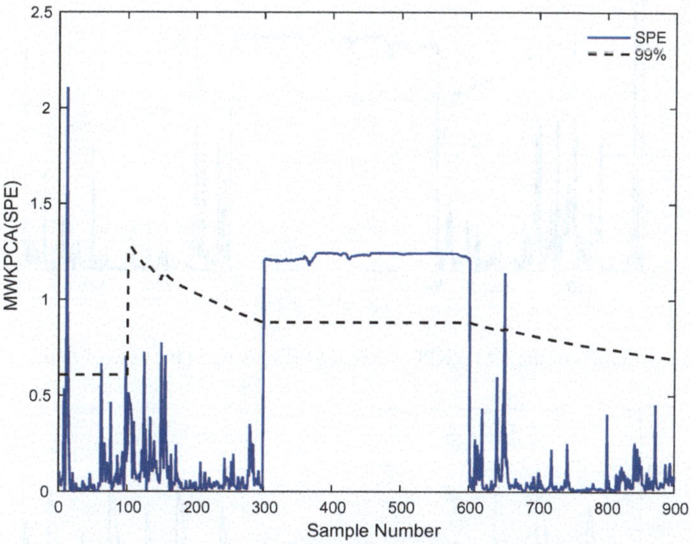

Fig. 2.6 Time evolution of the MWKPCA-based SPE statistic in the case of fault on variable x_2

Using the VMWKPCA method, the variation of the moving window size in the case of fault on variable x_2 is presented in Fig. 2.8. This figure shows that the moving window size remains constant between samples 300 and 600.

In order to ensure that meaningful comparison can be made, TC, FAR, and good detection rate (GDR) metrics are computed for both MWKPCA and VMWPCA methods. The GDR is computed by calculating percentage of faulty observations detected in the faulty region. The fault detection results are summarized in Table 2.2. These results show that the VMWKPCA reduces the TC compared to the MWKPCA technique.

After the presence of fault has been detected, it is important to identify the faulty variable. The partial VMWKPCA method is used to diagnose the provenance of this fault. This technique allows the structuring of the residuals by building a set of models, so that each model is sensitive to certain variables and insensitive to others. In this study, the models are built according to the following incident matrix.

Using partial VMWKPCA, we built 18 models of VMWKPCA. Each model is insensitive to one variable (sensors) as it is illustrated in Table 2.3. Figure 2.9 shows the evolution of the experimental signatures in the case of the fault on variable x_2. The

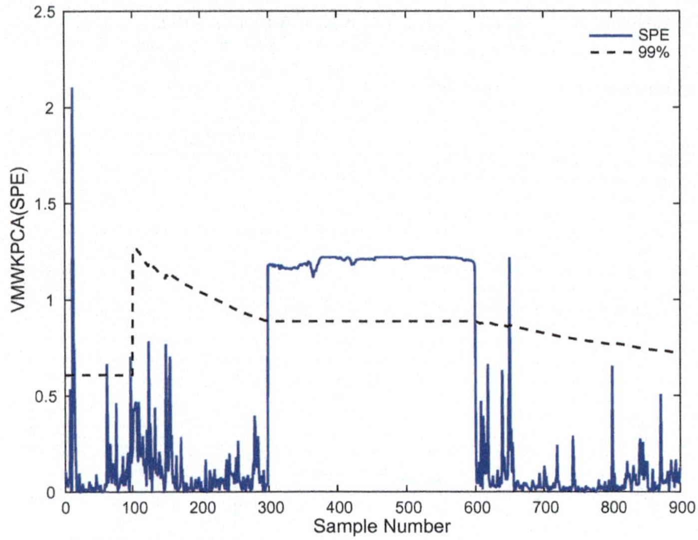

Fig. 2.7 Time evolution of the VMWKPCA-based SPE statistic in the case of fault on variable x_2

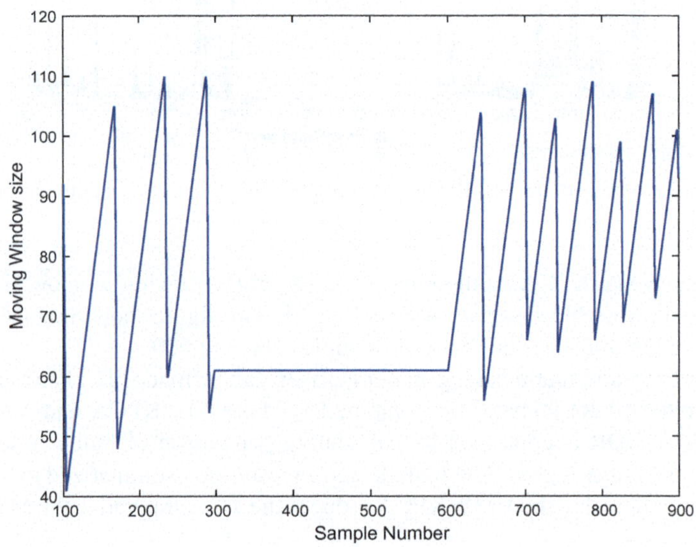

Fig. 2.8 Time evolution of the moving window size

experimental signature is obtained after codifying the SPE indices, where exceeding the threshold of detection is represented by 1 and less than the threshold is represented by 0. This gives the following experimental signature (1 0 1 1 1 1 1 1 1 1 1 1 1 1 1 1 1). This signature is identical to the second column of the theoretical table, which means that the variable x_2 is considered as the faulty one.

Table 2.2 Performances of MWKPCA and VMWKPCA methods in case of fault on the variable x_2

	MWKPCA	VMWKPCA
Average computation time (s)	0.0037	0.0011
False alarm rate (%)	2.0056	2.0056
Good detection rate (%)	100	100

Table 2.3 Incidence matrix

	d_1	d_2	d_3	d_4	d_5	d_6	d_7	d_8	d_9	d_{10}	d_{11}	d_{12}	d_{13}	d_{14}	d_{15}	d_{16}	d_{17}	d_{18}
r_1	0	1	1	1	1	1	1	1	1	1	1	1	1	1	1	1	1	1
r_2	1	0	1	1	1	1	1	1	1	1	1	1	1	1	1	1	1	1
r_3	1	1	0	1	1	1	1	1	1	1	1	1	1	1	1	1	1	1
r_4	1	1	1	0	1	1	1	1	1	1	1	1	1	1	1	1	1	1
r_5	1	1	1	1	0	1	1	1	1	1	1	1	1	1	1	1	1	1
r_6	1	1	1	1	1	0	1	1	1	1	1	1	1	1	1	1	1	1
r_7	1	1	1	1	1	1	0	1	1	1	1	1	1	1	1	1	1	1
r_8	1	1	1	1	1	1	1	0	1	1	1	1	1	1	1	1	1	1
r_9	1	1	1	1	1	1	1	1	0	1	1	1	1	1	1	1	1	1
r_{10}	1	1	1	1	1	1	1	1	1	0	1	1	1	1	1	1	1	1
r_{11}	1	1	1	1	1	1	1	1	1	1	0	1	1	1	1	1	1	1
r_{12}	1	1	1	1	1	1	1	1	1	1	1	0	1	1	1	1	1	1
r_{13}	1	1	1	1	1	1	1	1	1	1	1	1	0	1	1	1	1	1
r_{14}	1	1	1	1	1	1	1	1	1	1	1	1	1	0	1	1	1	1
r_{15}	1	1	1	1	1	1	1	1	1	1	1	1	1	1	0	1	1	1
r_{16}	1	1	1	1	1	1	1	1	1	1	1	1	1	1	1	0	1	1
r_{17}	1	1	1	1	1	1	1	1	1	1	1	1	1	1	1	1	0	1
r_{18}	0	1	1	1	1	1	1	1	1	1	1	1	1	1	1	1	1	0

2.6 Conclusion

In this study, we propose a new fault isolation method which is applicable to the process monitoring using kernel PCA. Recently, the fault detection method using kernel PCA has been developed. However, the fault isolation scheme suitable for kernel PCA monitoring has rarely been found. To isolate the faulty sensors, we develop new partial VMWKPCA method for fault isolation which is an extension of the partial PCA technique. The relevance of the proposed method was illustrated for fault isolation on air quality monitoring networks in Lorraine, France. The results demonstrated the effectiveness of the proposed partial VMWKPCA method for identifying the provenance of the fault.

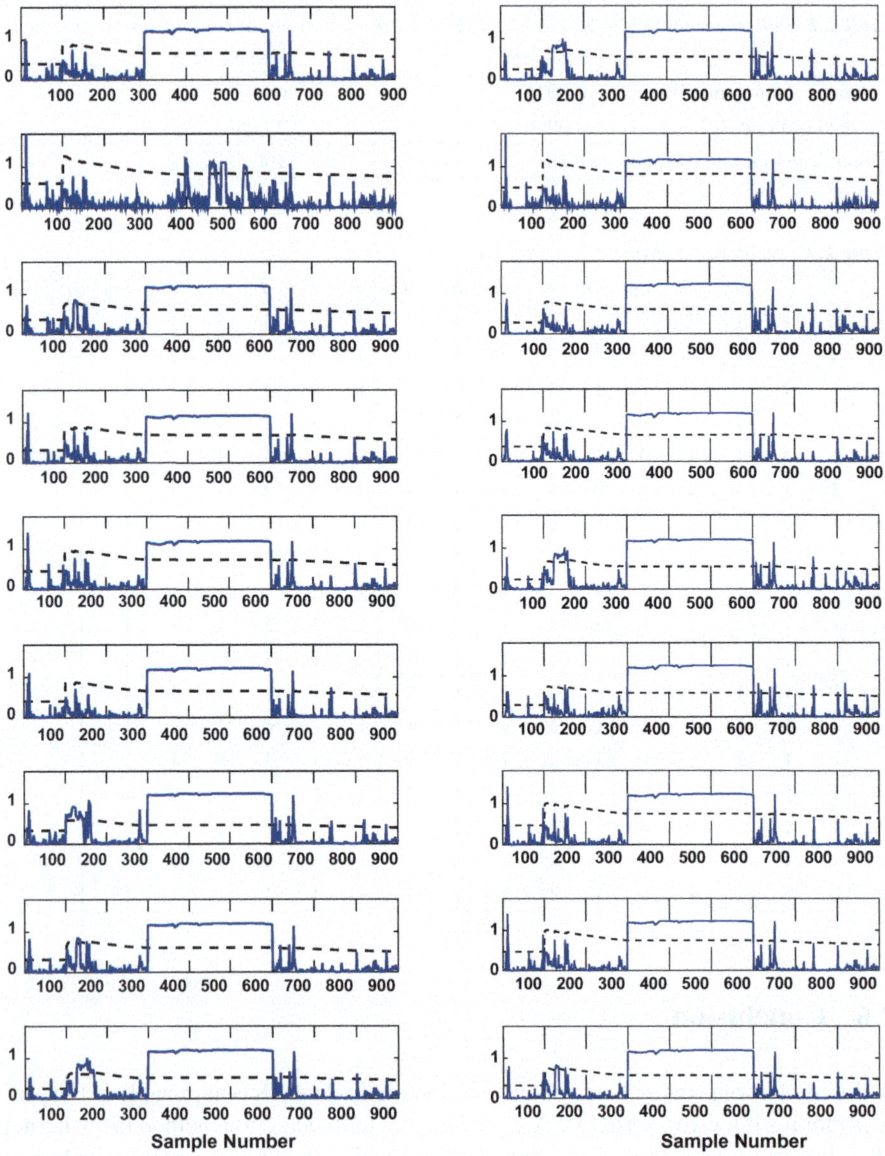

Fig. 2.9 Time evolution of SPEs corresponding to the second column of the different partial VMWKPCA models

References

Alcala, C. F., & Qin, S. J. (2010). Reconstruction-based contribution for process monitoring with kernel principal component analysis. *Industrial & Engineering Chemistry Research*, *49*(17), 7849–7857.

Cho, J.-H., Lee, J.-M., Choi, S. W., Lee, D., & Lee, I.-B. (2005). Fault identification for process monitoring using kernel principal component analysis. *Chemical Engineering Science*, *60*(1), 279–288.

Dong, D., & McAvoy, T. J. (1996). Nonlinear principal component analysis based on principal curves and neural networks. *Computers & Chemical Engineering*, *20*(1), 65–78.

Dunia, R., & Qin, S. J. (1998). Joint diagnosis of process and sensor faults using principal component analysis. *Control Engineering Practice*, *6*(4), 457–469.

Fazai, R., Taouali, O., Harkat, M. F., & Bouguila, N. (2016). A new fault detection method for nonlinear process monitoring. *The International Journal of Advanced Manufacturing Technology*, *87*(9–12), 3425–3436.

Fezai, R., Mansouri, M., Taouali, O., Harkat, M. F., & Bouguila, N. (2018). Online reduced kernel principal component analysis for process monitoring. *Journal of Process Control*, *61*, 1–11.

Gertler, J., & Cao, J. (2005). Design of optimal structured residuals from partial principal component models for fault diagnosis in linear systems. *Journal of Process Control*, *15*(5), 585–603.

Harkat, M.-F., Mourot, G., & Ragot, J. (2006). An improved PCA scheme for sensor FDI: Application to an air quality monitoring network. *Journal of Process Control*, *16*(6), 625–634.

Harkat, M.-F., Mourot, G., & Ragot, J. (2009). Multiple sensor fault detection and isolation of an air quality monitoring network using RBF-NLPCA model. *IFAC Proceedings Volumes*, *42*(8), 828–833.

Harkat, M.-F., Mansouri, M., Nounou, M., & Nounou, H. (2018). Enhanced data validation strategy of air quality monitoring network. *Environmental Research*, *160*, 183–194.

Jaffel, I., Taouali, O., Harkat, M. F., & Messaoud, H. (2016). Moving window KPCA with reduced complexity for nonlinear dynamic process monitoring. *ISA Transactions*, *64*, 184–192.

Jaffel, I., Taouali, O., Harkat, M. F., & Messaoud, H. (2017). Kernel principal component analysis with reduced complexity for nonlinear dynamic process monitoring. *The International Journal of Advanced Manufacturing Technology*, *88*(9–12), 3265–3279.

Joe Qin, S. (2003). Statistical process monitoring: Basics and beyond. *Journal of Chemometrics*, *17*(8–9), 480–502.

Kramer, M. A. (1991). Nonlinear principal component analysis using autoassociative neural networks. *AIChE Journal*, *37*(2), 233–243.

Lee, J.-M., Yoo, C., Choi, S. W., Vanrolleghem, P. A., & Lee, I.-B. (2004a). Nonlinear process monitoring using kernel principal component analysis. *Chemical Engineering Science*, *59*(1), 223–234.

Lee, J.-M., Yoo, C., & Lee, I.-B. (2004b). Fault detection of batch processes using multiway kernel principal component analysis. *Computers & Chemical Engineering*, *28*(9), 1837–1847.

Lee, J.-M., Qin, S. J., & Lee, I.-B. (2006). Fault detection and diagnosis based on modified independent component analysis. *AIChE Journal*, *52*(10), 3501–3514.

Liu, X., Kruger, U., Littler, T., Xie, L., & Wang, S. (2009). Moving window kernel PCA for adaptive monitoring of nonlinear processes. *Chemometrics and Intelligent Laboratory Systems*, *96*(2), 132–143.

Nguyen, V. H., & Golinval, J.-C. (2010). Fault detection based on kernel principal component analysis. *Engineering Structures*, *32*(11), 3683–3691.

Nomikos, P., & MacGregor, J. F. (1995). Multivariate SPC charts for monitoring batch processes. *Technometrics*, *37*(1), 41–59.

Schölkopf, B., Smola, A., & Müller, K.-R. (1998). Nonlinear component analysis as a kernel eigenvalue problem. *Neural Computation*, *10*(5), 1299–1319.

Sheriff, M. Z., Mansouri, M., Karim, M. N., Nounou, H., & Nounou, M. (2017). Fault detection using multiscale PCA-based moving window GLRT. *Journal of Process Control*, *54*, 47–64.

Taouali, O., Jaffel, I., Lahdhiri, H., Harkat, M. F., & Messaoud, H. (2016). New fault detection method based on reduced kernel principal component analysis (RKPCA). *The International Journal of Advanced Manufacturing Technology*, *85*(5–8), 1547–1552.

Widodo, A., Yang, B.-S., & Han, T. (2007). Combination of independent component analysis and support vector machines for intelligent faults diagnosis of induction motors. *Expert Systems with Applications*, *32*(2), 299–312.

Zhang, Y., Li, S., & Teng, Y. (2012). Dynamic processes monitoring using recursive kernel principal component analysis. *Chemical Engineering Science*, *72*, 78–86.

Chapter 3
Sensor Fault Detection and Estimation Based on UIO for LPV Time Delay Systems Using Descriptor Approach

Zina Bougatef, Nouceyba Abdelkrim, Abdel Aitouche and Mohamed Naceur Abdelkrim

Abstract This work investigates fault detection of linear parameter-varying (LPV) systems with varying time delay. An unknown input observer (UIO) is designed here to estimate faults and to generate residuals. Based on the descriptor approach, we elaborate an augmented system which considers faults as auxiliary. Only sensor faults are considered in this work and then estimated and detected using a delayed observer. The convergence of the observer is analyzed in terms of Lyapunov quadratic stability by generating linear matrix inequalities (LMIs). The work closes by applying the proposed approach to a delayed truck–trailer system with sensor faults.

Keywords Linear parameter-varying system · Sensor faults · Time delay · Unknown input observer · LMIs

3.1 Introduction

Sensor or actuator faults can cause deterioration of industrial and human systems which leads to very dangerous results for human operator especially (Isermann and Balle 1997; Rodrigues et al. 2007). Currently, the sensor fault diagnosis becomes an important requirement of industrial systems. Several works are interested by this topic

Z. Bougatef (✉) · M. N. Abdelkrim
Research Lab. MACS, National Engineering School of Gabes, University of Gabés, Gabés, Tunisia
e-mail: zinabougatef.enis@gmail.com

M. N. Abdelkrim
e-mail: nouceyba.naceur@laposte.net

N. Abdelkrim
Higher Institute of Industrial Systems of Gabés, University of Gabés, Gabés, Tunisia
e-mail: naceur.abdelkrim@enig.rnu.tn

A. Aitouche
Research CRISTAL Laboratory, Hautes Etudes d'Ingénieur, University of Lille 1, Lille, France
e-mail: abdel.aitouche@yncrea.fr

N. Derbel et al. (eds.), *Diagnosis, Fault Detection & Tolerant Control*, Studies in Systems, Decision and Control 269,
https://doi.org/10.1007/978-981-15-1746-4_3

(Staroswiecki et al. 2004, 1999; Rodrigues et al. 2013). Fault detection and isolation are two necessary steps in fault diagnosis. In fact, a great work has been considered to establish some methods, such as the model-based techniques (Knittel et al. 2003; Gertler 2017; Ding 2008). Let's talk about model-based approaches that use classic theories, and these approaches attract attention of researchers such as Chen and Patton (2012), Rodrigues et al. (2014). Descriptor (singular) systems are characterized by a general formulation and widely used in modeling of different systems, (e.g., Yeu et al. 2005) authors design an unknown input observer for fault detection and isolation of descriptor systems and in Taniguchi et al. (2000), Wang et al. (2004) authors develop fuzzy approach for nonlinear systems without and with time delay, respectively, using descriptor method. However, singular linear parameter-varying (LPV) systems are recently studied in fault diagnosis (Lopez-Estrada et al. 2014, 2015).

In the literature, there are many methods for fault detection and isolation (FDI) such as the analytical redundancy methods (Gertler 1992; Staroswiecki and Comtet-Varga 2001; Gertler 2013; Tellili et al. 2014; Bougatef et al. 2016) and the Kalman filter techniques (Wei et al. 2010; Foo et al. 2013). But unknown input observer (UIO) is recently used in fault diagnosis (Sotomayor and Odloak 2005; Marx et al. 2007; Chen and Patton 2012; Ahmadizadeh et al. 2014; Liu and Gao 2017). Especially for LPV systems, (Li et al. 2018) develop UIO for state estimation and fault detection using LPV model. Observers solve the problem of unmeasured states and/or unknown inputs like faults. Indeed, in the case of expensive sensors, the use of observers becomes an economic technique (Ichalal and Mammar 2015).

Time delay can appear obviously in many systems, and this phenomena complicate their design. The results available in this field are few. Besides, sensor fault diagnosis has been considered in Ichalal et al. (2009) and Tian and Yue (2013) using fuzzy descriptor systems, and in Chen et al. (2011), Zhai et al. (2014) based on the H_∞ approach, but this problem is not considered yet for singular LPV systems with time delay. In this chapter, the fault diagnosis problem of LPV systems with varying time delay by using the descriptor approach is elaborated. This method is based on considering the faults as an auxiliary state in order to facilitate fault estimation and detection, and then using a UIO for estimating the augmented states and generating residuals. The designed observer depends on the time delay created in the studied system. Moreover, the existence and the convergence of the UIO are obtained, respectively, by solving the linear matrix inequalities (LMIs) and using the Lyapunov theory. For these purposes, first, we represent the studied system as an LPV descriptor delayed system, and second, we develop the UIO. Finally, an algorithm for the UIO design will be formulated to calculate the observer gains.

This paper is organized as follows. Section 3.2 gives the studied (LPV) system under an augmented form and gives the system matrices. In Sect. 3.3, a UIO for delayed systems is designed. Section 3.4 deals with the algorithm which is used to compute the observer gains. Then, in Sect. 3.5, simulation results prove the effectiveness of the proposed approach. The conclusion is done in Sect. 3.6.

3.2 System Description

Consider the following delayed singular LPV system with sensor fault:

$$\begin{cases} E\dot{x}(t) = \sum_{i=0}^{n} \rho_i A_i x(t-\tau_i(t)) + \rho_i B_i u(t) \\ y(t) = Cx(t) + Df_s(t) \\ \tau_0(t) = 0 \\ 0 \leq \tau_i(t) \leq \tau_m \; ; i = 1 \ldots n \\ x(t) = \phi(t) \end{cases} \tag{3.1}$$

where $x(t) \in R^n$,$u(t) \in R^c$,$y(t) \in R^m$, and $f_s(t)$ denote, respectively, the state vector, the control input, the output vector, and the sensor fault vector. $A(\rho)$, $B(\rho)$ are the matrices of the time-bounded varying parameter ρ, such that

$$\begin{aligned} A(\rho) &= \sum_{i=0}^{n} \rho_i A_i \\ B(\rho) &= \sum_{i=0}^{n} \rho_i B_i \end{aligned} \tag{3.2}$$

where ρ_i are weights of the LPV subsystems satisfying

$$\sum_{i=0}^{n} \rho_i = 1, \quad 0 \leq \rho_i \leq 1 \tag{3.3}$$

and C is a matrix with appropriate dimension. $\tau_i(t)$ for $i = 0 \ldots n$ are the time-varying delays in which τ_m is the delays upper bound and $\tau_0(t) = 0$.

An augmented system is constructed as follows:

$$\begin{cases} \bar{E}\dot{\bar{x}}(t) = \sum_{i=0}^{n} \rho_i (\bar{A}_i \bar{x}(t-\tau_i(t)) + \bar{B}_i u(t) + \bar{D}_i x_s(t)) \\ y(t) = \bar{C}\bar{x}(t) \end{cases} \tag{3.4}$$

such that the augmented state vector is

$$\bar{x}(t) = \begin{bmatrix} x(t) \\ x_s(t) \end{bmatrix} \tag{3.5}$$

and

$$\overline{E} = \begin{bmatrix} E & 0 \\ 0 & 0 \end{bmatrix}, \overline{A}_i = \begin{bmatrix} A_i & 0 \\ 0 & -I_p \end{bmatrix}, \overline{B}_i = \begin{bmatrix} B_i \\ 0 \end{bmatrix}, \overline{C} = [C \; I_p], \overline{D} = \begin{bmatrix} 0 \\ I_p \end{bmatrix}$$

$$x_s(t) = Df_s(t)$$

The state vector $x(t)$ and fault $f_s(t)$ can be estimated if the estimation of augmented state $\bar{x}(t)$ is obtained. Then, the problem of fault estimation of system (3.1) is converted as a UIO design for the augmented system.

3.3 UIO Design

We present here a designed unknown input observer approach for delayed LPV systems. The existence and convergence conditions are formulated as LMIs terms.

The observer considered for the system (3.4) is given by

$$\begin{cases} \dot{z}(t) = \sum_{i=0}^{n} \rho_i (N_i z(t-\tau_i(t)) + L_i y(t-\tau_i(t)) + G_i u(t)) \\ \hat{\bar{x}}(t) = z(t) + H_2 y(t) \\ \hat{y}(t) = \bar{C}\hat{\bar{x}}(t) \\ r(t) = y(t) - \hat{y}(t) \\ z(t) = 0, \ -\tau_m \le t \le 0 \end{cases} \tag{3.6}$$

where $\hat{x}$ is the state estimation vector of x. H_2, N_i, L_i, and G_i are the observer matrices which should be determined and r(t) is called residual vector that is used to detect faults.

The augmented state estimation error is

$$\bar{e}(t) = \bar{x}(t) - \hat{\bar{x}}(t) \tag{3.7}$$

$$= \bar{x}(t) - z(t) - H_2\bar{C}\bar{x}(t) \tag{3.8}$$

$$= (I_n - H_2\bar{C})\bar{x}(t) - z(t) \tag{3.9}$$

If $H_1 \in R^{n \times n}$ can be calculated from the following condition:

$$H_1\bar{E} = I_n - H_2\bar{C} \tag{3.10}$$

then

$$\bar{e}(t) = H_1\bar{E}\bar{x}(t) - z(t) \tag{3.11}$$

and the error dynamic is given as

$$\dot{\bar{e}}(t) = H_1\bar{E}\dot{\bar{x}}(t) - \dot{z}(t) \tag{3.12}$$

After substituting (3.1) and (3.2) in (3.12), it becomes

$$\begin{aligned} \dot{\bar{e}}(t) = &\sum_{i=0}^{n} \rho_i H_1 \Big(\bar{A}_i \bar{x}(t-\tau_i(t)) + \bar{B}_i u(t) + \bar{D}_i x_s(t) \Big) \\ &- \sum_{i=0}^{n} \rho_i \Big(N_i z(t-\tau_i(t)) + L_i \bar{C}\bar{x}(t-\tau_i(t)) + G_i u(t) \Big) \end{aligned} \tag{3.13}$$

some manipulations result

$$\dot{\bar{e}}(t) = \sum_{i=0}^{n} \Big[\rho_i \; N_i \bar{e}\big(t - \tau_i(t)\big) + \big(H_1 \bar{A}_i - L_i \bar{C} - N_i H_1 \bar{E}\big)\bar{x}\big(t - \tau_i(t)\big) + \big(H_1 \bar{B}_i - G_i\big)u(t) + \big(H_1 \bar{D}_i\big)x_s(t) \Big] \tag{3.14}$$

Then, we have

$$H_1 \bar{E} = I_n - H_2 \bar{C} \tag{3.15}$$
$$H_1 \bar{A}_i = L_i \bar{C} + N_i H_1 \bar{E} \tag{3.16}$$
$$G_i = H_1 \bar{B}_i \tag{3.17}$$
$$H_1 \bar{D}_i = 0 \tag{3.18}$$

if the above conditions are satisfied and so we can rewrite the error dynamics as follows:

$$\begin{cases} \dot{\bar{e}}(t) = \sum_{i=0}^{n} \rho_i N_i \bar{e}(t - \tau_i(t)) \\ r(t) = \bar{C}\bar{e}(t) \end{cases} \tag{3.19}$$

Remark 1 In order to solve (3.15), we assume this condition:

$$rank \begin{bmatrix} E \\ C \end{bmatrix} = n \tag{3.20}$$

Lemma 1 *The delayed system described by (3.19) can be stable if there exist a symmetric definite positive matrices P and Q_i for $i = 1 \dots n$ which satisfy the following LMI:*

$$\begin{bmatrix} PN_0 + N_0^T P + Q_1 + \cdots + Q_n & PN_1 & PN_2 & \cdots & PN_n \\ \star & -(1-\lambda_1)Q_1 & 0 & \cdots & 0 \\ \star & \star & -(1-\lambda_2)Q_2 & \cdots & 0 \\ \vdots & \vdots & \vdots & \ddots & \vdots \\ \star & \star & \star & \cdots & -(1-\lambda_n)Q_n \end{bmatrix} < 0 \tag{3.21}$$

Proof Considering the following Lyapunov–Krasovskii function with P and Q_i $(i = 1 \dots n)$ are symmetric definite positive matrices:

$$V[e(t)] = e^T(t)Pe(t) + \sum_{i=1}^{n} \int_{t-\tau_i(t)}^{t} e^T(s)Q_i e(s)ds \tag{3.22}$$

Then, we can calculate the derivative of Lyapunov function as follows:

$$\begin{aligned}\dot{V}[e(t)] &= e^T(t)[PN_0 + N_0^T P]e(t) \\ &+ \sum_{i=1}^{n} \Big[\rho_i \big[e^T(t)PN_i e\big(t-\tau_i(t)\big) + e^T\big(t-\tau_i(t)\big)N_i^T Pe(t)\big] \\ &+ e^T(t)Q_i e(t) - [1-\dot{\tau}_i](t)e^T\big(t-\tau_i(t)\big)Q_i e\big(t-\tau_i(t)\big)\Big]\end{aligned} \tag{3.23}$$

By considering $\dot{\tau}_i \leq \lambda_i$ for $i = 1 \ldots n$, we can obtain

$$\dot{V}[e(t)] \leq \sum_{i=0}^{n} \begin{bmatrix} e(t) \\ e(t-\tau_1) \\ e(t-\tau_2) \\ \vdots \\ e(t-\tau_n) \end{bmatrix}^T \Pi \begin{bmatrix} e(t) \\ e(t-\tau_1) \\ e(t-\tau_2) \\ \vdots \\ e(t-\tau_n) \end{bmatrix} \tag{3.24}$$

where

$$\Pi = \begin{bmatrix} PN_0 + N_0^T P + Q_1 + \cdots + Q_n & PN_1 & PN_2 & \ldots & PN_n \\ \star & -(1-\lambda_1)Q_1 & 0 & \ldots & 0 \\ \star & \star & -(1-\lambda_2)Q_2 & \ldots & 0 \\ \star & \vdots & \vdots & \ddots & \vdots \\ \star & \star & \star & \star & -(1-\lambda_n)Q_n \end{bmatrix} \tag{3.25}$$

The dynamic error (3.19) is stable if $\dot{V}(e(t))$ is definite and negative (Wu et al. 2010).

3.4 Method of Calculating the Observer Gains

Matrices of the observer (3.6) are calculated such that (3.15), (3.16), (3.17), and (3.18) are satisfied, so we have

$$\underbrace{\begin{bmatrix} H_1 & H_2 \end{bmatrix}}_{H} \underbrace{\begin{bmatrix} \bar{E} & \bar{D} \\ \bar{C} & 0 \end{bmatrix}}_{Y} = \underbrace{\begin{bmatrix} I_n & 0 \end{bmatrix}}_{W} \tag{3.26}$$

Defined matrix $H = \begin{bmatrix} H_1 & H_2 \end{bmatrix}$ and $Y = \begin{bmatrix} \bar{E} & \bar{D} \\ \bar{C} & 0 \end{bmatrix}$.

Y is a known matrix but H is an unknown matrix and needs to be solved. Solvability of (3.24)can be obtained if

$$rank(Y) = n + rank(\bar{D}) \tag{3.27}$$

Then, the solution of (3.24) is

$$H = WY^{+} \tag{3.28}$$

where the pseudo-inverse matrix Y^{+} is of $(n \times (n+m))$ and W is of $(n \times (n + K_{f_s}))$ dimension:

$$Y^{+} = \left(Y^T Y\right)^{-1} Y^T \tag{3.29}$$

Equation (3.28) can be partitioned into two parts:

$$H = \left[\, H_1 \; H_2 \,\right] = W \left[\, Y_1^{+} \; Y_2^{+} \,\right] + K \left[\, V_1 \; V_2 \,\right] \tag{3.30}$$

where $Y_1 \in R^{n \times n}$, $Y_2 \in R^{n \times m}$, and $V = I_n - YY^{+}$.

For $i = 0 \ldots n$, we introduce the variable K_i, such that

$$K_i = L_i - N_i H_2 \tag{3.31}$$

Now, (3.14) will be given as follows:

$$N_i = H_1 A_i - K_i C \tag{3.32}$$

Substituting H_1 in (3.32) results

$$N_i = H_{10} A_i + K X_1 A_i - K_i C \tag{3.33}$$

where $H_{10} = WY_1^{+}$

Theorem 1 *If there exist positive symmetric definite matrices P and Q_i for $i = 1 \ldots n$ and matrices M and P_i for $i = 0 \ldots n$, such that*

$$\begin{bmatrix} \phi_0 & \phi_1 & \phi_2 & \cdots & \phi_n \\ \star & -(1-\lambda_1)Q_1 & 0 & \cdots & 0 \\ \star & \star & -(1-\lambda_2)Q_2 & \cdots & 0 \\ \vdots & \vdots & \vdots & \ddots & \vdots \\ \star & \star & \star & \cdots & -(1-\lambda_n)Q_n \end{bmatrix} < 0 \tag{3.34}$$

where

$$\begin{aligned} \phi_0 = {} & PH_{10}A_0 + MX_1A_0 - P_0C \\ & + (PH_{10}A_0 + MX_1A_0 - P_0C)^T + Q_1 + \cdots + Q_n \end{aligned} \tag{3.35}$$

$$\phi_i = PH_{10}A_i + MX_1A_i - P_iC \tag{3.36}$$

for $i = 1 \ldots n$. Then the observer (3.2) exists and the estimation error converges toward zero.

$$K = P^{-1}M \tag{3.37}$$

$$K_i = P^{-1}P_i, i = 0 \ldots n \tag{3.38}$$

Proof Replacing (3.33) in (3.21) and using Lemma 1, it can be demonstrated.

Algorithm
Step 1. Calculate $H_{10} = WY_1^+$ and X_1 from n first columns of X.
Step 2. Solve LMI (3.34) and obtain the matrices P, M, Q_i for $i = 1 \ldots n$ and P_i for $i = 0 \ldots n$.
Step 3. Compute K and K_i for $i = 0 \ldots n$ from Eqs. (3.35) and (3.37).
Step 4. Compute G, H_2, and N_i for $i = 0 \ldots n$, respectively, from Eqs. (3.16), (3.18) and (3.33).
Step 5. Compute L_i for $i = 0 \ldots n$ from Eq. (3.31).

3.5 Simulation Studies

In this section, an example of truck–trailer system is studied in order to prove the effectiveness of the designed observer. A dynamic model is considered:

$$\begin{aligned}
\dot{x}_1(t) &= -a\frac{v(t)\bar{t}}{L_t t_0}x_1(t) - (1-a)\frac{v(t)\bar{t}}{L_t t_0}x_1(t-\tau_1) + \frac{v(t)\bar{t}}{l t_0}u(t) \\
\dot{x}_2(t) &= a\frac{v(t)\bar{t}}{L_t t_0}x_1(t) + (1-a)\frac{v(t)\bar{t}}{L_t t_0}x_1(t-\tau_1) \\
\dot{x}_3(t) &= \frac{v(t)\bar{t}}{t_0}\sin\left[x_2(t) + a\frac{v(t)\bar{t}}{2L_l}x_1(t) + (1-a)\frac{v(t)\bar{t}}{L_t}x_1(t-\tau_1)\right]
\end{aligned} \tag{3.39}$$

where $x_1(t)$ is the difference angle between the truck and the trailer, $x_2(t)$ is the trailer angle, $x_3(t)$ is the vertical position of the trailer's rear end, and $u(t)$ is the steering angle.

The values of the truck–trailer parameters are taken as follows: $a = 0.7$, $l = 2.8$, $\bar{t} = 2$, $L_t = 5.5$, $t_0 = 0.5$, and $\tau_1 = 0.5$. It is assumed that the speed of backing up $v(t)$ varies according to $v(t) = \left[-1.2\ 0.6\right]$.

In reference to Jia et al. (2015), if $(x_2(t) + a\frac{v\bar{t}}{2L_t}x_1(t) + (1-a)\frac{v\bar{t}}{L_t}x_1(t-\tau_1))$ is about 0, then the system (3.39) can be represented by

$$\begin{cases} E\dot{x}(t) = A_0(\rho_0)x(t) + A_1(\rho_1)x(t-\tau_1) + Bu(t) \\ y(t) = Cx(t) + Df_s(t) \end{cases} \tag{3.40}$$

with

$$A_0(v(t)) = \begin{bmatrix} -a\frac{v(t)\bar{t}}{L_t t_0} & 0 & 0 \\ a\frac{v(t)\bar{t}}{L_t t_0} & 0 & 0 \\ -a\frac{v^2(t)t^2}{2L_t t_0} & \frac{v(t)\bar{t}(t)}{t_0} & 0 \end{bmatrix}, A_1(v(t)) = \begin{bmatrix} -(1-a)\frac{v(t)\bar{t}}{L_t t_0} & 0 & 0 \\ (1-a)\frac{v(t)\bar{t}}{L_t t_0} & 0 & 0 \\ (1-a)\frac{v^2(t)\bar{t}^2}{2L_t t_0} & 0 & 0 \end{bmatrix}$$

$$B(v(t)) = \begin{bmatrix} \frac{v(t)\bar{t}}{l t_0} \\ 0 \\ 0 \end{bmatrix}, E = \begin{bmatrix} 1 & 0 & 0 \\ 0 & 1 & 0 \\ 0 & 0 & 1 \end{bmatrix}, D = \begin{bmatrix} 1 & 0 & 0 \\ 0 & 1 & 0 \\ 0 & 0 & 1 \end{bmatrix} C = I_3$$

The weighting functions $\rho_i(v(t))$ are

$$\rho_0(v(t)) = \frac{v(t) - v_{\max}}{v_{\min} - v_{\max}} = \frac{v(t) + 1.2}{0.6}$$

$$\rho_1(v(t)) = \frac{v(t) - v_{\min}}{v_{\max} - v_{\min}} = \frac{v(t) + 0.6}{-0.6}$$

On solving (3.34) by using the MATLAB LMI toolbox, we obtain the Lyapunov matrices:

$$P = \begin{bmatrix} -0.9781 & 1.0243 & 0.1901 \\ 1.0243 & -0.0463 & 1.3589 \\ 0.1901 & 1.3589 & -0.0070 \end{bmatrix}$$

$$Q_1 = \begin{bmatrix} -2.4656 & 0.2332 & 0.2686 \\ 0.2332 & -0.6557 & -0.0132 \\ 0.2686 & -0.0132 & -3.0927 \end{bmatrix}$$

$$P_0 = \begin{bmatrix} 2.6898 & -0.1272 & -0.1465 \\ -0.1272 & 0.7153 & 0.0072 \\ -0.1465 & 0.0072 & 3.3739 \end{bmatrix}$$

$$P_1 = \begin{bmatrix} 2.6898 & -0.1272 & -0.1465 \\ -0.1272 & 0.7153 & 0.0072 \\ -0.1465 & 0.0072 & 3.3739 \end{bmatrix}$$

Then, the UIO gains are as follows:

$$H_1 = \begin{bmatrix} 1.1500 & 0.0408 & -0.0380 \\ 0.0362 & 1.1611 & 0.0035 \\ -0.0239 & -0.0293 & 1.1957 \end{bmatrix}, H_2 = \begin{bmatrix} -0.1500 & -0.0408 & 0.0380 \\ -0.0362 & -0.1611 & -0.0035 \\ 0.0239 & 0.0293 & -0.1957 \end{bmatrix}$$

$$K = \begin{bmatrix} 1.3000 & 0.0816 & -0.0759 \\ 0.0725 & 1.3221 & 0.0070 \\ -0.0478 & -0.0585 & 1.3913 \end{bmatrix} K_0 = \begin{bmatrix} -2.2205 & 0.1855 & 2.1324 \\ 0.2110 & -0.0187 & 2.1767 \\ 1.5875 & 0.3859 & -1.5281 \end{bmatrix}$$

$$K_1 = \begin{bmatrix} -2.2205 & 0.1855 & 2.1324 \\ 0.2110 & -0.0187 & 2.1767 \\ 1.5875 & 0.3859 & -1.5281 \end{bmatrix}, L_0 = \begin{bmatrix} -2.6908 & 0.0141 & 2.6563 \\ 0.2766 & -0.0511 & 2.5728 \\ 2.1402 & 1.3527 & -1.8923 \end{bmatrix}$$

$$L_1 = \begin{bmatrix} -2.6328 & 0.0528 & 2.6435 \\ 0.2266 & -0.0668 & 2.5853 \\ 1.8367 & 0.5469 & -1.8761 \end{bmatrix}, G = \begin{bmatrix} -1.6428 \\ -0.0518 \\ 0.0341 \end{bmatrix}$$

The vector of faults is

$$f_s(t) = [\, f_{s1}(t) \; f_{s2}(t) \; f_{s3}(t) \,]^T$$

Faults $f_{s1}(t)$, $f_{s2}(t)$, and $f_{s3}(t)$ affect, respectively, the first, the second, and the third sensors of the system.

The simulation results in the presence of sensor fault transient forms are given in Figs. 3.1, 3.2, 3.3, 3.4, and 3.5. The sensor fault estimation errors are given by

$$\begin{aligned} e_1(t) &= f_{s1}(t) - \hat{f}_{s1}(t) \\ e_2(t) &= f_{s2}(t) - \hat{f}_{s2}(t) \\ e_3(t) &= f_{s3}(t) - \hat{f}_{s3}(t) \end{aligned}$$

- Case 1: Fault-free case

We will illustrate, first, estimated states and the estimation errors in fault-free case.

- Case 2: Faulty case: Transient form

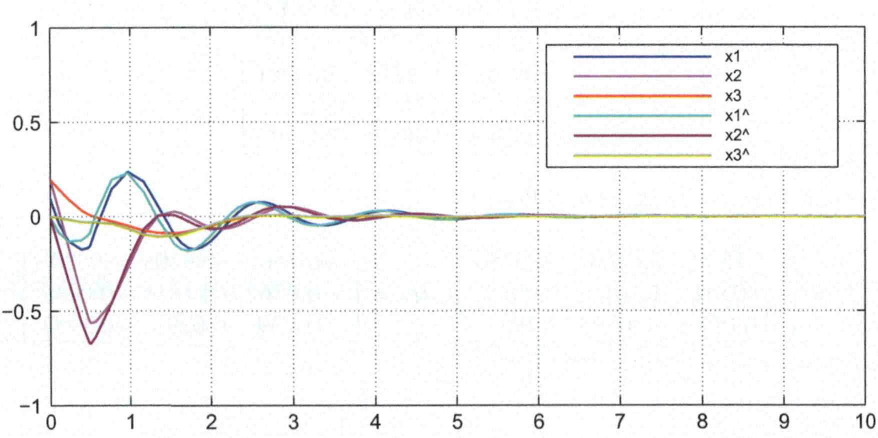

Fig. 3.1 The time evolution of real and estimated states $x_1(t)$, $x_2(t)$ *and* $x_3(t)$ in fault-free case

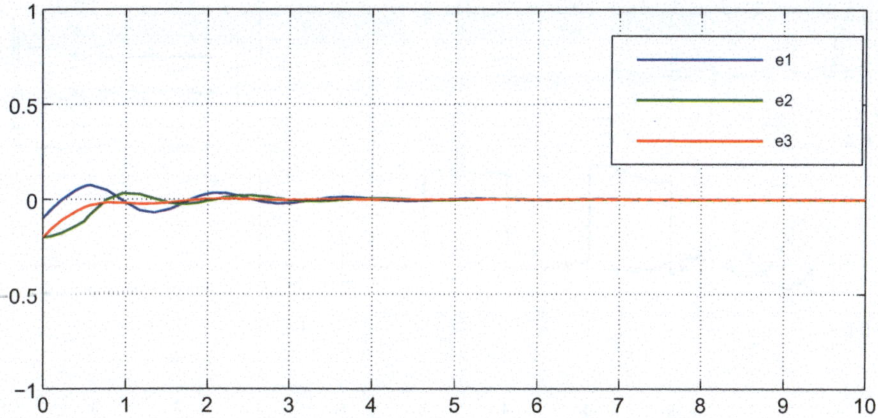

Fig. 3.2 The time evolution of the estimation errors $e_1(t)$, $e_2(t)$, and $e_3(t)$

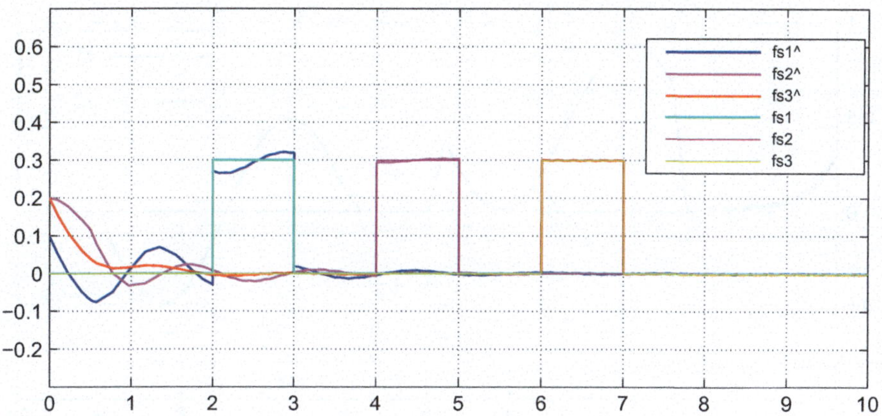

Fig. 3.3 The time evolution of real and estimated faults

Then, transient form faults occur on first, second, and third sensors, respectively, at different simulation times.

Figures 3.1, 3.2, and 3.3 show that based on the descriptor approach, the designed UIO can estimate both the state of the delayed LPV system and the faults (auxiliary states). Moreover, we consider from the time evolution of residuals (see Fig. 3.4) that the faults are successfully detected and also isolated.

- Case 3: Faulty case: Sine form

Now, the first and second sensors are considered and the sensor fault matrix is selected as

$$D = \begin{bmatrix} 1 & 0 \\ 0 & 1 \\ 0 & 0 \end{bmatrix}$$

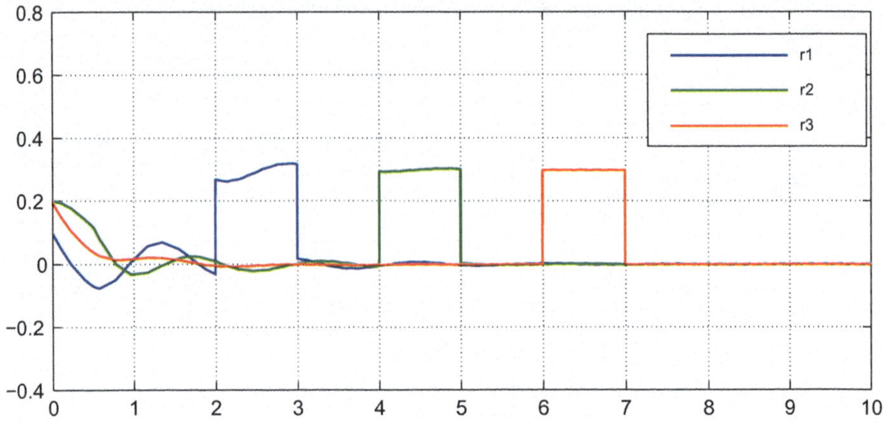

Fig. 3.4 The time evolution of the residual $r_1(t)$, $r_2(t)$, and $r_3(t)$ in the case of faults

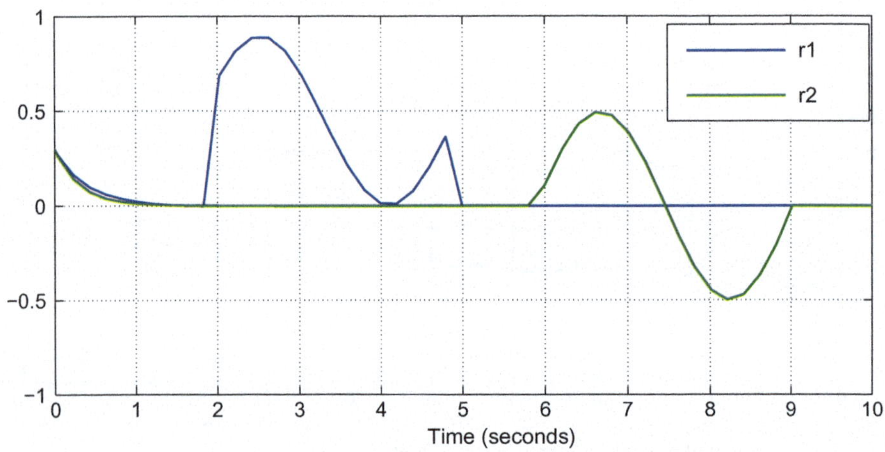

Fig. 3.5 The time evolution of the residuals $r_1(t)$ and $r_2(t)$ in the case of faults

The sensor fault signals will have a sine form as given below:

$$\begin{cases} f_1(t) = \sin(2t + 0.5) + 1,\ 2 \leq t \leq 5 \\ \qquad 0,\ otherwise \\ f_2(t) = \sin(2t + 0.5),\ 6 \leq t \leq 9 \\ \qquad 0,\ otherwise \end{cases} \tag{3.41}$$

The simulation results in this case are given as follows.

Figure 3.5 represents the time evolution of residuals which indicate the presence of sensor faults having sine forms.

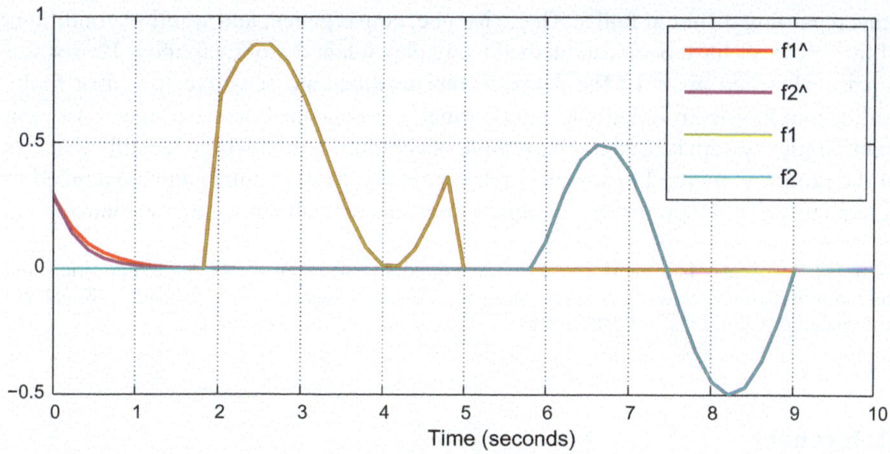

Fig. 3.6 The time evolution of real and estimated faults

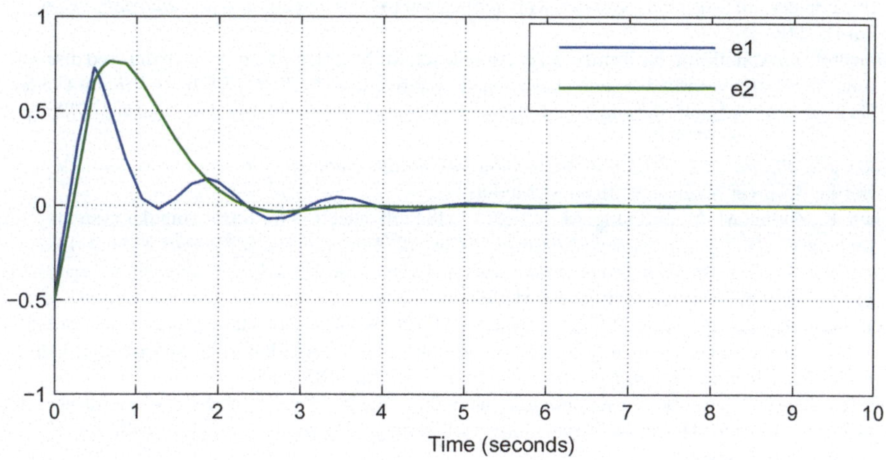

Fig. 3.7 The time evolution of the estimation errors $e_1(t)$ and $e_2(t)$

Figures 3.5, 3.6, and 3.7 demonstrate that our proposed approach is efficient for different sensor fault forms. The sensor faults are not only well detected but also perfectly estimated using the UIO and the technique of residual generation.

3.6 Conclusion

In this chapter, we use the descriptor approach applied to the delayed LPV systems in the case of sensor faults, and this approach is efficient in fault estimation and detection. Then, we design the UIO-based fault diagnosis for singular delayed LPV systems regarding different faults. The existence, convergence, and stability conditions of this observer have been established using dependent delay Lyapunov–Krasovskii functional which gives LMIs. As well, the residuals are sensitive to sensor faults and decoupling from unknown inputs. Finally, a delayed linear parameter-varying truck–trailer system is used as an example of simulation to illustrate the effectiveness of the proposed method in various fault scenarios. These results can be extended in active fault-tolerant control by compensating sensor faults in state law control.

Acknowledgements This work has been supported by the Ministry of High Education and Research of Tunisia, with the cooperation of the Center of Research in Informatics, Signal and Automatics of Lille, France (CRISTAL).

References

Ahmadizadeh, S., Zarei, J., & Karimi, H. R. (2014). Robust unknown input observer design for linear uncertain time delay systems with application to fault detection. *Asian Journal of Control*, *16*(4), 1006–1019.
Bougatef, Z., Abdelkrim, N., Tellili, A., & Abdelkrim, M. N. (2016). Diagnosis of delayed discrete-time singularly perturbed systems based on slow subsystem. In *2016 17th International Conference on Sciences and Techniques of Automatic Control and Computer Engineering (STA)* (pp. 239–243). Sousse: IEEE.
Chen, J. & Patton, R. J. (2012). *Robust model-based fault diagnosis for dynamic systems* (Vol. 3). Berlin: Springer Science & Business Media.
Chen, L., Zhong, M.-Y., & Zhang, M.-Y. (2011). H8 fault detection for linear singular systems with time-varying delay. *International Journal of Control, Automation and Systems*, *9*(1), 9–14.
Ding, S. X. (2008). *Model-based fault diagnosis techniques: design schemes, algorithms, and tools*. Berlin: Springer Science & Business Media.
Foo, G. H. B., Zhang, X., & Vilathgamuwa, D. M. (2013). A sensor fault detection and isolation method in interior permanent-magnet synchronous motor drives based on an extended kalman filter. *IEEE Transacions on Industrial Electronics*, *60*(8), 3485–3495.
Gertler, J. (1992). Analytical redundancy methods in fault detection and isolation: survey and synthesis. In *Fault Detection, Supervision and Safety for Technical Processes 1991* (pp. 9–21). Amsterdam: Elsevier.
Gertler, J. (2013). *Fault detection and diagnosis*. Berlin: Springer.
Gertler, J. (2017). *Fault detection and diagnosis in engineering systems*. Abingdon: Routledge.
Ichalal, D., & Mammar, S. (2015). On unknown input observers for lpv systems. *IEEE Transactions on Industrial Electronics*, *62*(9), 5870–5880.
Ichalal, D., Marx, B., Ragot, J., & Maquin, D. (2009). Fault diagnosis for takagi-sugeno nonlinear systems. *IFAC Proceedings Volumes*, *42*(8), 504–509.
Isermann, R., & Balle, P. (1997). Trends in the application of model-based fault detection and diagnosis of technical processes. *Control Engineering Practice*, *5*(5), 709–719.
Jia, Q., Chen, W., Zhang, Y., & Li, H. (2015). Fault reconstruction and fault-tolerant control via learning observers in takagi-sugeno fuzzy descriptor systems with time delays. *IEEE Transactions on Industrial Electronics*, *62*(6), 3885–3895.
Knittel, D., Laroche, E., Gigan, D., & Koc, H. (2003). Tension control for winding systems with two-degrees-of-freedom h/sub/spl infin//controllers. *IEEE Transactions on Industry Applications*, *39*(1), 113–120.

Li, S., Wang, H., Aitouche, A., Tian, Y., & Christov, N. (2018). Active fault tolerance control of a wind turbine system using an unknown input observer with an actuator fault. *International Journal of Applied Mathematics and Computer Science*, *28*(1), 69–81.

Liu, X., & Gao, Z. (2017). Robust finite-time fault estimation for stochastic nonlinear systems with brownian motions. *Journal of the Franklin Institute*, *354*(6), 2500–2523.

Lopez-Estrada, F.-R., Ponsart, J.-C., Astorga-Zaragoza, C.-M., Camas-Anzueto, J.-L., & Theilliol, D. (2015). Robust sensor fault estimation for descriptor-LPV systems with unmeasurable gain scheduling functions: Application to an anaerobic bioreactor. *International Journal of Applied Mathematics and Computer Science*, *25*(2), 233–244.

Lopez-Estrada, F.-R., Ponsart, J.-C., Theilliol, D., Astorga-Zaragoza, C., & Aberkane, S. (2014). Fault diagnosis based on robust observer for descriptor-LPV systems with unmeasurable scheduling functions. *IFAC Proceedings Volumes*, *47*(3), 1079–1084.

Marx, B., Koenig, D., & Ragot, J. (2007). Design of observers for takagi-sugeno descriptor systems with unknown inputs and application to fault diagnosis. *IET Control Theory & Applications*, *1*(5), 1487–1495.

Rodrigues, M., Hamdi, H., Braiek, N. B., & Theilliol, D. (2014). Observer based fault tolerant control design for a class of lpv descriptor systems. *Journal of the Franklin Institute*, *351*(6), 3104–3125.

Rodrigues, M., Sahnoun, M., Theilliol, D., & Ponsart, J.-C. (2013). Sensor fault detection and isolation filter for polytopic lpv systems: A winding machine application. *Journal of Process Control*, *23*(6), 805–816.

Rodrigues, M., Theilliol, D., Aberkane, S., & Sauter, D. (2007). Fault tolerant control design for polytopic LPV systems. *International Journal of Applied Mathematics and Computer Science*, *17*(1), 27–37.

Sotomayor, O. A., & Odloak, D. (2005). Observer-based fault diagnosis in chemical plants. *Chemical Engineering Journal*, *112*(1–3), 93–108.

Staroswiecki, M., & Comtet-Varga, G. (2001). Analytical redundancy relations for fault detection and isolation in algebraic dynamic systems. *Automatica*, *37*(5), 687–699.

Staroswiecki, M., Hoblos, G., & Aitouche, A. (1999). Fault tolerance analysis of sensor systems. In *Proceedings of the 38th IEEE Conference on Decision and Control* (Vol. 4, pp. 3581–3586).

Staroswiecki, M., Hoblos, G., & Aitouche, A. (2004). Sensor network design for fault tolerant estimation. *International Journal Of Adaptive Control and Signal Processing*, *18*(1), 55–72.

Taniguchi, T., Tanaka, K., & Wang, H. O. (2000). Fuzzy descriptor systems and nonlinear model following control. *IEEE Transactions on Fuzzy Systems*, *8*(4), 442–452.

Tellili, A., Abdelkrim, N., Jaouadi, B., & Abdelkrim, M. N. (2014). Diagnosis of discrete-time singularly perturbed systems based on slow subsystem. *acta mechanica et automatica*, *8*(4), 175–180.

Tian, E. & Yue, D. (2013). Reliable h8 filter design for t-s fuzzy model-based networked control systems with random sensor failure. *International Journal of Robust and Nonlinear Control*, *23*(1), 15–32.

Wang, Y., Sun, Z.-Q., & Sun, F.-C. (2004). Robust fuzzy control of a class of nonlinear descriptor systems with time-varying delay. *International Journal of Control, Automation, and Systems*, *2*(1), 76–82.

Wei, X., Verhaegen, M., & van Engelen, T. (2010). Sensor fault detection and isolation for wind turbines based on subspace identification and kalman filter techniques. *International Journal of Adaptive Control and Signal Processing*, *24*(8), 687–707.

Wu, M., He, Y., & She, J.-H. (2010). *Stability analysis and robust control of time-delay systems*. Berlin: Springer.

Yeu, T.-K., Kim, H.-S., & Kawaji, S. (2005). Fault detection, isolation and reconstruction for descriptor systems. *Asian Journal of Control*, *7*(4), 356–367.

Zhai, D., Zhang, Q.-L., & Li, J.-H. (2014). Fault detection for singular multiple time delay systems with application to electrical circuit. *Journal of the Franklin Institute*, *351*(12), 5411–5436.

Chapter 4
Fault Diagnosis of Linear Switched Systems Based on Hybrid Observer

Héla Gara and Kamel Ben Saad

Abstract This chapter presents a fault detection and localization method for switched linear systems which are a class of hybrid dynamical systems involving interaction between continuous dynamics and discrete dynamics. Fault diagnosis is an important task that may effectively improve reliability and safety of industrial systems by reducing shutdown numbers. The proposed diagnosis method is based on two hybrid observer modules. The first module is used to identify the current operating mode and the second one is a bank of Luenberger observer which aims to detect and localize the fault. In order to prove its efficiency and validity, the proposed method was applied to detect two types of faults that could affect a boost converter: a sensor fault and a system parameter fault.

Keywords Hybrid observer · Switched linear systems · Boost converter · Faults detection and localization

4.1 Introduction

Automated processes aim to increase productivity and to reduce the cost of equipment maintenance. However, the complexity of production processes implies increasingly a lot of needs in terms of availability, performance, and safety constraints. Thus, it is necessary to have a diagnosis algorithm to detect and to localize the defective elements in a dynamic system that can affect its normal operations. Indeed, the diagnosis procedure is necessary, not only to improve the performance and the productivity of the systems, but also to limit the consequences of failures that can be catastrophic for industrial installations and harmful for machine operators.

H. Gara (✉) · K. B. Saad
LARA Laboratory, University of Tunis El Manar
National Engineering School of Tunis, Tunis, Tunisia
e-mail: gara.hela2009@yahoo.fr

K. B. Saad
e-mail: kamel.bensaad@enit.utm.tn

N. Derbel et al. (eds.), *Diagnosis, Fault Detection & Tolerant Control*,
Studies in Systems, Decision and Control 269,
https://doi.org/10.1007/978-981-15-1746-4_4

Several works have addressed the theme of diagnosis allowing the implementation, in the industrial environment, of some modern approaches resulting from the automatic control field and the artificial intelligence domain. Recently, the field of industrial process and supervision focuses on the diagnosis problem of many systems that involve interactions between continuous dynamics and discrete dynamics. These systems are called hybrid dynamical systems.

In this chapter, we study the problems of the diagnosis of some faults that can affect a class of hybrid systems. A diagnosis method based on a hybrid observer is proposed in order to give the information on the appearance or not of a fault.

4.2 Motivation, Related Work, and Objectives

Diagnosis is involved in many parts of modern industry. This explains the great interest of the automation community on the diagnosis of dynamical systems. It is obvious that a faulty system can also cause significant damage to industrial system, to human, and to the environment.

Many methods have been developed over the past 30 years to diagnose the two types of systems: continuous systems and discrete event systems. Whatever the developed approaches and the type of considered system, diagnosis is always built by the same principle of comparison between actual behaviors and theoretical behaviors. The model-based methods use the normal system behavior model of the system to predict its theoretical evolution.

On the other side, if the objectives are more ambitious, and the faults must be precisely localized and identified, it is necessary to have a model of the system behaviors in the presence of faults. Many systems are not purely continuous or discrete but combine the two aspects. These systems are known commonly as hybrid dynamic systems (HDS) or hybrid systems. There are several classes of HDS. Among these classes, we can cite switched linear systems that evolve following modes of functioning according to imposed discrete events.

Recently, hybrid systems have been the subject of major research works concerning modeling, simulation, and synthesis of proposed controller (Branicky 1996; Labinaz et al. 1996; Antsaklis et al. 1993; Ilraith et al. 1999). Several research works extended known and classical methods for fault detection and localization to this class of hybrid systems.

Thus, with regards to the extension of the methods resulting from the continuous systems, we find methods based on observers (Gertler 1993; Iserman 1984; Willsky 1976) and non-model-based diagnosis methods (Patton et al. 2000). The problems of hybrid observer design and fault detection methods for HDS have been studied using various techniques. Indeed, the role of a hybrid observer (HO) is to estimate both states: the continuous state and the discrete state of the HDS. The fundamental problem during the synthesis of this type of observer concerns the determination of the convergence conditions of the HO and by taking into account the switching phenomenon (Balluchi et al. 2002; Pettersson 2006).

A bank of Luenberger observers was proposed in Belkhiat et al. (2012), Takrouni et al. (2016), Gara and Ben Saad (2018b) to guarantee the sensor fault detection. In Belkhiat et al. (2011), an observer has been designed to ensure fault detection when the knowledge of the discrete mode is unknown. The diagnosis method based on sliding mode observer was proposed in Mejri et al. (2017) to detect actuator faults for a class of hybrid system. The authors in El Mekki and Ben Saad (2016) used a fault diagnosis based on a super-twisting sliding mode observer.

This chapter is organized as follows. In Sect. 4.3, different classes of hybrid dynamical systems are presented. Section 4.4 includes a state of the art of some diagnosis methods. Section 4.5 presents the proposed diagnosis method based on a HO. The simulation results of the proposed diagnosis method applied to a DC–DC boost converter are presented in Sect. 4.6. Finally, the conclusion is given in Sect. 4.7.

4.3 Hybrid Dynamical Systems

In this section, we will focus on the modeling of hybrid systems for the purpose of diagnosis.

4.3.1 Presentation of Hybrid Dynamical System

A hybrid dynamic system is composed of a continuous dynamic system, a discrete event system, and an interface managing the interactions between them as shown in Fig. 4.1. The discrete state and the continuous one are described by Nerode and Kohn (1993).

- **Discrete dynamics**: The discrete dynamics of a HDS is associated to a discrete event system. The transitions between the discrete states are consequence of the occurrence of discrete events which express the influence of the continuous state on the discrete one (Antsaklis et al. 1997).

  These events can be classified into two types:

  - Controlled events that are external events, corresponding to discrete automatic controls or discrete manual controls.
  - Autonomous or spontaneous events that are internal events are produced when the continuous state vector crosses some boundaries of the state space.

- **Continuous dynamics**: The continuous dynamics of a hybrid dynamical system can be modeled by state equations which are generally nonlinear.
- **Interface between dynamics**: The interface expresses the interactions between the continuous part and the discrete part of the HDS. The discrete state determines the specific continuous dynamics of the continuous part (Branicky 1995).

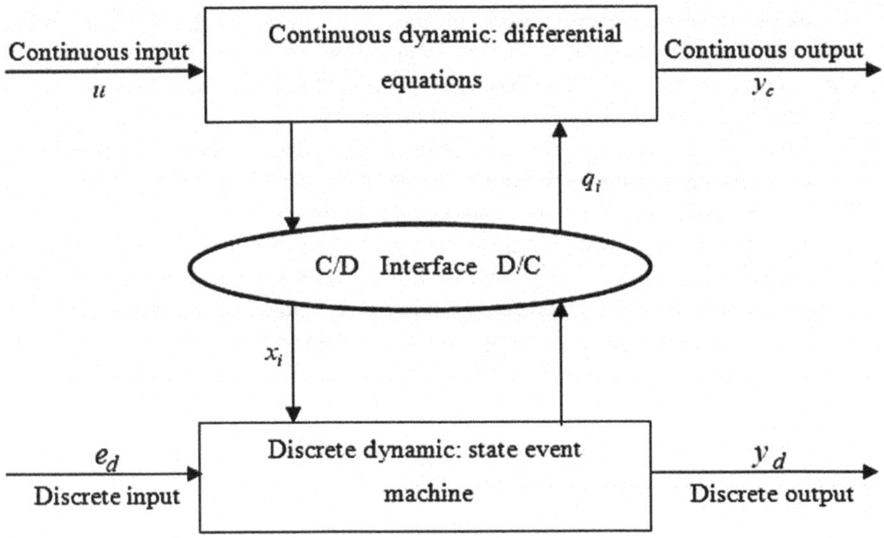

Fig. 4.1 Hybrid dynamical system: interaction between the continuous and discrete part

4.3.2 *Different Classes of Hybrid Systems*

In automation control theory researchers defined several classes of HDS. We will briefly mention some classes (Chua and Dang 1986; Johansson 2003; Haddad et al. 2006; Van der Schaft and Schumacher 2000):

(a) Switched linear systems are characterized by continuous dynamics and react with a set of linear subsystems which are controlled by a discrete event system indicating the active subsystem. Each subsystem is called mode. The discrete dynamic is reacted by a switching law which may depend either on spontaneous switching or on forced switching. As an example, we can mention the electronic circuits controlled by elements with discrete characteristics (switch, diode, thyristor, transistor, etc.).
(b) Linear jump systems are composed of a continuous state and a discrete one controlled by unknown external inputs. The dynamic of these states is reacted by an autonomous system modeled by a set of differential equations without input.
(c) Piecewise constant affine linear systems are characterized by a hybrid state which is composed of a discrete state and a continuous one. Continuous dynamics evolve through a piecewise linear system and discrete dynamics consequence of the invariant discrete transitions.

In the following, we will focus on the class of the switched linear systems, Fig. 4.2.

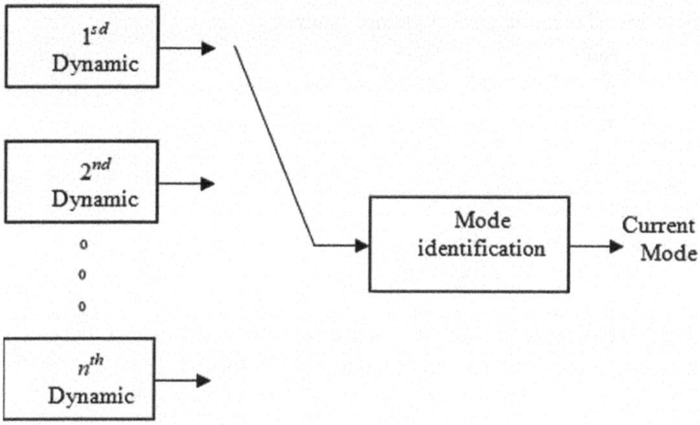

Fig. 4.2 Switched linear system

4.4 System Diagnosis

4.4.1 Definition

In this part, we will cite definitions of some diagnosis keywords and terms.

- Normal system functioning: A system is in a normal functioning state when its variables and characteristic parameters remain near their nominal values. Otherwise, the system is faulty.
- Fault: A fault is any unacceptable difference between a nominal behavior and a measured behavior of a process. This difference is ideally zero in the absence of fault (Isermann 2006).
- Failure: A failure is a permanent interruption of the ability of a process to perform a required function with a performance defined in its technical specifications. A fault appears when there is a difference between the theoretical and measured characteristics. This difference is the cause of a failure or a breakdown that causes the system shutdown (Isermann and Ballé 1997). A breakdown results always from a failure and consequence from a fault:

$$\text{Fault} \longrightarrow \text{Failure} \longrightarrow \text{breakdown}.$$

- Fault detection: The detection of a fault consists in deciding whether the system is in a normal functioning state or not.
- Localization of a fault: It consists in the determination of the faulty elements after the fault detection (Patton et al. 2000).
- Diagnosis: It is the identification of the probable cause of a failure using logical reasoning based on observations.

Table 4.1 Signature table for the case of multiple faults

	Fault f_1	Fault f_2	...	Fault f_j
Residual r_1	1	0	...	0
Residual r_2	0	1	...	0
$\vdots$	$\vdots$	$\vdots$	$\vdots$	$\vdots$
Residual r_j	0	0	...	1

- Residual: Also called fault indicator, it expresses the difference between the measured value and the calculated one provided by a model.

4.4.2 *Different Stages of the Diagnosis of a System*

The diagnosis of an industrial system requires a number of steps summarized in Fig. 4.3 and detailed later (Isermann 2006).

The diagnosis of faults involves three essential steps: fault detection, fault localization, and decision-making.

Fault detection step consists in the generation of signals indicating the appearance of one or several faults.

Fault localization step: This step consists in the determination of the faulty elements. Indeed, in the case of multiple faults, the residual vector must have some properties allowing a unique characterization of each fault. Two techniques ensure a proper residual evaluation (Patton et al. 2000):

- the construction of structured residuals and
- the construction of directional residuals.

The structured residuals are the residuals sensitive to a single fault and insensitive to others. These residuals are generated according to a signature table, as shown in Table 4.1.

Thus, a nonzero residual indicates the appearance of the fault for which it is sensitive.

The directional residuals are constructed in the presence of a given fault. The residual vector is oriented following a precise direction in the residual space.

Figure 4.4 presents, in the space of parity (plane (r_1, r_2, r_3)), the different directions, which can be taken as the vector of parity, according to the faults.

Decision step: The diagnosis logic consists in taking a decision resulting from the residual evolution.

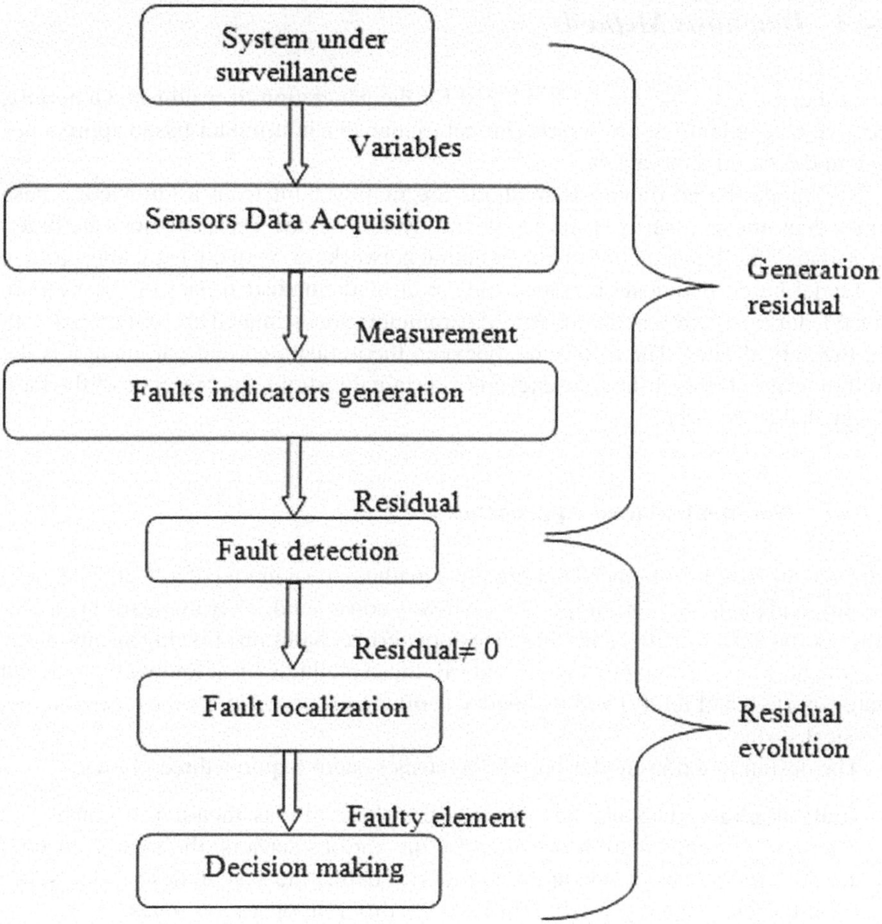

Fig. 4.3 Different steps of diagnosis

Fig. 4.4 Directional residuals

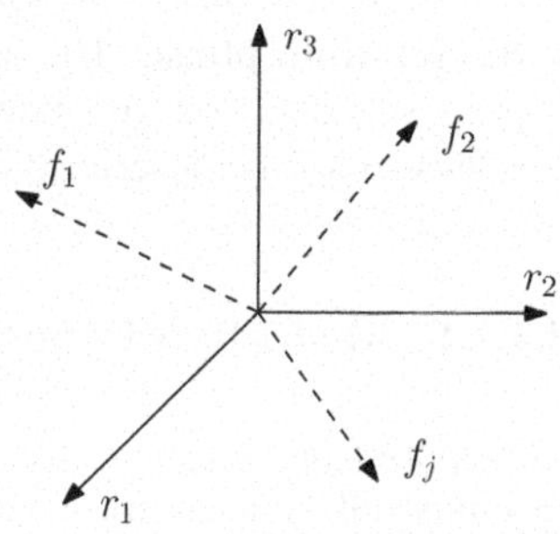

4.4.3 Diagnosis Methods

Several approaches have been developed for the generation of residuals. Generally, they are grouped into two principles branches, namely, non-model-based approaches and model-based approaches.

Non-model-based diagnosis methods are mainly built from a knowledge base which does not necessarily require a specific system model. These methods are based on artificial intelligence or on artificial neural networks or on fuzzy logic approaches.

Model-based diagnosis methods rely on a mathematical model of the system. Indeed, during operation, the measured parameters are estimated and compared with the theoretical ones. The difference between theoretical and real parameters is the fault indicator. If this difference exceeds a certain threshold, the presence of the fault is signaled to the user.

4.4.3.1 Non-model-based Approaches

Non-model-based methods do not require a mathematical model of a system because the physical phenomenon complexity is already considered. They use learning procedures derived from artificial intelligence or pattern recognition allowing an automatic classification of the known classes. Indeed, these methods must know all operating states (normal and faulty) and its history is often unacceptable for the case of a real physical system.

The design of a non-model-based diagnosis system requires three phases:

- **Analysis phase**: The analysis phase is the phase of classification. It consists in examining the information provided by the various sensors, the system history, and the experiences of operators, in order to define the system description space as well as the classes that will represent the different operation states.
- **Choice phase**: This phase consists in building a decision rule that will establish boundaries between the different classes. The decision rule will assign a new observation to one of the known classes.
- **Exploitation phase**: The aim is to implement the diagnosis system in order to propose a decision for a new observation made. In fact, this phase depends on the relevance of the description space and the performances of the decision rule.

4.4.3.2 Model-Based Approaches

The approaches based on mathematical models for fault detection and localization in industrial systems were introduced in the early 70s. Several studies have been carried out on these methods (Willsky 1976; Chow and Willsky 1984; Basseville 1988; Gertler 1993; Isermann and Ballé 1997).

These methods are alternatives to physical or analytical redundancy, which consists in using multiple actuators, sensors, processors, and software to measure and/or control a particular variable.

The model-based methods have a major advantage over the non-model-based methods because the information about the normal behavior of the system is taken into account via a reference model.

Among the used approaches for the generation of residuals, we find

- the parity space approach,
- the observer-based approach, and
- the parametric estimation approach.

Parity space approach
This method relies on the coherence between the process model and the measurements provided by the sensors. This coherence produces fault indicators which are represented by the analytical redundancy relationships. The principle of redundancy is to determine the value of the system variables by different ways and analyses. If the difference between the measured value and the calculated ones, which is called residual, is zero, then the measurements are consistent with the model and the system is declared flawless. A nonzero residual indicates the appearance of a fault.

Observer-based approach
The basic principle of residual generation using observers is to estimate the outputs by using measured inputs and outputs. The difference between the estimated outputs and the measured ones, also called the "estimation errors on the outputs," generates the residual vector, (Fig. 4.5). These residuals must provide reliable indicators of the behavior of the process. This method has been applied to solve several problems of detection and localization of sensor and actuator faults. In the presence of a fault, the sensitive residuals must be nonzero in order to be able to detect and localize it.

Parametric estimation approach
The principle of parametric estimation is to estimate, continuously and in real time, the system parameters using input and output measurements. Then, the residual vector is the difference between these estimated values and the nominal ones obtained for a normal operating state of the process. So, we can use the parametric estimation approach for diagnosis purposes. Figure 4.6 presents the principle of

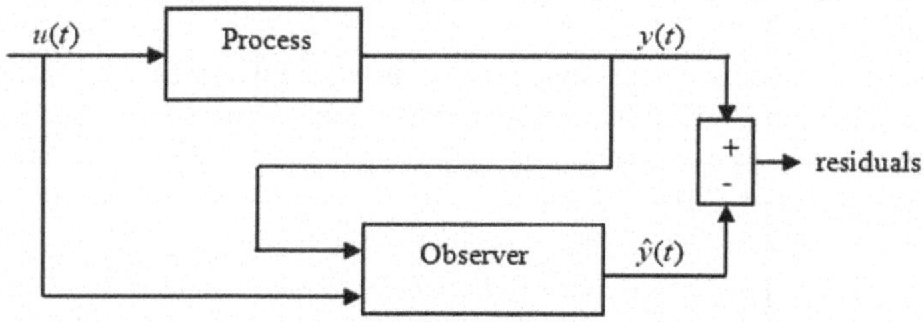

Fig. 4.5 Principle of residual generation using an observer

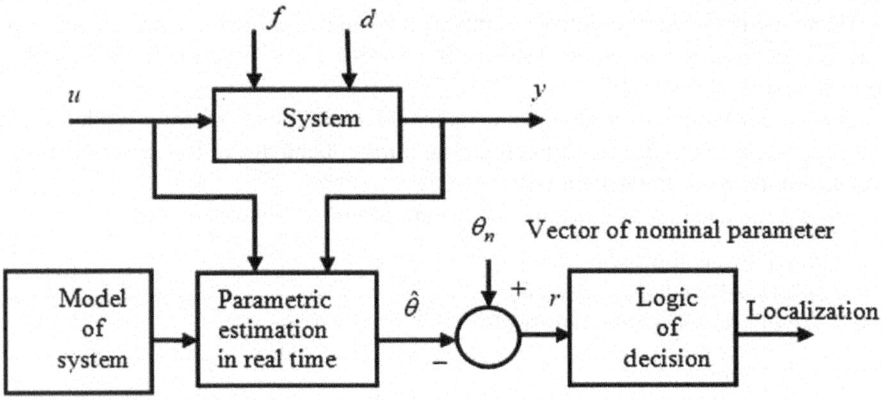

Fig. 4.6 Principle of residual generation by parametric estimation

residuals generation by real-time estimation of the system parameters and denotes the vector of the parameters resulting from the real-time estimation and the corresponding nominal values, respectively.

4.5 Proposed Diagnosis Approach

This section deals with the synthesis of HO for the fault detection of a class of switched linear systems. The HO is synthesized to generate residuals allowing the detection of faults affecting the continuous part of switched linear systems (Belkhiat et al. 2011; El Mekki and Ben Saad 2017; Gara and Ben Saad 2018a).

4.5.1 Switched Linear System Model

The switched linear systems are a subclass of HDS. They are characterized by a hybrid dynamic involving a continuous and discrete dynamics. The continuous state evolution is governed by a linear dynamical subsystem which is controlled by a discrete event system.

When a switched transition occurs, the continuous state is initialized automatically to a new value which is given by an active linear subsystem.

A switched linear system composed of N linear subsystems can be described as follows (Takrouni et al. 2016):

$$(S_q) \begin{cases} \dot{x}(t) = A_q x(t) + B_q u(t) + F_{xq} f(t) + D_q d(t) \\ y(t) = C_q x(t) + F_{yq} f(t) \end{cases} \tag{4.1}$$

for $q = 1 \ldots, N$ and where:
q: the index of the active mode at the instant t,
$x \in \mathbb{R}^n$: the state vector,
$u \in \mathbb{R}^m$: the input vector,
$y \in \mathbb{R}^p$: the measurement vector,
$d \in \mathbb{R}^m$: the unknown input vector, and
$f \in \mathbb{R}^\ell$: the detected fault.

In the following, a fault detecting approach based on a hybrid observer is presented.

4.5.2 Fault Detection Based on a Hybrid Observer

The aim of the proposed diagnosis method is to estimate accurately the continuous state by using the measured inputs and outputs of the system. This estimation requires to know, for each time, the current mode of functioning.

Then, the designed hybrid observer is composed of two modules. The first one identifies the active mode and the second one is a linear observer which estimates the system state for each mode.

In this work, a bank of Luenberger observer is used. The estimated output $\widehat{y}_q(t)$ is compared to the measured one $y_q(t)$ to generate the vector of residuals noted $r_q(t)$ (Fig. 4.7).

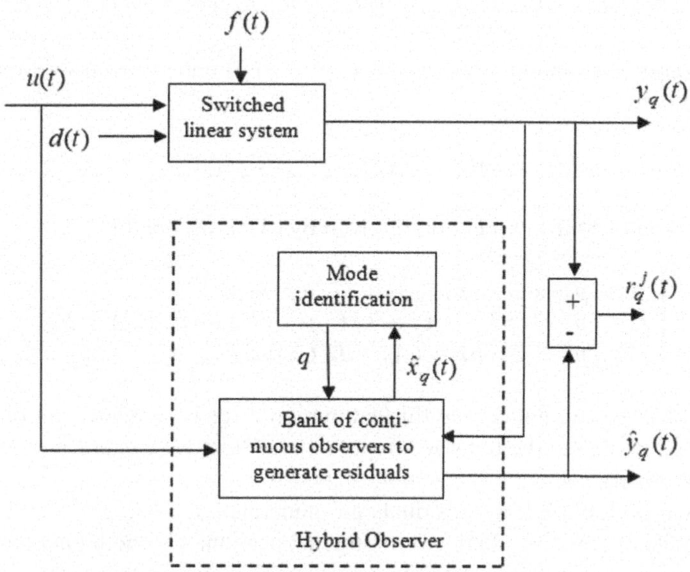

Fig. 4.7 Proposed diagnosis approach principle

Along this work, we consider these two following assumptions:

Assumption 1 The hybrid dynamic system does not allow any measurable discrete output: $y_d = \{\}$.

Assumption 2 All pairs (A_q, C_q) are observable.

A Luenberger observer for the switched linear system (4.1) can be defined as follows:

$$(O_q)\begin{cases} \dot{\widehat{x}}(t) = A_q\widehat{x}(t) + B_q u(t) + L_q[y_j(t) - \widehat{y}_j(t)] \\ \widehat{y}(t) = C_q\widehat{x}(t) \\ r_q^j(t) = y_j(t) - \widehat{y}_j(t) \end{cases} \quad (4.2)$$

where $\widehat{x} \in \mathbb{R}^n$ is the state estimation vector, $\widehat{y}_j$ is the jth component of the estimated output vector, $r_q^j(t)$ is the component of the residuals vector, and L_q is the Luenberger observer gain calculated for each identified mode.

$r_q^j(t)$is the estimation error on the outputs obtained from the Luenberger observer.

The dynamic of the estimation error of the state, defined by $e_{xq}(t) = x(t) - \widehat{x}(t)$, can be written as follows (Patton et al. 2000):

$$\dot{e}_{xq} = (A_q - L_q C_q)e_{xq} + (F_{xq} - L_q F_{yq})f(t) + D_{xq}d(t) \quad (4.3)$$

Therefore, using Laplace transform, the dynamic of the state estimation error can be reformulated as follows:

$$e_{xq}(p) = [pI - (A_q - L_q C_q)]^{-1}[(F_{xq} - L_q F_{yq})f(p) + D_{xq}d(p)] \quad (4.4)$$

Then, the output estimation error $e_{yq} = y(p) - \widehat{y}(p)$ can be expressed by the following expression:

$$e_{yq}(p) = C_q[x(p) - \widehat{x}(p)] + F_{yq}(p)f(p) = C_q e_{xq}(p) + F_{yq}(p)f(p) \quad (4.5)$$

By replacing in (4.5) the expression of given by (4.4), we obtain

$$\begin{cases} e_{yq}(p) = G_{fq}(p)f(p) + G_{dq}(p)d(p) \\ G_{fq}(p) = [pI - (A_q - L_q C_q)]^{-1}(F_{xq} - L_q F_{yq}) + F_{yq} \\ G_{dq}(p) = C_q[pI - (A_q - L_q C_q)]^{-1}D_{xq} \end{cases} \quad (4.6)$$

From (4.6), we can notice that this estimation error is sensitive, not only to the faults $f(p)$, but also to the unknown input $d(p)$. Then, we can use it to generate a vector of residuals.

As shown in Table 4.2, the residuals are generated according to the theoretical signature table of the sensor faults by taking into account all functioning mode of the switched linear system. Thus, the jth on-zero residual is sensitive to the appearance of the jth fault. Then, the system is declared faulty.

Table 4.2 Signatures table for the localization of multiple faults

	f_1	f_2	...	f_j
r_1	1	0	...	0
r_2	0	1	...	0
$\vdots$	$\vdots$	$\vdots$	$\vdots$	$\vdots$
r_j	0	0	...	1

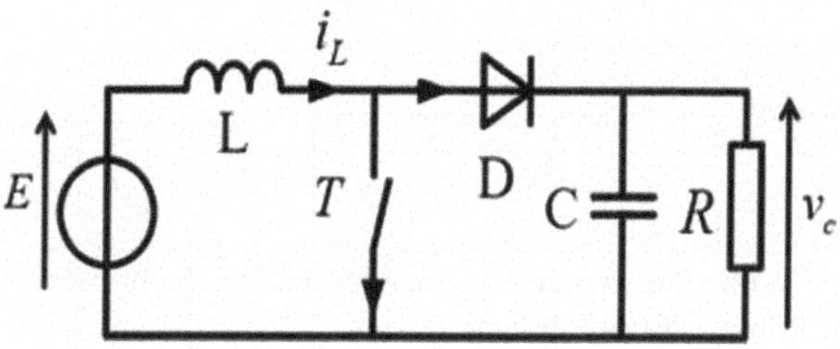

Fig. 4.8 Structure of the studied boost converter

4.6 Case Study

4.6.1 *Studied System*

To illustrate the performance of the proposed diagnosis method based on HO, we choose a boost converter as an example of switched linear system.

The structure of the studied converter, presented in Fig. 4.8, is composed of an inductance L, a capacitor C, a switch T, a diode D, and a load resistance R.

The boost is a DC–DC converter that produces an output voltage v_c higher than the input one E. The relation between v_c and E is described by the following equation:

$$v_c = \frac{1}{1-\alpha} E \tag{4.7}$$

where α is the control signal duty cycle (Sreekumar and Agarwal 2008).

In this work, we have considered the case of a boost converter operating in continuous conduction mode where the current through the inductor i_L never falls to zero during the commutation cycle. Then, two modes of functioning can be identified according to the switch state T.

During the conduction period, the first mode (mode 1) corresponds to the closed state of T. The mathematical model of the studied converter during Mode 1 can be described as follows:

$$\begin{cases} E = L\dfrac{di_L}{dt} \\ C\dfrac{dv_c}{dt} + \dfrac{v_c}{R} = 0 \end{cases} \Longrightarrow \begin{cases} \dfrac{di_L}{dt} = \dfrac{1}{L}E \\ \dfrac{dv_c}{dt} = -\dfrac{1}{RC}v_c \end{cases} \tag{4.8}$$

By considering the state vector $x^T = [i_L \; , \; v_c]^T$, the state-space representation of the studied converter corresponding to the mode 1 can be expressed as follows:

$$(S_1) \quad \begin{cases} \dot{x}(t) = A_1 x(t) + B_1 u(t) \\ y(t) = C_1 x(t) \end{cases} \tag{4.9}$$

where

$$A_1 = \begin{bmatrix} 0 & 1 \\ 0 & -\frac{1}{RC} \end{bmatrix}, \quad B_1 = \begin{bmatrix} \frac{1}{L} \\ 0 \end{bmatrix}, \quad C_1 = \begin{bmatrix} 1 & 0 \\ 0 & 1 \end{bmatrix}$$

When the switch T is open (Mode 2), the boost converter can be described by the following differential equations:

$$\begin{cases} L\dfrac{di_L}{dt} = E - v_c \\ C\dfrac{dv_c}{dt} + \dfrac{v_c}{R} = i_L \end{cases} \Longrightarrow \begin{cases} \dfrac{di_L}{dt} = \dfrac{1}{L}E - \dfrac{1}{L}v_c \\ \dfrac{dv_c}{dt} = \dfrac{1}{C}i_c - \dfrac{1}{RC}v_c \end{cases} \tag{4.10}$$

The state-space representation for Mode 2 is expressed as follows:

$$(S_2) \quad \begin{cases} \dot{x}(t) = A_2 x(t) + B_2 u(t) \\ y(t) = C_2 x(t) \end{cases} \tag{4.11}$$

where

$$A_2 = \begin{bmatrix} 0 & -\frac{1}{L} \\ \frac{1}{C} & -\frac{1}{RC} \end{bmatrix}, \quad B_2 = \begin{bmatrix} \frac{1}{L} \\ 0 \end{bmatrix}, \quad C_2 = \begin{bmatrix} 1 & 0 \\ 0 & 1 \end{bmatrix}$$

The parameters of the studied converter are given in Table 4.3.

Therefore, matrices of the model converter of two modes of functioning are calculated as follows:

Mode 1: $q = 1$

$$A_1 = \begin{bmatrix} 0 & 1 \\ 0 & -303 \end{bmatrix}, \quad B_1 = \begin{bmatrix} 1000 \\ 0 \end{bmatrix}, \quad C_1 = \begin{bmatrix} 1 & 0 \\ 0 & 1 \end{bmatrix}, \quad F_{y1} = \begin{bmatrix} 1 & 0 \\ 0 & 1 \end{bmatrix}$$

Mode 2: $q = 2$

$$A_2 = \begin{bmatrix} 0 & -\frac{1}{L} \\ \frac{1}{C} & -\frac{1}{RC} \end{bmatrix}, \quad B_2 = \begin{bmatrix} \frac{1}{L} \\ 0 \end{bmatrix}, \quad C_2 = \begin{bmatrix} 1 & 0 \\ 0 & 1 \end{bmatrix}$$

Table 4.3 Parameters of the studied boost

Parameters	Input voltage	Resistance	Inductance	Capacitor	Switching period	Duty cycle
Values	5 V	100 Ω	0.001 H	3.3×10^{-6} F	40 ms	0.5

4.6.2 Implementation of the Hybrid Observer

In our case study, a bank of Luenberger observers is used to localize the sensor faults. Then, each mode q has its Luenberger observer which is sensitive to the two outputs of the boost converter i_L and v_c.

To ensure the convergence of each observer, it is necessary to determine the gain matrices L_q.

Indeed, each mode of the converter has two orders, and then two pole values are determined to realize the pole placement. Therefore, according to the dynamics of each mode, we choose

$$p_1 = -3 \times 10^5\ ,\quad p_2 = -2 \times 10^5$$

So, the obtained observer gains are

$$L_1 = \begin{bmatrix} 300000 & 0 \\ 0 & 199700 \end{bmatrix},\quad L_2 = \begin{bmatrix} 300000 & -1000 \\ 30300 & 199700 \end{bmatrix}$$

The state observer for the first mode is defined as follows:

$$(O_1)\ \begin{cases} \begin{bmatrix} \dot{\hat{x}}_1 \\ \dot{\hat{x}}_2 \end{bmatrix} = 10^5 \begin{bmatrix} -3 & 0 \\ 0 & -2 \end{bmatrix} \begin{bmatrix} \hat{x}_1 \\ \hat{x}_2 \end{bmatrix} + \begin{bmatrix} 1000 & 300000 & 0 \\ 0 & 0 & 199700 \end{bmatrix} \begin{bmatrix} u \\ y_1 \\ y_2 \end{bmatrix} \\ \begin{bmatrix} \hat{y}_1 \\ \hat{y}_2 \end{bmatrix} = \begin{bmatrix} 1 & 0 \\ 0 & 1 \end{bmatrix} \begin{bmatrix} \hat{x}_1 \\ \hat{x}_2 \end{bmatrix} \end{cases} \tag{4.12}$$

So, the state observer for the second mode is given by the following expression:

$$(O_2)\ \begin{cases} \begin{bmatrix} \dot{\hat{x}}_1 \\ \dot{\hat{x}}_2 \end{bmatrix} = 10^5 \begin{bmatrix} -3 & 0 \\ 0 & -2 \end{bmatrix} \begin{bmatrix} \hat{x}_1 \\ \hat{x}_2 \end{bmatrix} + \begin{bmatrix} 1000 & 300000 & -1000 \\ 0 & 30300 & 199700 \end{bmatrix} \begin{bmatrix} u \\ y_1 \\ y_2 \end{bmatrix} \\ \begin{bmatrix} \hat{y}_1 \\ \hat{y}_2 \end{bmatrix} = \begin{bmatrix} 1 & 0 \\ 0 & 1 \end{bmatrix} \begin{bmatrix} \hat{x}_1 \\ \hat{x}_2 \end{bmatrix} \end{cases} \tag{4.13}$$

The two matrices $G_{fq}(p)$ are defined as follows:

Table 4.4 Fault signature table

	f_{i_L}	f_{v_c}
r_1^1	1	0
r_1^2	0	1
r_2^1	1	1
r_2^2	1	1

$$G_{f1}(p)=\begin{bmatrix} \frac{p}{p+300000} & 0 \\ 0 & \frac{p+303}{p+200000} \end{bmatrix},\quad G_{f2}(p)=\begin{bmatrix} \frac{p}{p+300000} & \frac{1000}{p+300000} \\ -\frac{30300}{p+200000} & \frac{p+303}{p+200000} \end{bmatrix}$$

Then, the error estimations of the first and the second modes are expressed as follows:

$$r_1(p) = e_{y1}(p) = \begin{bmatrix} \frac{p}{p+300000} & 0 \\ 0 & \frac{p+303}{p+200000} \end{bmatrix} f(p) \tag{4.14}$$

$$r_2(p) = e_{y2}(p) = \begin{bmatrix} \frac{p}{p+300000} & \frac{1000}{p+300000} \\ -\frac{30300}{p+200000} & \frac{p+303}{p+200000} \end{bmatrix} f(p) \tag{4.15}$$

The table of theoretical signatures associated to r_q^j is given in Table 4.4.

It is clear that for the first mode, the residuals r_1^1 and r_1^2 can localize the faults. However, for the second mode, the residuals r_2^1 and r_2^2 have identical signatures. Thus, fault localization is impossible for this mode.

4.6.3 Simulation Results

The principle of the residual generation based on the proposed HO is shown by the block diagram of Fig. 4.9.

Simulations are performed under MATLAB/SIMULINK over the time interval [0, 0.1s].

Figures 4.10, 4.11, and 4.12 present the measured outputs of the studied boost converter (i_L, v_c), the estimated outputs provided by the proposed hybrid observer, and the estimation error (e_{y1}, e_{y2}), respectively.

Sensor faults

Let us consider the sensor faults f_{i_L} and f_{v_c} occurring successively at $t = 0.05$s. The faults are additive and modeled by steps which represent a variation of the nominal values of the sensors. The residuals of each mode are shown in Figs. 4.13 and 4.14.

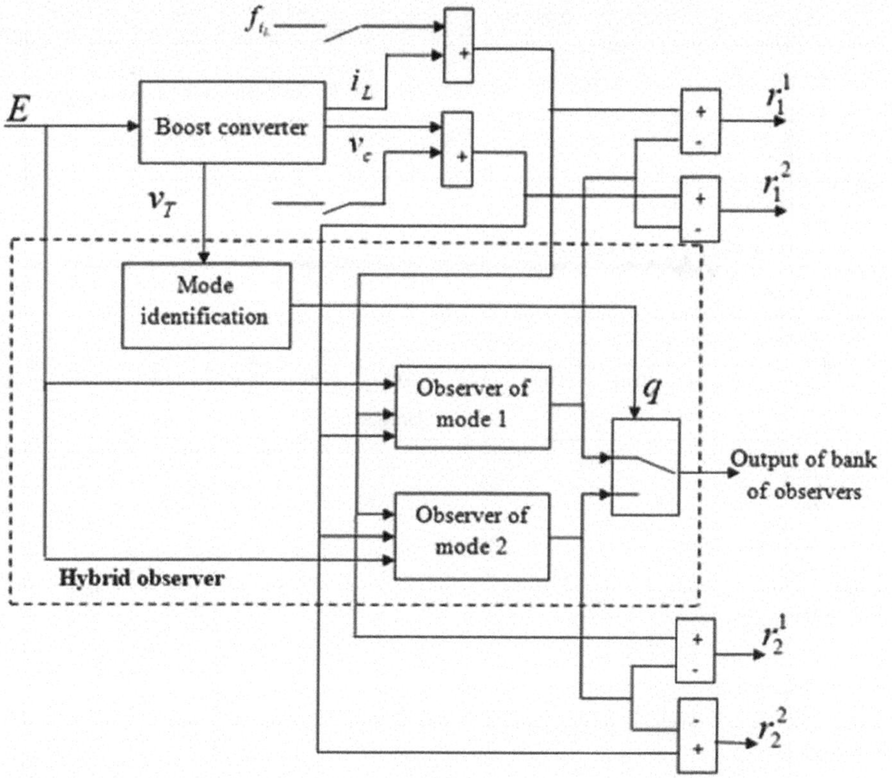

Fig. 4.9 Bank of residual principle

It is clear that the detected faults, at time, evolve according to the signature table presented in Table 4.4.

When the sensor fault occurs, each sensitive component of the residual vector deviates significantly from zero ensuring its detection. Then, this residual generator is sensitive to the considered sensor fault.

Parameter faults

Now, we consider that the fault consists in the variation of one converter parameter. For example, at $t = 0.05$s, the value of the resistance R decreases from 100Ω to 9Ω. The simulated output voltage of the boost converter is shown in Fig. 4.15.

Figures 4.16 and 4.17 show the structured residuals for each mode in the presence of the considered parameter variation fault.

From the simulation results, we can notice that the sensitive residuals vary suddenly when the parameter fault occurs and the insensitive residuals remain to converge to zero. For the two modes of functioning, the signatures of these residuals are not identical. Thus, they are well structured. Therefore, through these residual signature, we can localize the parameter fault for the case of the studied system.

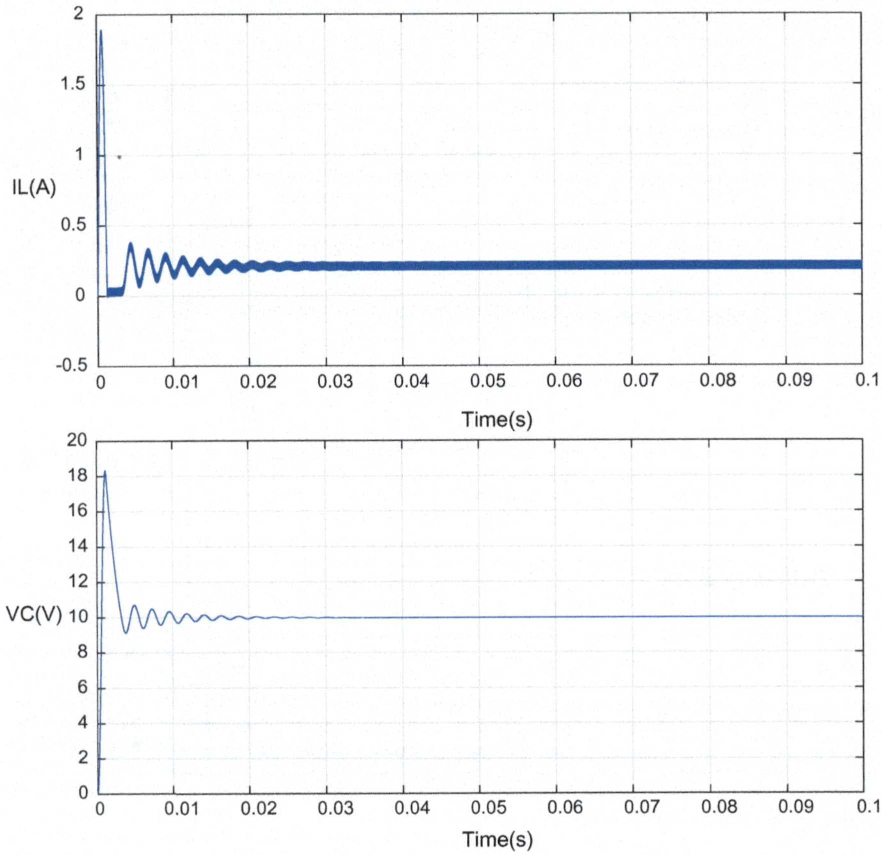

Fig. 4.10 Outputs of boost converter

From the obtained simulation results, we can deduce that the proposed diagnosis method proves clearly its validity to detect and to localize a resistance load value change.

4.7 Conclusion

In this chapter, a general overview of hybrid dynamical systems was presented. In fact, hybrid systems involve an interaction between continuous and discrete dynamics. A brief description of different diagnosis methods is presented. Two approaches are cited: model-based approaches and non-model-based approaches. The model-based diagnosis methods use mathematical relationships between the inputs and outputs of the system. We can cite parity space, observer and parametric estimation.

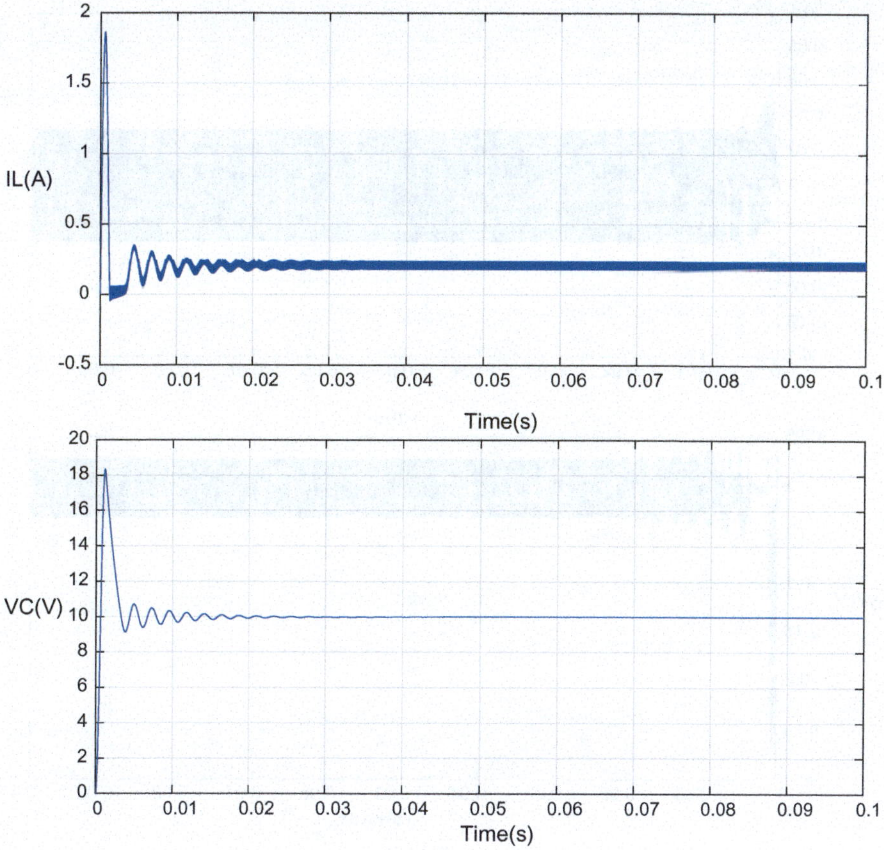

Fig. 4.11 Estimated outputs of the hybrid observer

However, the non-model-based approaches rely mainly on a knowledge base and on the recognition of the signatures already observed. They do not necessarily require an accurate model of the system. These methods can be based on artificial intelligence approaches or on artificial neural network approaches or on fuzzy logic approaches.

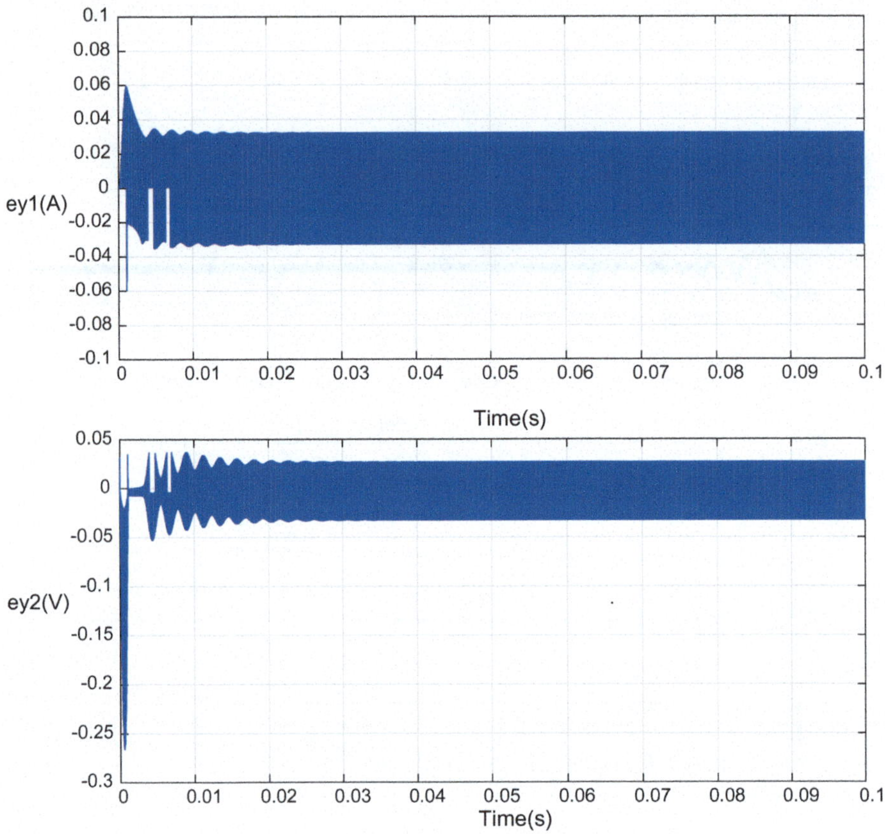

Fig. 4.12 Estimated error

To diagnose the switched linear system, which is a class of HDS, we have proposed a hybrid observer. The proposed HO uses two modules. The first one identifies the active mode and the second one is a bank of Luenberger observers used to estimate the active mode state vector. A bank of residuals is generated to detect the fault. Finally, we applied the proposed approach to a boost converter. The simulation results show the efficiency of the designed hybrid observer to detect and to localize both sensor faults and parameter faults.

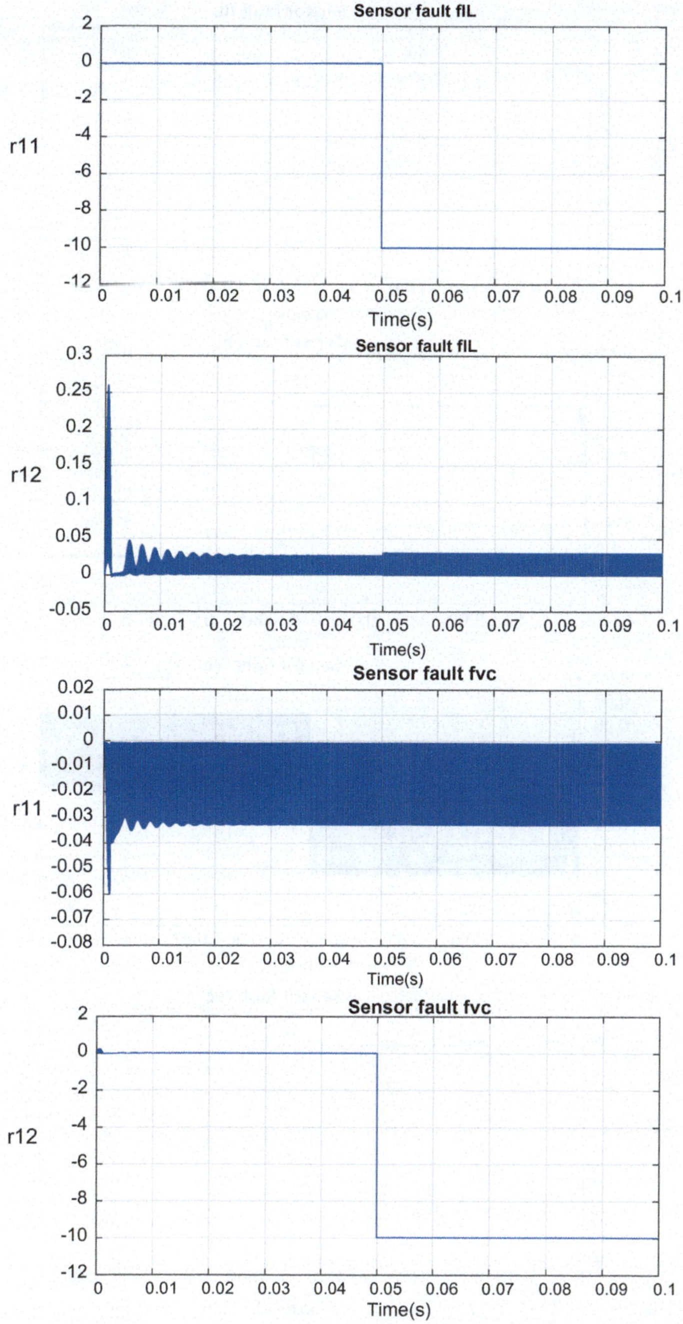

Fig. 4.13 Evolution of the residuals (r_1^1, r_1^2) of the first mode at the presence of faults f_{i_L} and f_{v_c}

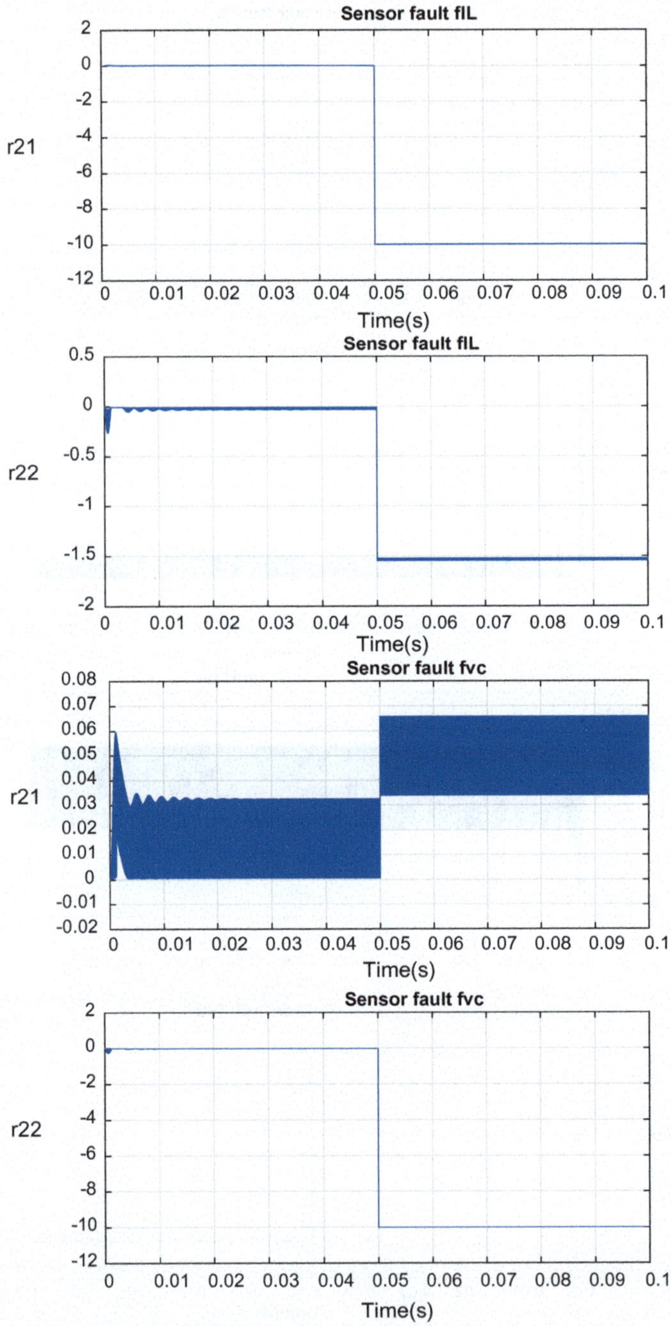

Fig. 4.14 Evolution of the residuals (r_2^1, r_2^2) of the second mode at the presence of faults f_{i_L} and f_{v_c}

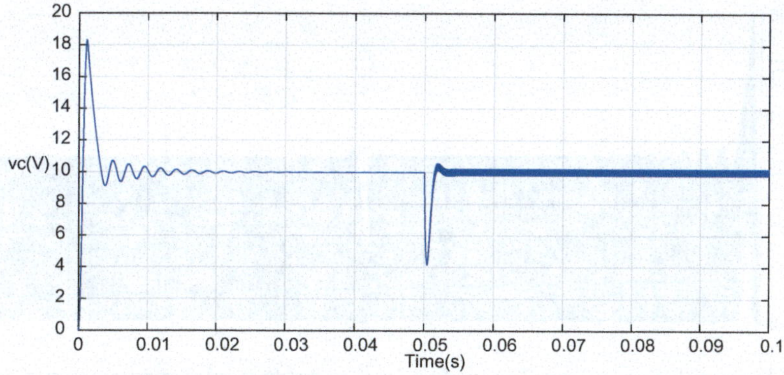

Fig. 4.15 Output voltage of boost converter for the case of the variation from 100 to 9Ω

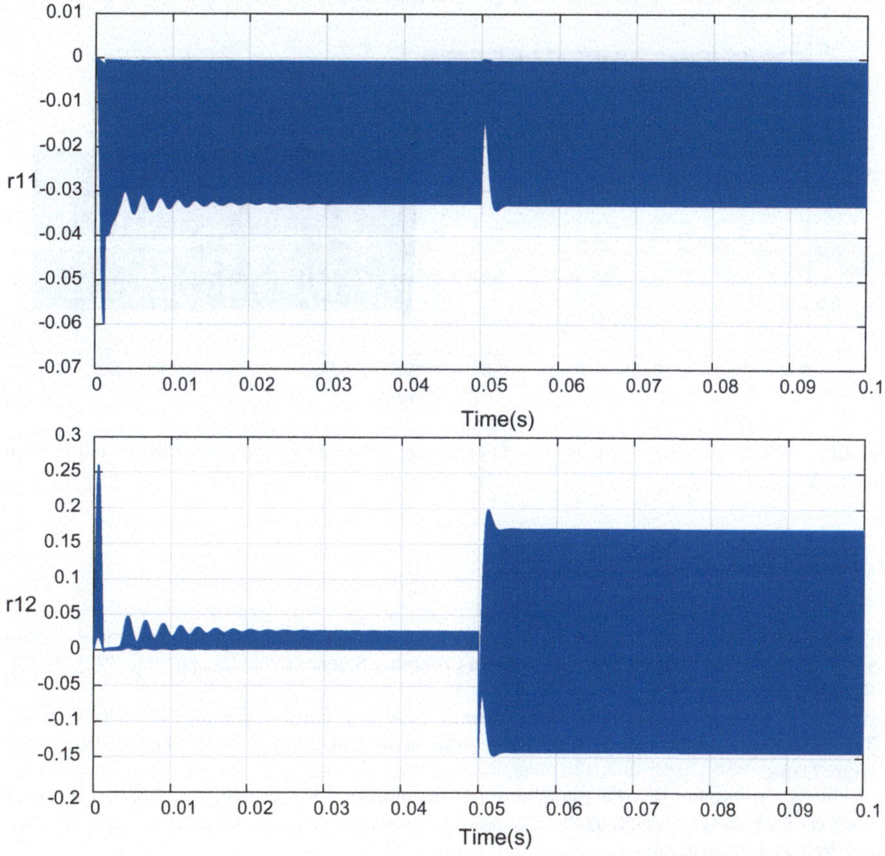

Fig. 4.16 Evolution of the residuals (r_1^1, r_1^2) of the first mode for the case of the variation from 100 to 9Ω

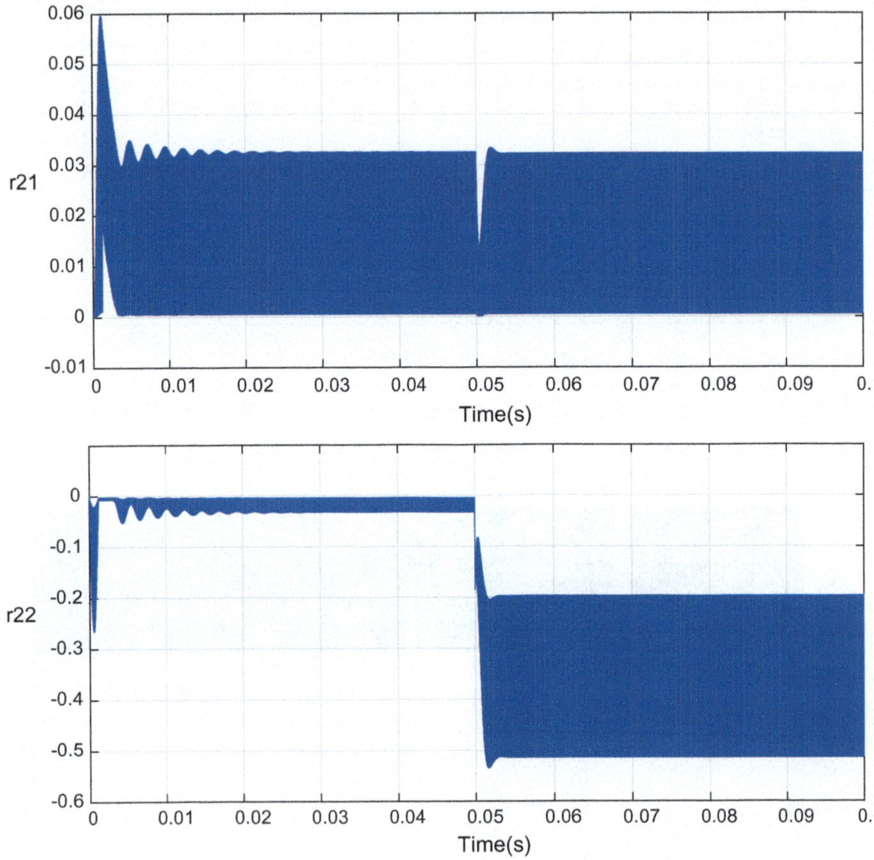

Fig. 4.17 Evolution of the residuals (r_2^1, r_2^2) of the second mode for the case of the variation from 100 to 9Ω

References

Antsaklis, P., Stiver, J., & Lemmon, M. (1993). Hybrid system modeling and autonomous control systems. In *Hybrid Systems*. Lecture Notes in Computer Science (Vol. 736, pp. 366–392), Berlin: Springer.

Antsaklis, P., Koutsoukosand, X.D., & Zaytoon, J. (1997). Control of complex system: A survey. Technical Report of the ISIS G' On hybrid group at the University of Notre Dame ISIS-97-017, Notre Dame.

Balluchi, A., Benvenuti, L., Di Benedetto, M., Sangiovanni-Vincentelli, A. (2002). Design of observer for hybrid system. In *Hybrid Systems: Computation and Control (HSCC'02)* (Vol. 2289, pp. 76–89). Berlin: Springer.

Basseville, M. (1988). Detecting changes in signals and systems: A survey. *Automatica*, *24*(3), 309–326.

Belkhiat, D. E. C., Messai, N. & Manamanni, N. (2011). Design of a robust fault detection based observer for linear switched systems with external disturbances. *Nonlinear Analysis: Hybrid Systems, 5*, 206–219.

Belkhiat, D.E.C., Manamanni, N., Messai, N., & Djemai, M. (2012). Fault detection & isolation for a class of hybrid systems: A dedicated switched robust observer scheme. In *20th Mediterranean Conference on Control and Automation (MED)* (pp. 978–983). Barcelona.

Branicky, M. (1996). General hybrid dynamical systems: modeling, analysis, and control. In *Lecture Note: Hybrid System III* (Vol. 4066, pp. 187–200). Berlin: Springer.

Branicky, M. (1995). *Studies in hybrid systems: modeling, analysis, and control*. Ph.D. thesis, Massachusetts Institute of Technology, Cambridge.

Chow, E. Y., & Willsky, A. S. (1984). Analytical redundancy and the design of robust failure detection systems. *IEEE Transactions on Automatic Control, 29*, 603–614.

Chua, L. O., & Dang, A. C. (1986). Canonical piecewise-linear modeling. *IEEE Transactions Circuit and System, 33*(5), 511–525.

El Mekki, A., & Ben Saad, K. (2016). A BLDC fault diagnosis approach based on a super-twisting sliding mode observer. In *4th International Conference on Control Engineering and Information Technology (CEIT-2016) Tunisia, Hammamet, 16–18 December*.

El Mekki, A., & Ben Saad, K. (2017). Fault diagnosis of buck converter driven DC motor based on a hybrid unknown input observer. In *Conférence Internationale en Automatique & Traitement de Signal (ATS-2017) Proceedings of the Engineering and Technology—PET* (Vol. 22, pp. 103–107). Tunisia.

Gara, H., & Ben Saad, K. (2018a). Design of a fault detection hybrid observer for linear switched systems. In *15th International Multi-conference on Signals, Systems, and Devices SSD'2018, Hammamet*.

Gara, H., & Ben Saad, K. (2018b). Fault detection for linear switched systems based on a bank of Luenberger observers. In *2nd International Conference on Advanced Systems and Electrical Technologies IC-ASET 2018, Hammamet*.

Gertler, J. (1993). Analytical redundancy methods in failure detection and isolation. *Control Theory and Advanced Technology, 9*(1), 259–285.

Haddad, W., Chellaboina, V. & Nersesov, S. G. (2006). *Impulsive and hybrid dynamical systems. Princeton series in applied mathematics*. Princeton: Princeton University Press.

Ilraith, S., Biswas, G., Clancy, D., & Guota, V. (1999). Towards diagnosing hybrid systems. In *10th International Workshop on Principles of Diagnosis, LochAwe, Ecosse* (pp. 193–203).

Iserman, R. (1984). Process fault diagnosis based on modeling and estimation methods: A survey. *Automatica, 20*(4), 387–404.

Isermann, R., & Ballé, P. (1997). Trends in the application of model-based fault detection and diagnosis of technical process. *Control Engineering Practice, 5*(5), 709–719.

Isermann, R. (2006). *Fault diagnosis systems*. Berlin: Springer.

Johansson, M. (2003). *Piecewise linear control systems: A computational approach*. Berlin: Springer.

Labinaz, G., Bayoumi, M., & Rudie, K. (1996). Modeling and control of hybrid system: A survey. In *13th Triennial World Congress*. San Francisco, USA.

Mejri, S., Dhahri, S., & Sellami, A. (2017). Actuator faults reconstruction for a class of dynamical hybrid system. In *ICCAD'17, Hamammet, Tunisia, January*.

Nerode, A., & Kohn, W. (1993). *Models for hybrid systems: automata, topologies, controllability, observability*. Lecture Notes in Computer Science, Hybrid Systems I (Vol. 736, pp. 317–356), Berlin: Springer.

Patton, R. J., Frank, P. M., & Clark, R. N. (2000). *Issues of fault diagnosis for dynamic systems*. Berlin: Springer.

Pettersson, S. (2006). Designing switched observers for switched systems using multiple Lyapunov functions and dwell-time switching. In *2nd IFAC Conference on Analysis and Design of Hybrid Systems*. Alghero, Italy.

Sreekumar, C., & Agarwal, V. (2008). A hybrid control algorithm for voltage regulation in DC-DC boost converter. *IEEE Transaction*, *55*(6), 2530–2538.

Takrouni, A., Zanzouri, N., & Labidi, I. (2016). Robust diagnosis for hybrid noise systems. *International Journal Modelling, Identification and Control*, *25*(3), 177.

Van der Schaft, A., & Schumacher, H. (2000). *An introduction to hybrid dynamical systems*. Berlin: Springer.

Willsky, A. S. (1976). A survey of design methods for failure detection in dynamic systems. *Automatica*, *12*(6), 601–611.

Chapter 5
Neutral Time-Delay System: Diagnosis and Prognosis Using UIO Observer

M. Bettaher, A. Elhsoumi and S. Bel Hadj Ali Naoui

Abstract The present work presents a diagnosis approach design and proposes a new prognosis method for a class of neutral time-delay system. An unknown input observer (UIO) is designed to achieve actuator fault detection. Another UIO with a new form is used to generate an algebraic method of system prognosis. The two observers are applied to the transmission line system as an example of a neutral time-delay system. Simulation results reveal that the application of the proposed observers for the transmission line system approach providing Fault detection is achieved by the two methods. The diagnosis approach detection occurs at the same time of fault appearance. However, the prognosis approach detects the fault before its happening which can give a chance to correct system behavior early with minimal losses.

Keywords Neutral time-delay system · Diagnosis · Prognosis · UIO
Fault detection

5.1 Introduction

Industrial companies are now facing substantial competition in the market to fulfill the demands of their customers in terms of service, quality, and product performance. Maintenance is a key to in this race of providing quality service in maintaining the

M. Bettaher (✉) · A. Elhsoumi · S. Bel Hadj Ali Naoui
MACS Laboratory: Modeling, Analysis and Control of Systems, National Engineering School of Gabes (ENIG), University of Gabés, Gabés, Tunisia
e-mail: bettaher.marwa@yahoo.fr

A. Elhsoumi
e-mail: hsoumi.aicha@yahoo.fr

S. Bel Hadj Ali Naoui
e-mail: saloua.belhadjali@enit.rnu.tn

S. Bel Hadj Ali Naoui
Preparatory Institute for Engineering Studies—El Manar, El Manar University, Tunis, Tunisia

N. Derbel et al. (eds.), *Diagnosis, Fault Detection & Tolerant Control*,
Studies in Systems, Decision and Control 269,
https://doi.org/10.1007/978-981-15-1746-4_5

viability and the dependability of the operation. The presence of delays can introduce undesirable effect; some defects may occur in the system behavior in the actuator, sensor or in the model which make systems behavior, unstable.

Therefore, diagnosis and prognosis are the essential objectives of research to make model control easier. In the past, maintenance strategies based on the diagnosis were used to facilitate corrective maintenance by detecting, locating, and identifying faults in a system. Overall, there are two diagnostic approaches: the quantitative approach using mainly analytical models (parity space, observers, parameter estimation, etc.) (Isermann 1997, 2005; Yang and Saif 1996; Toscano 2005; Kratz et al. 1998; Bettaher et al. 2017; Wang et al. 2009) and the qualitative approach using methods without analytical models (based essentially on Artificial intelligence) (Piechowiak 2003; Basseville and Cordier 1996).

Yet, in some critical situations the prediction of failures is even more desirable than their diagnosis because prognosis allows us to estimate the remaining life of a system from its current state to a failed state. The prognosis methods can be classified according to three approaches:prognosis based on a physical model, prognosis guided by data, and prognosis based on feedback data (Ribot 2009; Schutz 2009; Gucik-Derigny 2011; Adams and Nataraju 2002).

The development of diagnosis and prognosis approaches remains limited to some works. To this end, our research aims at the developing of a diagnosis and prognosis approach based on observer with unknown inputs for neutral time-delay systems.

In this chapter aims at developing a prognosis and diagnostic approach based on observer with unknown inputs for neutral time delay systems.

This work is organized as follows: In Sect. 5.2, an algorithm for obtaining unknown input observer (UIO) to be used in the diagnosis strategy is developed. The Sect. 5.3, presents the proposed degradation prognosis method to evaluate the state of cumulative degradation and its dynamics. The Sect. 5.4, illustrates the relevance of the diagnosis and the prognosis method applied to a transmission line system. Discussions about diagnosis and prognosis works are developed in Sect. 5.5. Finally, conclusions are given in the last section.

5.2 Diagnosis of a Linear Neutral Delay System

The diagnosis is to determined the type, the size, the place, and the time of occurrence of a fault, it follows the fault detection and includes its isolation and identification. Consider the following neutral time-delay system.

$$\begin{cases} \dot{x}(t) = A_1 x(t) + A_2 x(t-h) + A_d \dot{x}(t-d) + Bu(t) + Ek(t) + Ff(t) \\ y(t) = Cx(t) \\ x(\theta) = \rho(\theta), \forall \theta \in [-l, 0] \end{cases} \tag{5.1}$$

where $x \in \mathbb{R}^n$ is the state vector, u $\in \mathbb{R}^m$ is the input vector, $y(t) \in \mathbb{R}^P$ is the output vector and $f(t) \in \mathbb{R}^r$ represents the unknown actuator fault and $k(t) \in \mathbb{R}^q$ is the disturbances; $h > 0$ and $d > 0$ are respectively delays on the state and its derivative.

A_1, A_2, A_d, E, F, B and C are constant matrices and matrix F is of full column rank, i.e., rank $(F) = r$. $l = \max\{h, d\}$ and $\rho(t) \in \mathbb{R}^n$ is the continuous function of the initial condition in [–l, 0].

To detect the actuator faults the unknown input observer developed by Elhsoumi et al. (2013) can be used with the following structure:

$$\begin{cases} \dot{z}(t) = \widehat{A}_1 z(t) + \widehat{A}_2 z(t-h) + \widehat{A}_d \dot{z}(t-d) \\ \qquad +D_1 y(t) + D_2 y(t-h) + D_d \dot{y}(t-d) + Bu(t) \\ \widehat{x} = z(t) + My(t) \\ z(\theta) = \rho(\theta), \forall \theta \in [-l, 0] \end{cases} \tag{5.2}$$

with $z(t) \in \mathbb{R}^g$ is the vector estimate $g \leq$ n, $\widehat{A}_1$, $\widehat{A}_2$, $\widehat{A}_d$, D_1, D_2, D_d, M, and B are unknown matrices to determine in order that $\widehat{x}$ converge to $x(t)$. $\rho(t) \in \mathbb{R}^n$ is a function of initial condition vector continuous on $[-l, 0]$. The expression of state error is obtained by:

$$e(t) = x(t) - \widehat{x}(t) \tag{5.3}$$

The residual vector is $r(t) = Ce(t)$.

To achieve the fault detection, residual behavior should be depend only to faults and achieve the following conditions:

$$\begin{cases} r(t) = 0 \quad \text{if} \quad f = 0 \quad \text{and} \quad \forall x, u, d \\ r \neq 0 \quad \text{if} \quad f \neq 0 \quad \text{and} \quad \forall x, u, d \end{cases} \tag{5.4}$$

The matrices that describe the observer developed by Elhsoumi are obtained by the following equation (Elhsoumi et al. 2013):

$$(\widehat{A}_1 \;\; \widehat{A}_2 \;\; \widehat{A}_d \;\; D_1 \;\; D_2 \;\; D_d \;\; \widehat{B} \;\; M) \bigcup = \bigvee \tag{5.5}$$

with:

$$M = E[(CE)^T CE]^{-1}(CE)^T$$

$$\bigcup = \begin{pmatrix} I - MC & \cdots & 0 & 0 & 0 & 0 \\ \vdots & \ddots & 0 & 0 & 0 & 0 \\ 0 & 0 & I - MC & 0 & 0 & 0 \\ 0 & 0 & 0 & I - MC & 0 & 0 \\ C & \cdots & 0 & 0 & 0 & 0 \\ \vdots & \ddots & \vdots & \vdots & \vdots & 0 \\ 0 & \cdots & C & 0 & \vdots & 0 \\ 0 & \cdots & 0 & C & 0 & 0 \\ 0 & \cdots & 0 & 0 & I & 0 \\ 0 & \cdots & 0 & 0 & 0 & CE \end{pmatrix}. \tag{5.6}$$

and

$$\bigvee = ((I - MC)A_1 \ (I - MC)A_2 \ (I - MC)A_d \ (I - MC)B \ E) \tag{5.7}$$

The author in Elhsoumi et al. (2013) proposed a generalization of asymptotic condition observer to estimate unknown input in the case of neutral time-delay system. To obtain a solution of system (5.5), it is necessary that the decoupling condition unknown inputs: $\text{rank}(EC) = \text{rank}(E)$ is verified; This condition is usually encountered in the synthesis of UIO in the case of neutral time-delay systems.

5.3 Prognosis for a Class of Neutral Time-Delay System

5.3.1 Model Class for Prognosis

Nowadays, the Management Prognosis Health (MPH) is trying to increase degradation prognosis methodologies to estimate the dynamics of system degradation state and predict the Life Time Remaining (LTR). For this reason, a conventional general model class is used by Gucik-Derigny (2011):

$$\begin{cases} \dot{x} = w(\kappa, u) \\ y = q(\kappa, u) \\ \kappa \in v, u \in U, y \in \delta \\ w(.,.) : v \times U \to v, y(.,.) : v \times U \to \delta \end{cases} \tag{5.8}$$

$\kappa = [x, \dot{x}, \alpha]^T$ is associated with the state x and degradation α. f is a function which is at least class C^1. v is an open set of $x \in \mathbb{R}^n$, U is an open set of $\mathbb{R}^m$ regarding environmental and load conditions. δ is an open set of $\mathbb{R}^p$, $t \in \mathbb{R}^+$ is the time variable. In order to describe the behavior of this system, a subclass of the model (5.8) is used:

$$\begin{cases} \dot{x} = w(x, \dot{x}, \lambda(\alpha), u); x(t_0) = x_0 \\ \dot{\alpha} = \varepsilon s(x, \dot{x}, \alpha), \alpha(t_0) = \alpha_0 \\ y = q(x, \dot{x}, \alpha, u) \end{cases} \tag{5.9}$$

$x \in \mathbb{R}^n$ is a state vector which has a fast dynamic behavior and $\dot{x}$ is the derivative of the states. $\lambda \in \mathbb{R}^r$ is the vector of parameters, function of α.

$\alpha \in \mathbb{R}^n$ is the set of states of degradation associated with slow dynamics, $u \in \mathbb{R}^m$ is the vector composed of the environmental conditions and solicitation. The ratio of separation of fast time scales and slow dynamics is described by $0 < \varepsilon << 1$. $y \in \mathbb{R}^p$ is the output vector.

Degradation is defined as a process of alteration caused by changes in the structural properties of the system, irreversibly. It is synonymous with a slope of functionality when it reaches a critical threshold. Therefore, $\dot{\alpha} = \varepsilon s(x, \dot{x}, \alpha)$, w, s, q are assumed as being of continuously differentiable functions of their arguments $x, \alpha, u, \dot{x}$.

The principal contribution of this paper is to introduce models based prognostic methodology to rebuild dynamics accumulation of the degradation state, using only known system input and output measures. The method is based first on the fact that the deterioration will be brought to an unknown input.

5.3.2 System Description and Problem Formulation

a. UIO based degradation prognosis:

`Assumption 1`: the output function is not a function of the degradation state which consider a model class in the form of system (5.9).

$$\begin{cases} \dot{x}(t) = A_1(\alpha)x(t) + A_2(\alpha)x(t-l) + A_d(\alpha)\dot{x}(t-l) + B(\alpha)u(t) \\ y(t) = Cx(t) \end{cases} \tag{5.10}$$

$x \in \mathbb{R}^n$, $y \in \mathbb{R}^p$, $u \in \mathbb{R}^m$, $\alpha \in \mathbb{R}^q$ and the matrices A_1, A_2, A_d, B, and C are defined with appropriate dimensions.

`Assumption 2`: In the structural properties the changes caused by degradation in the structural properties are supposed to be modeled with linear structure as follows:

$$\begin{cases} A_1(\alpha) = A_{1_0} + \sum_{i=1}^{q} A_{1_i}\alpha_i\ , \quad A_2(\alpha) = A_{2_0} + \sum_{i=1}^{q} A_{2_i}\alpha_i \\ A_d(\alpha) = A_{d_0} + \sum_{i=1}^{q} A_{d_i}\alpha_i\ , \quad B(\alpha) = B_0 + \sum_{i=1}^{q} B_i\alpha_i \end{cases} \tag{5.11}$$

Bye making (5.11) in (5.10) and by considering hypothesis 2, the system obtained is as follows:

$$\begin{cases} \dot{x} = A_{1_0}x(t) + A_{2_0}x(t-l) + A_{d_0}\dot{x}(t-l) + B_0u(t) + \Omega(x, \dot{x}, u, \alpha) \\ y = Cx(t) \end{cases} \tag{5.12}$$

`Assumption 3`: We consider a matrix J such as $\text{rank}(J) \leq p$. In addition, $\text{rank}(J) = q \leq p$ and $\Omega = J\Omega_1$

$$\begin{cases} \dot{x} = A_{1_0}x(t) + A_{2_0}x(t-l) + A_{d_0}\dot{x}(t-l) + B_0u(t) + J\Omega_1(x, \dot{x}, u, \alpha) \\ y = Cx(t) \end{cases} \tag{5.13}$$

In order to design an observer for linear systems with neutral delay, a methodology is developed by Elhsoumi et al. (2013) with T_1 is a non-singular transformation matrix $x = T_1\overline{x}$. $T_1 = [N\ J_1]$ with $N \in M_{(n,n-q)}$, $J_1 \in M_{n,q} \in \text{span}\{J_i,\ i = 1 \ldots n\}$, J_i $(i = 1 \ldots n)$ represent column vectors of matrix J.

By applying the state T_1 transformation, we obtain:

$$\begin{cases} \dot{\overline{x}} = \overline{A}_{1_0}\overline{x}(t) + \overline{A}_{2_0}\overline{x}(t-l) + \overline{A}_{d_0}\dot{\overline{x}}(t-l) + \overline{B}_0 u + T_1^{-1} J \Omega_1(\overline{x}, \dot{\overline{x}}, u, \alpha) \\ y = \overline{C}\overline{x} \end{cases} \tag{5.14}$$

with the following matrices:

$$\begin{cases} \overline{A}_{1_0} = T_1^{-1} A_{1_0} T_1 = \begin{bmatrix} \overline{A}_{1_{11}} & \overline{A}_{1_{12}} \\ \overline{A}_{1_{21}} & \overline{A}_{1_{22}} \end{bmatrix} \\ \overline{A}_{2_0} = T_1^{-1} A_{2_0} T_1 = \begin{bmatrix} \overline{A}_{2_{11}} & \overline{A}_{2_{12}} \\ \overline{A}_{2_{21}} & \overline{A}_{2_{22}} \end{bmatrix} \\ \overline{B}_0 = T_1^{-1} B_0 = \begin{bmatrix} \overline{B}_1 \\ \overline{B}_2 \end{bmatrix} \\ \overline{A}_{d_0} = T_1^{-1} A_{d_0} T_1 = \begin{bmatrix} \overline{A}_{d_{11}} & \overline{A}_{d_{12}} \\ \overline{A}_{d_{21}} & \overline{A}_{d_{22}} \end{bmatrix} \\ \overline{C} = C T_1 \end{cases}$$

The state vector is decomposed into two parts, one part $\overline{x}_1$ that is not affected by the degradation and a second part $\overline{x}_2$ affected by the degradation.

$$\overline{x} = [\overline{x}_1 \ \overline{x}_2]^T$$

$$\overline{x}_1 \in R^{n-q}, \overline{x}_2 \in R^q$$

O^{-1} is a non-singular transformation matrix for operations on the J column and rank $(O) = n$ such that

$$\begin{cases} \dot{\overline{x}} = \overline{A}_{1_0}\overline{x} + \overline{A}_{2_0}\overline{x}(t-l) + \overline{A}_{d_0}\dot{\overline{x}}(t-l) \\ +\overline{B}u + T_1^{-1} J O^{-1} O \Omega_1(\overline{x}, \dot{\overline{x}}, u, \alpha) \\ \overline{y} = \overline{C}\overline{x} \end{cases} \tag{5.15}$$

with

$$\overline{J} = T_1^{-1} J O^{-1} = \begin{pmatrix} 0_{n-q,n-q} & 0_{n-q,q} \\ 0_{q,n-q} & I_{q,q} \end{pmatrix}$$

$$\overline{J} v(\overline{x}, \dot{\overline{x}}, u, \alpha) = O \Omega_1(\overline{x}, \dot{\overline{x}}, u, \alpha)$$

System (5.14) can be defined in the following form

$$\begin{cases} \dot{\overline{x}} = \overline{A}_{1_0}\overline{x} + \overline{A}_{2_0}\overline{x}(t-l) + \overline{A}_d\dot{\overline{x}}(t-l) + \overline{B}u + \overline{J}v(\overline{x}, \ \dot{\overline{x}}, u, \alpha) \\ y = \overline{C}\overline{x} \end{cases} \tag{5.16}$$

Hence can be considered with following form the system without faults and unknown input:

$$(I_{n-q}\ 0)\dot{\overline{x}} = (\overline{A}_{1_{11}}\overline{A}_{1_{12}})\overline{x} + (\overline{A}_{2_{11}}\overline{A}_{2_{12}})\overline{x}(t-l) + (\overline{A}_{d_{11}}\overline{A}_{d_{12}})\dot{\overline{x}}(t-l) + \overline{B}_1 u \quad (5.17)$$

`Assumption` 4: A transformation of the output is performed under the assumption that the CJ_1 matrix is full rank column. Using the output transformation $\overline{y} = [\overline{y}_1\ \overline{y}_2] = U^{-1}y$ with the following matrices U = $[CJ_1\ Q]$ with the matrices Q$\in M_{p,(p-q)}$, $U_1 \in M_{q,p}$, $U_2 \in M_{p-q,p}$
Then

$$\overline{y}_1 = U_1 y = U_1 C N \overline{x}_1 + \overline{x}_2 \quad (5.18)$$

$$\overline{y}_2 = U_2 y = U_2 C N \overline{x}_1 \quad (5.19)$$

Using expression (5.18), the following expression can be deduced:

$$\overline{x}_2 = U_1 y - U_1 C N \overline{x}_1 \quad (5.20)$$

Inserting (5.20) in (5.16), and using the output term (5.19) which does not include the behavior of degradation and retaining only the state $\overline{x}_1$ of (5.16); The obtained system after transformation is written as:

$$\begin{cases} \dot{\overline{x}}_1 = \widetilde{A}_{1'}\overline{x}_1 + \widetilde{A}_{2'}\overline{x}_l + \widetilde{A}_{d'}\dot{\overline{x}}_{d_1} + \overline{B}_1 u + J_1 y_1 + J_2 y_{l_1} + J_d \dot{y}_d \\ \overline{y} = \widetilde{C}_1 \overline{x}_1 \end{cases} \quad (5.21)$$

Can be defined the matrices as follows:

$$\begin{cases} \widetilde{A}_{1'} = \overline{A}_{1_{11}} - \overline{A}_{1_{12}} U_1 C N \\ \widetilde{A}_{2'} = \overline{A}_{2_{11}} - \overline{A}_{2_{12}} U_1 C N \\ \widetilde{A}_{d'} = \overline{A}_{d_{11}} - \overline{A}_{d_{12}} U_1 C N \\ J_1 = \overline{A}_{1_{12}} U_1 \\ J_2 = \overline{A}_{2_{12}} U_1 \\ J_d = \overline{A}_{d_{12}} U_1 \\ \widetilde{C}_1 = U_2 C N \end{cases}$$

Under the condition that system (5.20) is observable, an unknown input observer can be built for the healthy system:

$$\begin{cases} \dot{\overline{z}}(t) = \widehat{A}_{1'}\overline{z} + \widehat{A}_{2'}\overline{z}_l + \widehat{A}_{d'}\dot{\overline{z}}_d + \widehat{Z}_1\overline{y}_1 + \widehat{Z}_2\overline{y}_l + \widehat{Z}_d\dot{\overline{y}}_d + \widehat{B}_1 u + \widehat{J}_1 y_1 + \widehat{J}_2 y_l + \widehat{J}_d \dot{y}_d \\ \widehat{\overline{x}}_1 = \overline{z}(t) + M_1\overline{y} \\ \overline{z}(\theta) = \alpha(\theta), \forall \in [-l, 0] \end{cases} \quad (5.22)$$

With $z \in \mathbb{R}^g$ is the state estimate vector (g <n-q) and $\widehat{A_1}$, $\widehat{A_2}$, $\widehat{A_d}$, $\widehat{Z_1}$, $\widehat{Z_2}$, $\widehat{Z_d}$, $\widehat{B_1}$, $\widehat{J_1}$, $\widehat{J_2}$, $\widehat{J_d}$ and $\widehat{M_1}$ are unknown matrices to be determined such that $\widehat{\overline{x}}$ converges to $\overline{x}$.

b. Reconstruction of the degradation state and unknown input: Considering expression (5.18) and remembering that the state transformation T_1 has been used in expression (5.14), the state vector is defined as

$$\widehat{x} = T_1 \widehat{\overline{x}} = \begin{pmatrix} \widehat{\overline{x}}_1 \\ \widehat{\overline{x}}_2 \end{pmatrix} = T_1 \begin{pmatrix} \overline{z}(t) + M_1 \overline{y} \\ U_1 y - U_1 C N \widehat{\overline{x}}_1 \end{pmatrix} \tag{5.23}$$

$\widehat{x}, \widehat{\overline{x}}, \widehat{\overline{x}}_1, \widehat{\overline{x}}_2$ Are respectively the estimates of x, $\overline{x}, \overline{x}_1, \overline{x}_2$

According to Eq. (5.21), we obtain by Elhsoumi et al. (2013):

$$\begin{pmatrix} \dot{\widehat{\overline{x}}}_1 \\ \dot{\widehat{\overline{x}}}_2 \end{pmatrix} = \begin{pmatrix} \overline{A}_{1_{11}} & \overline{A}_{1_{12}} \\ \overline{A}_{1_{21}} & \overline{A}_{1_{22}} \end{pmatrix} \begin{pmatrix} \widehat{\overline{x}}_1 \\ \widehat{\overline{x}}_2 \end{pmatrix} + \begin{pmatrix} \overline{A}_{2_{11}} & \overline{A}_{2_{12}} \\ \overline{A}_{2_{21}} & \overline{A}_{2_{22}} \end{pmatrix} \begin{pmatrix} \widehat{\overline{x}}_1(t-l) \\ \widehat{\overline{x}}_2(t-l) \end{pmatrix}$$
$$+ \begin{pmatrix} \overline{A}_{d_{11}} & \overline{A}_{d_{12}} \\ \overline{A}_{d_{21}} & \overline{A}_{d_{22}} \end{pmatrix} \begin{pmatrix} \dot{\widehat{\overline{x}}}_1(t-l) \\ \dot{\widehat{\overline{x}}}_2(t-l) \end{pmatrix} + \begin{pmatrix} \overline{B}_1 \\ \overline{B}_2 \end{pmatrix} u + \begin{pmatrix} 0 \\ I_q \end{pmatrix} \widehat{v} \tag{5.24}$$

The two preceding equations gives:

$$\dot{\widehat{\overline{x}}}_2 = \left(\overline{A}_{1_{21}} \; \overline{A}_{1_{22}}\right) \widehat{\overline{x}}_2 + \left(\overline{A}_{2_{21}} \; \overline{A}_{2_{22}}\right) \widehat{\overline{x}}_2(t-l) + \left(\overline{A}_{d_{21}} \; \overline{A}_{d_{22}}\right) \dot{\widehat{\overline{x}}}_2(t-l) + \overline{B}_2 u + \widehat{v} \tag{5.25}$$

Therefore, the vector estimate of unknown input is expressed by:

$$\begin{aligned} \widehat{v} = \dot{\widehat{\overline{x}}}_2 &- (\overline{A}_{1_{21}} \overline{A}_{1_{22}}) \begin{pmatrix} \overline{z}_1 + M_1 U_2 y_1 \\ U_1(I - CNM_1U_2) y_1 + U_1 C N \overline{z}_1 \end{pmatrix} \\ &- (\overline{A}_{2_{21}} \overline{A}_{2_{22}}) \begin{pmatrix} \overline{z}_2 + M_1 U_2 y_2 \\ U_1(I - CNM_1U_2) y_2 + U_1 C N \overline{z}_2 \end{pmatrix} \\ &- (\overline{A}_{d_{21}} \overline{A}_{d_{22}}) \begin{pmatrix} \overline{z}_d + M_1 U_2 y_d \\ U_1(I - CNM_1U_2) y_d + U_1 C N \overline{z}_d \end{pmatrix} - \overline{B}_2 u \end{aligned} \tag{5.26}$$

The expression of the vector estimate of the unknown input is obtained by:

$$\begin{aligned} \widehat{v} = U_1 \dot{y} &+ (\overline{A}_{1_{22}} U_1 C N - U_1 C N \widehat{A}_{1'} - \overline{A}_{1_{21}}) \overline{z}_1 \\ &+ (\overline{A}_{2_{22}} U_1 C N - U_1 C N \widehat{A}_{2'} - \overline{A}_{2_{21}}) \overline{z}_2 \\ &- (\overline{A}_{1_{22}} U_1 + U_1 C N \widehat{J}_1) y_1 - (\overline{A}_{2_{22}} U_1 + U_1 C N \widehat{J}_2) y_2 \\ &- (\overline{A}_{d_{22}} U_1 + U_1 C N \widehat{J}_d) \dot{y}_d \\ &(\overline{A}_{d_{22}} U_1 C N - U_1 C N \widehat{A}_{d'} - \overline{A}_{1_{21}}) \dot{\overline{z}}_d - (U_1 C N \overline{B}_1 + \overline{B}_2) u \end{aligned} \tag{5.27}$$

Mathematically, the unknown input is built. Then, an expression of the degradation state based on the unknown input can be established. Associated with (5.15), matrix $\overline{J}$ is developed for this purpose.

$$\begin{aligned} \overline{J} &= O\Omega_1(\overline{x}, \dot{\overline{x}}, u, \alpha) \\ \Leftrightarrow & \begin{pmatrix} 0_{n-q,n-q} & 0_{n-q,q} \\ 0_{q,n-q} & I_{q,q} \end{pmatrix} \begin{pmatrix} v_{n-q} \\ v_q \end{pmatrix} \\ &= \begin{pmatrix} O\Omega_{1_{n-q,n-q}}(\overline{x}, \dot{\overline{x}}, u, \alpha) & O\Omega_{1_{n-q,q}}(\overline{x}, \dot{\overline{x}}, u, \alpha) \\ O\Omega_{1_{q,n-q}}(\overline{x}, \dot{\overline{x}}, u, \alpha) & O\Omega_{1_{q,q}}(\overline{x}, \dot{\overline{x}}, u, \alpha) \end{pmatrix} \end{aligned} \tag{5.28}$$

Then we have:

$$O\Omega_{1_{q,q}}(\overline{x}, \dot{\overline{x}}, u, \phi) = v_q \tag{5.29}$$

Finally, we can deduce the following result. Proposal: J is a solution of Eq. (5.29) with u is known input and $\overline{x}$, $\overline{x}(t-l)$ and $\dot{\overline{x}}(t-l)$, v_q are estimated.

5.4 Transmission Line Diagnosis and Prognosis

In this part, the example of the transmission line model is presented to demonstrate the effectiveness and flexibility of the proposed observer based diagnostic and prognosis approach and to compare the results of the two approaches.

5.4.1 *Diagnostic Study*

The transmission line system characterized by constant impedance can be considered as linear neutral delayed system form (5.1) described in Elhsoumi et al. (2013)

$$\begin{cases} \dot{x}(t) = -x(t) - 0.5x(t-l) + 0.5\dot{x}(t-l) \\ \quad + u(t) + 0.25d(t) + f(t) \\ y(t) = Cx(t), C = 1 \end{cases} \tag{5.30}$$

Now, the system is supposed to be affected by the following fault:

$$f(t) = \begin{cases} 1, & \text{if} \quad 4 \leqslant t \leqslant 5 \\ 0, & \text{otherwise} \end{cases}$$

The simulation result for faulty system is presented in Fig. 5.1.

The figures above show that the system is stable and the residual is insensitive to disturbances and sensitive only to a fault. Thus, the observer achieves the detection stage. Figure 5.1 shows that the residual variation occurs at time 4; it is the same period of fault appearance. Therefore, we can conclude that the presence of a fault at any time can be detected by the designed observer and the residual is usually sensitive only to faults and insensitive to disturbances.

5.4.2 *Prognosis Study*

To illustrate the relevance of the proposed prognosis methodology a simulation is performed on the transmission line system.

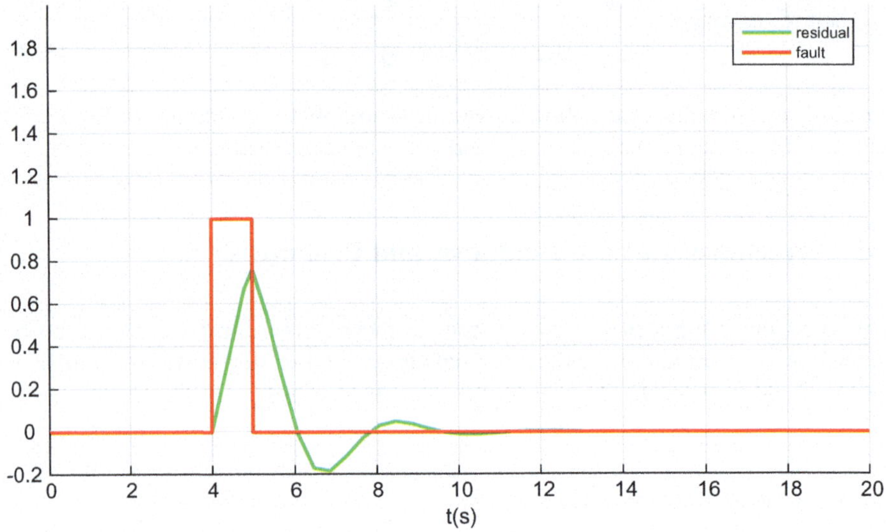

Fig. 5.1 Residual for faulty system

The model equation is given by:

$$\begin{cases} \dot{x}(t) = A_1(\alpha)x(t) + A_2(\alpha)x(t-l) \\ \quad + A_d(\alpha)\dot{x}(t-l) + B(\alpha)u(t) \\ y(t) = Cx(t) \end{cases} \tag{5.31}$$

With

$$\begin{cases} A_1 = \frac{m-\frac{1}{z'}}{c_l} \\ A_2 = -\frac{(z'-R_0)(m+\frac{1}{z'})}{c_l(z'+R_0)} \\ A_d = \frac{z'-R_0}{z'+R_0} \\ z' = z + \alpha \\ z = \sqrt{\frac{L}{c}} \end{cases} \tag{5.32}$$

A transmission line is constituted by one or more conductors carrying an electrical signal from a source (transmitter) to a receiving load. If the signal conveyed meets a breakdown of characteristic impedance $z = \sqrt{\frac{L}{C}}$ part of the transmission line, is reflected towards the emitter causing a deformation of its conductance. Therefore, degradation primarily affects the physical parameters when the crack changes. An image of the degradation is taken into account for our study. It is assumed that z' is at a critical level when $z' = 3z$. Rewrite (5.31) with $z' = z + \alpha$ (proposition), we get:

$$\begin{cases} \dot{x}(t) = \left(\frac{m}{c_l} - \frac{1}{(z+\alpha)c_l}\right)x(t) \\ \quad + \frac{((z+\alpha) - R_0)(m(z+\alpha)+1)}{c_l(z+\alpha)(z+\alpha+R_0)}x(t-l) \\ \quad + \left(1 - \frac{2R_0}{z+\alpha+R_0}\right)\dot{x}(t-l) + Bu(t) \\ y(t) = Cx(t) \end{cases} \tag{5.33}$$

The general expression of cumulative degradation is defined as follows:

$$\dot{\alpha} = \varepsilon s(x, \alpha) \tag{5.34}$$

The cumulative degradation (5.31) is estimated here by developing the methodology introduced above. The non-singular matrix of state transformation T_1 is defined by $T_1 = 1\ O^{-1}$ non-singular transformation matrix is written as: $O^{-1} = 1$. After applying the two previous transformations, system (5.30) can be written as:

$$\begin{cases} \dot{x}(t) = \left(\frac{m}{c_l} - \frac{1}{(z+\alpha)c_l}\right)x(t) \\ \quad + \frac{((z+\alpha) - R_0)(m(z+\alpha)+1)}{c_l(z+\alpha)(z+\alpha+R_0)})x(t-l) \\ \quad + \left(1 - \frac{2R_0}{z+\alpha+R_0}\right)\dot{x}(t-l) + Bu(t) \\ y(t) = Cx(t) \end{cases} \tag{5.35}$$

Then, it is assumed that rank (J) $\leqslant p$, also there exists J such that rank (J) = q = 1 $\leqslant p$ and $\Omega = J\Omega_1$

$$\dot{x} = \frac{m}{c_l}x - \dot{x}(t-l) + D\Delta_1(x, \dot{x}, u) \tag{5.36}$$

Non-singular U^{-1} output transformation is expressed as follows: $U^{-1} = 1$.

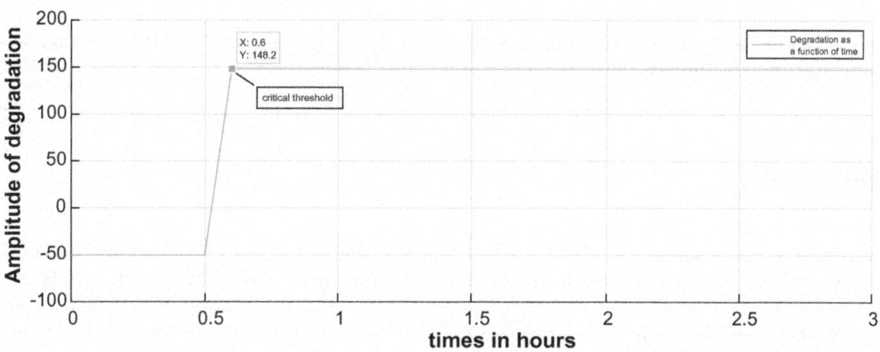

Fig. 5.2 Degradation as a function of time

After the calculation of the unknown input and the degradation state based on (5.26), (5.27) and (5.28), the curve of the degradation is illustrated in Fig. 5.2.

The remaining Time of Life is the time difference between the Time of appearance of failure and the present time of reference when the system evolves. Therefore, the remaining time in this system is equal to 0.6 s.

5.5 Discussion

The studies on fault diagnosis for delay systems (retarded type or neutral type) are very few and the FDI methods used are mainly based on model-based approaches (e.g., state observers Zhang et al. 2009; parity equations Zhang et al. 2002; geometric approach Meskin and Khorasani 2009 or online identification algorithms Jiang and Zhou 2005). However, the fault diagnosis problem for neutral time-delay systems has not been completely addressed in the above references. In recent years, only a few results on FDI of neutral systems with time-delay have been developed. This problem has only been investigated, for the first time, in Ren (2006) where an H_∞ filter approach is used for robust fault detection. Model-based prognosis methodology is applied recently for a class of nonlinear damage models (Gucik-Derigny 2011).

5.6 Conclusion

In this chapter, a new prognosis approach for observer design problem of a class of neutral time-delay systems is addressed. The main contribution of the present work lies in the use of an unknown input observer for the neutral time-delay system prognosis. The comparison between transmission line system diagnosis and prognosis results shows that faults are predicted with the prognosis approach before 0.6 s of its appearance. However, a fault is detected at the same time of its appearance with the diagnosis approach. Consequently, future work can investigate the application of a prognosis approach to other forms of neutral time-delay systems namely continuous or noncontinuous systems.

Acknowledgements This work was supported by the Ministry of the Higher Education and Scientific Research in Tunisia.

References

Adams, D. E., & Nataraju, M. (2002). A nonlinear dynamical systems framework for structural diagnosis and prognosis. *International Journal of Engineering Science*, *40*(17), 1919–1941.

Basseville, M. & Cordier, M.-O. (1996). *Surveillance et diagnostic de systèmes dynamiques: approches complémentaires du traitement de signal et de l'intelligence artificielle*. Ph.D. thesis, INRIA.

Bettaher, M., Elhsoumi, A., & Bel Hadj Ali Naoui, S. (2017). Neutral system with variable time-delay: Stability analysis and diagnosis. In *2017 18th International Conference on Sciences and Techniques of Automatic Control and Computer Engineering (STA)* (pp. 67–72). Monastir: IEEE.

Elhsoumi, A., Bel Hadj Ali, S., El Harabi, R., & Abdelkrim, M. N. (2013). Uio-based fault estimation for a class of neutral time-delay system. In *2013 14th International Conference on Sciences and Techniques of Automatic Control and Computer Engineering (STA)* (pp. 251–258). Piscataway: IEEE.

Gucik-Derigny, D. (2011). *Contribution au pronostic des systèmes à base de modèles: théorie et application*. Ph.D. thesis, L'Université paul cézanne aixmarseille III.

Isermann, R. (1997). Supervision, fault-detection and fault-diagnosis methods an introduction. *Control Engineering Practice*, *5*(5), 639–652.

Isermann, R. (2005). Model-based fault-detection and diagnosis-status and applications. *Annual Reviews in Control*, *29*(1), 71–85.

Jiang, C., & Zhou, D. (2005). Fault detection and identification for uncertain linear time-delay systems. *Computers & Chemical Engineering*, *30*(2), 228–242.

Kratz, F., Nuninger, W., & Ploix, S. (1998). Fault detection for time-delay systems: A parity space approach. In *Proceedings of the 1998 American Control Conference* (Vol. 4, pp. 2009–2011). Philadelphia: IEEE.

Meskin, N., & Khorasani, K. (2009). Robust fault detection and isolation of time-delay systems using a geometric approach. *Automatica*, *45*(6), 1567–1573.

Piechowiak, S. (2003). *Intelligence artificielle et diagnostic*. Saint-Denis: Editions Techniques Ingénieur.

Ren, J. (2006). LMI-based fault detection filter design for a class of neutral system with time delay in states. In *The Sixth World Congress on Intelligent Control and Automation, 2006. WCICA 2006* (Vol. 2, pp. 5581–5585). Dalian: IEEE.

Ribot, P. (2009). *Vers l'intégration diagnostic/pronostic pour la maintenance des systèmes complexes*. Ph.D. thesis, Université Paul Sabatier-Toulouse III, Toulouse.

Schutz, J. (2009). *Contribution à l'optimisation des plans d'exploitation et de maintenance, selon une approche basée sur le pronostic: application au domaine naval*. PhD thesis, L'Universit e Paul Verlaine, Metz.

Toscano, R. (2005). *Commande et diagnostic des systèmes dynamiques*. Paris: Elipses.

Wang, H.-Y., Tian, Z.-H., Shi, S.-J., & Weng, Z.-X. (2009). A robust fault detection scheme for a class of multi time delay systems. *Journal of Shanghai Jiaotong University (Science)*, *14*(3), 286–289

Yang, H. & Saif, M. (1996). Observer design and fault diagnosis for retarded dynamical systems. In *Proceedings of the 35th IEEE Conference on Decision and Control, 1996* (Vol. 2, pp. 1149–1154). Kobe: IEEE.

Zhang, K., Jiang, B., & Shumsky, A. (2009). A new criterion of fault estimation for neutral delay systems using adaptive observer. *Acta Automatica Sinica*, *35*(1), 85–91.

Zhang, P., Ding, S., Wang, G., & Zhou, D. (2002). Fault detection for multirate sampled-data systems with time delays. *International Journal of Control*, *75*(18), 1457–1471.

Chapter 6
Intrinsic Mode Function Selection and Statistical Information Analysis for Bearing Ball Fault Detection

Zahra Mezni, Claude Delpha, Demba Diallo and Ahmed Braham

Abstract Early fault detection and diagnosis is becoming one of the most challenging areas of research in systems health monitoring and management. Indeed, in this area, Condition-Based Maintenance (CBM) is becoming the standard as it has technical and economical benefits particularly in electromechanical systems. However, CBM requires efficient fault monitoring information. In rotating machinery, bearing faults are very common and their occurrence modifies the spectral content of vibration signals. However, in the case of ball faults, the spectrum's modification is weak making this fault detection tedious. Therefore, we propose here an efficient method to detect incipient ball faults. This method is based on the Empirical Mode Decomposition (EMD) that splits the original signal into different components called Intrinsic Mode Functions (IMFs). At first, we reduce the number of IMFs by keeping only the most energized. At second, using the Kullback–Leibler Divergence (KLD), we eliminate the less robust to load variations. Finally, the most sensitive IMFs are then selected and used for ball fault detection. To extract the fault features, the first four statistical moments of the retained components are computed and analyzed. Using experimental dataset from the Case Western Reserve University, we show that this methodology is efficient: from the initial 18 components, only 7 are retained and

Z. Mezni · A. Braham
Laboratoire Matériaux, Mesures et Applications, INSAT, Université de Carthage, Carthage, Tunisia
e-mail: mezni.zahra@gmail.com

A. Braham
e-mail: Ahmed.Braham@insat.rnu.tn

C. Delpha (✉)
Laboratoire des Signaux et Systèmes (L2S), UMR 8506 (CNRS, CentraleSupélec, Université Paris Sud, Université Paris Saclay), Université Paris Sud - Université Paris Saclay, 91192 Gif sur Yvette, France
e-mail: Claude.Delpha@l2s.centralesupelec.fr

D. Diallo
Group of Electrical Engineering Paris (GeePs), UMR 8507 (CNRS, CentraleSupélec, Univ. Paris Sud, Université Paris Saclay, Sorbonne Universités), Université Paris Sud - Université Paris Saclay, 91192 Gif sur Yvette, France
e-mail: Demba.Diallo@geeps.centralesupelec.fr

N. Derbel et al. (eds.), *Diagnosis, Fault Detection & Tolerant Control*, Studies in Systems, Decision and Control 269,
https://doi.org/10.1007/978-981-15-1746-4_6

the performance evaluation using Receiver Operating Curves (ROC) shows that with the variance of 3 IMFs, the incipient ball fault (0.007 in.) is detected with no false alarms.

Keywords Empirical mode decomposition · Intrinsic mode function selection · Fault detection and diagnosis · Statistical moment analysis · Kullback–Leibler divergence

6.1 Introduction

Nowadays, rotating machinery is growing substantially regarding the industrial field with the rapid development of science and technology. Therefore, diagnosing the machine, detecting the faults occurrence at an early stage and monitoring their evolution are the common targets for many researches to avoid downtime and increase reliability. In this context, several signal processing techniques have been extensively studied and widely applied in fault diagnosis including acoustic emission, motor current signature analysis (MCSA) (Abid et al. 2016), thermography, performance monitoring, and vibration analysis (Scheffer and Girdhar 2004) which is undoubtedly one of the most techniques to detect, apply maintenance operations to reduce, or eliminate recurring mechanical defects in rotating machinery. A major advantage of the latter method is that it can focus on solving problems before becoming too serious and cause unscheduled downtime as well as it can detect misalignment and unbalance before the bearings' deterioration especially that specific examinations revealed that the defects occurring in those components are dominant (40–90% from large to small machines) (Riera-Guasp et al. 2015). Hence, adopting and developing the suitable signal processing techniques are the challenging tasks to exploit all the information present in the response signals.

Fault diagnosis is based on three main phases, fault detection, fault isolation, and fault identification. The most crucial stage is the fault detection which is used to detect and identify any kind of abnormalities or malfunction in the system by specifying the time when the fault occurs as early as possible. This task remains vital to ensure the system normal operation especially under the time-varying operational conditions and the intrinsic dynamic nature which often lead to nonstationary signals. Such nonstationary signals are rich of information of machinery health conditions. Relevant features can be extracted from these signals for fault detection and diagnosis (Feng et al. 2013a). The classical signal processing methods are based on the assumptions of stationarity and linearity and can only provide analyses in terms of the statistical average in the time or frequency domain (Feng et al. 2013b). Such methods cannot stipulate any information regarding either the time evolution of the signal characteristics or the occurrence of time localized events (Flandrin 1989). To overcome this drawback, researches have been directed to the development of several advanced methods which take into consideration the real physical behavior of the system.

Over the last 20 years, many vibration-based fault diagnosis tools have been widely used especially the Short-Time Fourier Transform (STFT) (Satish 1998; Cusidó et al. 2008). Although STFT overcomes the disadvantages of FFT-based methods in processing nonstationary signals, it is also suitable for quasi-stationary signals instead of real nonstationary ones (Mohanty et al. 2017). Compared with STFT, wavelet (Staszewski and Tomlinson 1994; Yesilyurt and of Manchester. School of Engineering 1997) has been introduced as an improvement for analyzing nonstationary signals. This technique can be categorized into Peng and Chu (2004) Continuous Wavelet Transform (CWT), Discrete Wavelet Transform (DWT), and Wavelet Packet Transform (WPT) (Satish 1998; Yan et al. 2014). The performances of this technique depend on the choice of the mother wavelet function (Peng et al. 2005).

Recently, the Empirical Mode Decomposition (EMD) (Huang et al. 1998; Lei et al. 2013; Wu and Huang 2009), the core of the Hilbert–Huang Transform (HHT), has been applied to data analysis in various fields such as mechanical engineering (Feng et al. 2013b). It is a powerful and advantageous technique dealing with nonlinear and nonstationary signals for fault detection and diagnosis. Cai et al. (2011) improved a fault feature extraction method of rolling envelope spectrum based on EMD and spectral kurtosis. Cheng et al. (2008) introduced the EMD and partial Hilbert energy spectrum for the rolling bearing fault diagnosis. Zhou et al. (2011) used the demodulation method based on EMD and adaptive morphological filtering to separate the fault information and extract the fault feature frequency. Zhang et al. (2011) proposed a bearing fault diagnosis method based on EMD and singular value of the differential spectrum using vibration signals with a strong background noise.

6.2 Problem Statement

As it is known, the EMD's basic task is to decompose in the frequency domain the signal into a set of complete and almost orthogonal components named Intrinsic Mode Function (IMF) (Lei et al. 2013). Each IMF represents one part of the spectral content of the original signal in a narrow band of frequencies. Mechanical faults such as bearing and gearbox faults modify the power spectral density of the vibration signals (Harmouche et al. 2015a; Taylor 1994; Collacott 1979). Therefore, the fault information should be retrieved from the spectral content of the signal, meaning from the IMFs. However, not all the extracted IMFs are sensitive to the fault presence and/or level. Accordingly, getting rid of irrelevant features in the original feature set and retaining the most sensitive ones will significantly improve the detection accuracy. In this case study, we propose a method to select the most relevant IMFs for detecting incipient ball faults.

In this work, we first propose to reduce the number of IMFs by eliminating those who have the lower signal-to-noise ratio. Second, the Kullback–Leibler Divergence (KLD) (Anderson and Haas 2011) is then used to evaluate the sensitivity of the IMFs to the load variations. The KLD is well known for its ability to detect slight modifications (Silva and Narayanan 2006; Afgani et al. 2008; Harmouche et al. 2014; Delpha et al. 2017a, b). After these steps (EMD computation, reduction of the

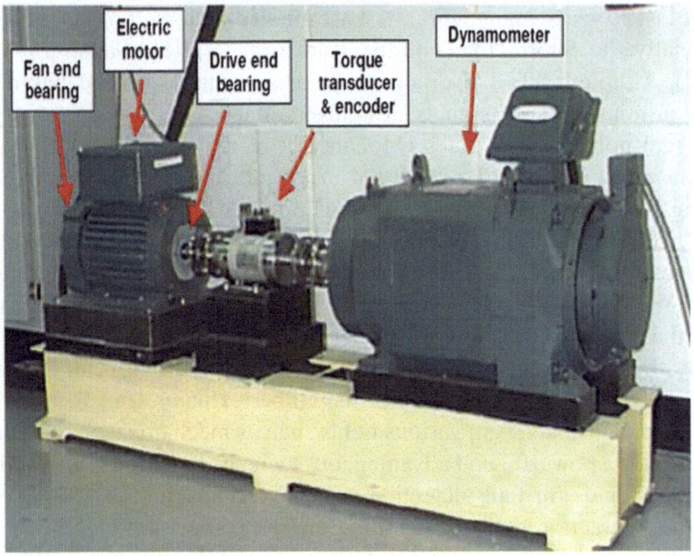

Fig. 6.1 CWRU bearing test bed

number of IMFs, and robustness to load variation), we analyze the characteristics of the remaining IMFs to evaluate their capability to detect the faults with the best performances.

For this study, the experimental vibration data is obtained from the Case Western Reserve University (CWRU) Bearing Data Center (Cwru 2016). The experiments were conducted using a 2 hp Reliance Electric motor as shown in Fig. 6.1. Motor bearings were seeded with faults using electro-discharge machining. Hereby, SKF deep groove ball bearings with three fault diameters ranging from 0.007 to 0.021 in. are treated. The geometrical specifications of the bearings are depicted in Table 6.1. Vibration signals are extracted at a sampling frequency of 12 kHz during 10 s using an accelerometer mounted at the 12 o'clock position (at both the drive end and fan end of the motor housing) and attached to the housing with magnetic bases. This vibration data has been recorded for motor loads from 0 to 3 hp (motor speeds of 1797 to 1720 rpm). In the following, the different load conditions will be denoted as

- L_0 : 0% of the nominal load;
- L_1 : 50% of the nominal load;
- L_2 : 100% of the nominal load;
- L_3 : 150% of the nominal load;

6.2.1 Related Work

This work is based on the use of the EMD as a processing technique to decompose the original signal into narrow frequency band signals named Intrinsic Mode

Table 6.1 Bearing information

Size(inches)						
Bearing location	Reference	Inside diameter	Outside diameter	Thickness	Ball diameter	Pitch diameter
Drive end	6205-2RS JEM SKF	0.9843	2.0472	0.5906	0.3126	1.537
Fan end	6203-2RS JEM SKF	0.6693	1.5748	0.4724	0.2656	1.122

Functions (IMFs). These components have already proved that they have profitable information describing the rolling elements state. In fact, one can notice the growing trend of works using this technique over the last decades, but the majority of these researches focus on spectral analysis by applying demodulation techniques categorized as mono/multidimensional approaches in order to highlight the time–frequency representation of the IMFs (Bouchikhi et al. 2011) and then identifying the corresponding frequency signatures. Moreover, obtaining relevant extracted information on the bearing fault is highly linked to the knowledge of the bearing structural information (size, format, ...). This strategy has been used in several recent works dealing with bearing fault diagnosis (Ali et al. 2015; Li et al. 2017; Elbouchikhi et al. 2017). In the time domain, very few solutions are proposed. The goal here is the efficient fault detection especially in the case of ball faults which are very small and most difficult to detect (Harmouche et al. 2015a). The major contribution of this work lies in the evaluation of the ball fault detection using the most relevant selected IMFs in the time domain. Statistical features are used to prove the efficiency of the selected IMFs. The kurtosis, which is extensively used in the detection and diagnosis of mechanical systems (Bensaad et al. 2017; Laha 2017), and the variance are used. The fault detection performances are evaluated based on probabilistic properties.

The rest of this chapter is organized as follows. In the next section, the signal preprocessing and feature extraction will be presented. Section 6.4 introduces the feature analysis for the ball fault detection. Section 6.5 is concerned with the fault detection and diagnosis. Finally, a conclusion in Sect. 6.6 reviews the main points discussed in this work.

6.3 Preprocessing

Fault detection and diagnosis (FDD) methods have been extensively used as process to detect, isolate, and identify faults in electrical machines. Generally, FDD methods are mostly classified as model-based methods which require sufficient and prior system knowledge as well as available accurate model, and model-driven-based methods which in contrast to the model-based approaches require only a sufficient historical process data that captures most of the process behavior (Venkatasubramanian et al.

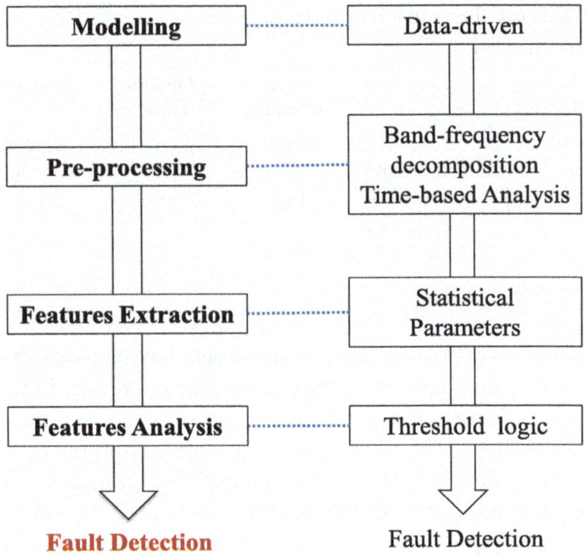

Fig. 6.2 Flowchart of the fault detection and diagnosis methodology

2003a, b). A schematic of FDD is illustrated in Fig. 6.2. Since no explicit models are required, this methodology is more attractive to practical applications with complex systems. As a result, many FDD methods have been developed such as in Harmouche et al. (2015a). Harmouche et al. have proposed a new approach combining the Principal Component Analysis (PCA) and the Linear Discriminant Analysis (LDA) as a robust technique based on the global spectrum for improving the fault diagnosis of ball bearings. Also, an ever increasing demand for the FDD in nuclear power plants (Ma and Jiang 2011) is necessary to improve safety and reliability. Among the techniques adopted for this purpose there are Wavelet Transform (WT) in Qian (2001) and Independent Component Analysis (ICA) in Hyvarinen (1999).

For this work, we adopt EMD for the preprocessing and feature extraction steps. Thus, we first introduce the basics of EMD and detail the use case.

6.3.1 Ensemble Mode Decomposition Basics

Empirical Mode Decomposition (EMD) was introduced by Huang et al. (1998). Basically, this method decomposes a nonstationary signal $x(t)$ into a finite number of modes called IMF_i $(i = 1, 2, \ldots, N)$ and a final residue function $r(t)$ representing the central tendency of the signal $x(t)$. The procedure main steps are detailed in Table 6.2; we obtain

$$x(t) = \sum_{i=1}^{N} IMF_i(t) + r(t) \tag{6.1}$$

Table 6.2 The EMD algorithm main steps

(Step-1) Initialize : $r_0 = x(t)$, and $i = 1$
(Step-2) Extract the ith IMF
2-a) Initialize : $h_{i(k-1)} = r_{i-1}, k = 1$
2-b) Extract the local extremum (maxima and minima) of $h_{i(k-1)}$
2-c) Interpolate the local extremum by cubic spline lines to form upper and lower envelops of $h_{i(k-1)}$
2-d) Calculate the mean $m_{i(k-1)}$ of the upper and lower envelops of $h_{i(k-1)}$
2-e) Let $h_{ik} = h_{i(k-1)} - m_{i(k-1)}$
2-f) If h_{ik} is an IMF:
then set $IMF_i = h_{ik}$,
else jump to step 2-b with $k = k + 1$ until : $SD = \frac{\sum \lvert h_{ik} - h_{i(k-1)} \rvert^2}{\sum h_{i(k-1)}^2} < \alpha$
(Step-3) Define the reminder residue $r_i = r_{i-1} - IMF_i$
(Step-4) If r_i still has at least 2 extremum
then jump to Step-2 with $i = i + 1$,
else the decomposition process is finished and r_i is the residue of the signal

The IMFs include different bands ranging from high to low frequencies (Rai and Upadhyay 2017). They have been defined as a class of function that must satisfy two conditions:

1. In the dataset, the number of extrema and the number of minima of zero-crossings must be equal or differ by one at most;
2. At any point, the local average of the upper and lower envelopes is zero.

The decomposition is carried out as described in Table 6.2.

Finally, to extract an IMF, the EMD uses a sifting process conditioned with the criterion SD where α is an arbitrary value between 0.2 and 0.3 (Hawley et al. 2010).

EMD was applied in various scientific areas, for example, nuclear physics (Montesinos et al. 2003), process control (Srinivasan et al. 2007; Luo et al. 2012), surface engineering (Zhang et al. 2014), biomedical signal processing (Huang et al. 1999), and so on. The number of publications using EMD has increased steadily over the past decade especially to fault diagnosis of rotating machinery. On this topic, many publications between theory and applications are investigated every year.

A general review on the application of EMD in fault diagnosis in terms of bearings leads to a variety of achievements (Duan et al. 2016; Abdelkader et al. 2016). Xu et al. (2009) have employed EMD to analyze the vibration signals of accelerated life tests of bearings and investigated the evolving trend of a bearing life cycle. Tsao et al. (2010) executed the envelope analysis on the selected IMFs which contain bearing fault information, and experimental results demonstrated the efficiency of their method in bearing fault detection (Lei et al. 2013).

Although EMD has shown potential and performance for bearing damage detection at a much earlier stage of fault development, it has several issues as well including end effects, extremum interpolation, and sifting stop criterion (Lei et al. 2013). Hence, many authors have proposed theoretical analyses to improve the EMD (Yang et al. 2008; Xuan et al. 2007; Hawley et al. 2010; Kopsinis and McLaughlin 2008). The Ensemble Empirical Mode Decomposition (EEMD) (Amirat et al. 2013; Xue et al. 2015; Singh et al. 2017) is another alternative that has been developed to increase the robustness of EMD against nuisances. In this method, the IMFs are obtained from an averaging process of several realizations of EMD with an additive noise. However, this method, besides being more complex and time-consuming than the EMD, also suffers at least from two limitations: the setting of the number of realizations and the noise level. Therefore, for sake of simplicity, the classical EMD will be used in the rest of this work.

6.3.2 Application to Vibration Signals

After performing EMD on a signal, we obtain a finite set of IMFs which include local features of the original signal. Among these IMFs some contain information useful for the fault identification but others are useless for diagnosing the faults. Therefore, a selection of the most relevant components is necessary, and the insensitive ones should be eliminated to construct a more accurate fault detection procedure.

In this context, we propose the following method described in Fig. 6.3 for identifying the sensitive modes:

1. We add an AWGN signal with SNR = 60 dB to the original signal;
2. We compute the EMD on the resulting noisy signal;
3. For each IMF, we compute the signal-to-noise ratio;
4. We repeat operations 1–3 for different load conditions.

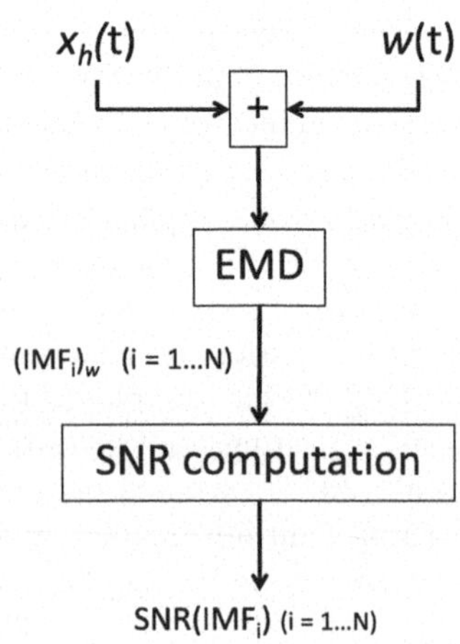

Fig. 6.3 SNR computation of the IMFs

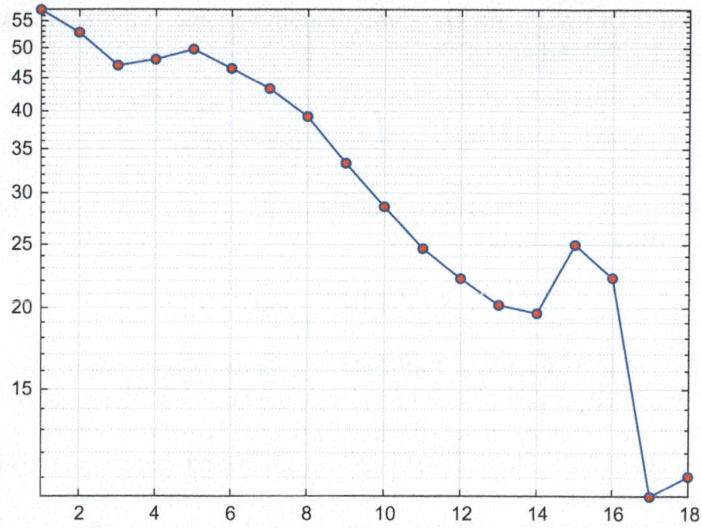

Fig. 6.4 Signal-to-noise ratio of the IMFs

Indeed, as the variance of the noise is always constant, then a decrease in the signal-to-noise ratio means that the energy of the corresponding IMF also decreases. The SNRs for all the IMF (SNR_{IMF}) are displayed in Fig. 6.4.

We consider the limit of relevance for the SNR_{IMF} as half of its full-scale range. In this case, the limit is 35 dB. Therefore, we can consider that only the first 8 IMFs are significant and will be retained for the evaluation in the fault detection process. It means that only the first eight IMFs are the most relevant ones as they contain most of the information nested in the original signal.

6.4 Feature Extraction

We propose in this section to evaluate the load variation effect on the IMFs obtained in the previous paragraph from the EMD. The objective is to eliminate any IMF which is sensitive to the load variation and can induce false alarms when looking at fault occurrence. This study is based on the Kullback–Leibler Divergence (KLD).

6.4.1 Kullback–Leibler Divergence Basics

The Kullback–Leibler divergence (Kullback and Leibler 1951) (also denoted as information divergence, relative entropy, or I-divergence) is one of the most frequently used information-theoretic divergence measures. Considering two Probability Density Function (PDF) $P(z)$ and $Q(z)$ of a random variable z, the Kullback–Leibler Information (KLI) is defined as

$$KLI(P||Q) = \int_{-\infty}^{+\infty} P(z) \log \frac{P(z)}{Q(z)} dz \quad (6.2)$$

The KL Divergence (KLD) is then defined as the symmetric version of the KL Information (Basseville 1989):

$$KLD = KLI(P||Q) + KLI(Q||P) \quad (6.3)$$

In the case of fault diagnosis, Kullback–Leibler divergence has received considerable attention in detecting incipiently developing faults and abnormal situations in vibration systems. More accurately, Delpha et al. (2017a, b), Youssef et al. (2016b, a), Harmouche et al. (2016, 2015b, 2014, 2013, 2012), and Zeng et al. (2014) proved that the KL divergence is more sensitive in detecting incipient faults than conventional scores, residual-based statistics, and the statistical local approach (Xie et al. 2015). Moreover, owing to the higher sensitivity in detecting faults at an earlier stage of development, the application of the KL divergence to fault detection and diagnosis has also been extended to dynamic systems (Bittencourt et al. 2014; Liu et al. 2013). However, to use the KLD, we must be sure that the PDF in healthy and faulty conditions share the same sample space (Kullback and Leibler 1951; Cover and Thomas 2005).

6.4.2 *Effect of the Load Variation on the IMFs*

In the following, we are going to compute the KLD for the first eight IMFs retained from the preprocessing step (Sect. 6.3).

The idea consists in the following main operations:

1. Decomposition of the healthy signal using EMD;
2. Compute the Probability Density Function (PDF) of each IMF;
3. Repeat operations (1) and (2) to generate 200 realizations of the healthy signal. For this operation, a very low-level Additive Gaussian Noise (AWGN) denoted as $w(t)$ will be considered. This noise level leading to an $SNR = 60\,\text{dB}$ has for main interest to create different realizations of the signal with the same properties than the original one.
4. Compute the EMD of each resulting signal;
5. Extract IMFs for each signal and then

    a. Compute the resulting PDF,
    b. Evaluate the KLD for the 200 realizations, and
    c. Compute the KLD mean value;

Fig. 6.5 Load effect evaluation based on KLD

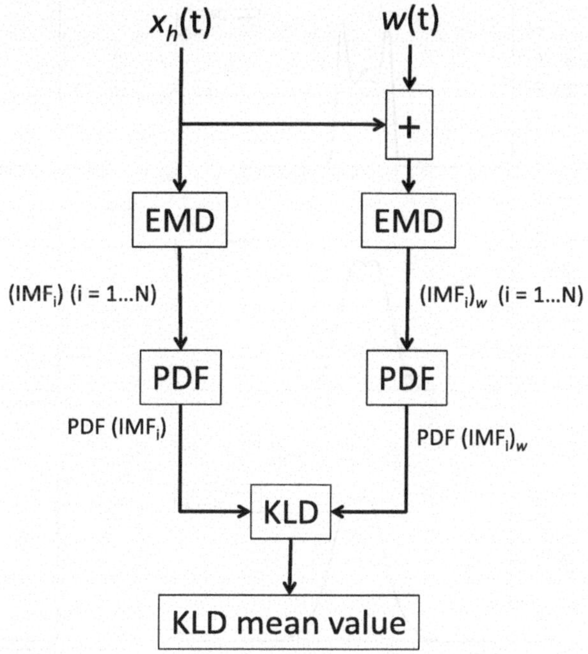

6. Repeat the operations (1)–(5) for the different loads.

The methodology is summarized in the flowchart displayed in Fig. 6.5.

Hereafter, with no loss of generality, we will only present the results concerning the no-load case (L_0).

As Fig. 6.6 shows that the PDFs of each IMF clearly share the same sample space, we can compute the KLD.

To evaluate the effect of the load variation, we compute the mean value of the KLD for 200 realizations. To compare the effect for each IMF, we calculate criteria denoted as Δ_R that measures the maximum deviation of the mean value from the no-load case taken as the reference.

$$\Delta_R = \frac{\Delta_{max}\{< KLD >_{IMF_i(L_j:j\neq 0)} \; / < KLD >_{IMF_i(L_0)}\}}{< KLD >_{IMF_i(L_0)}} \tag{6.4}$$

The results are summarized in Table 6.3.

The evolution of the relative maximum deviation is plotted in Fig. 6.7. It clearly shows that IMF_1 is the most sensitive to the load variation.

As a consequence to avoid confusion between load variation and fault effects, the IMF_1 cannot be considered as a reliable feature for fault detection. Therefore, in the following only the IMF_2–IMF_8 will be used to evaluate the fault detection.

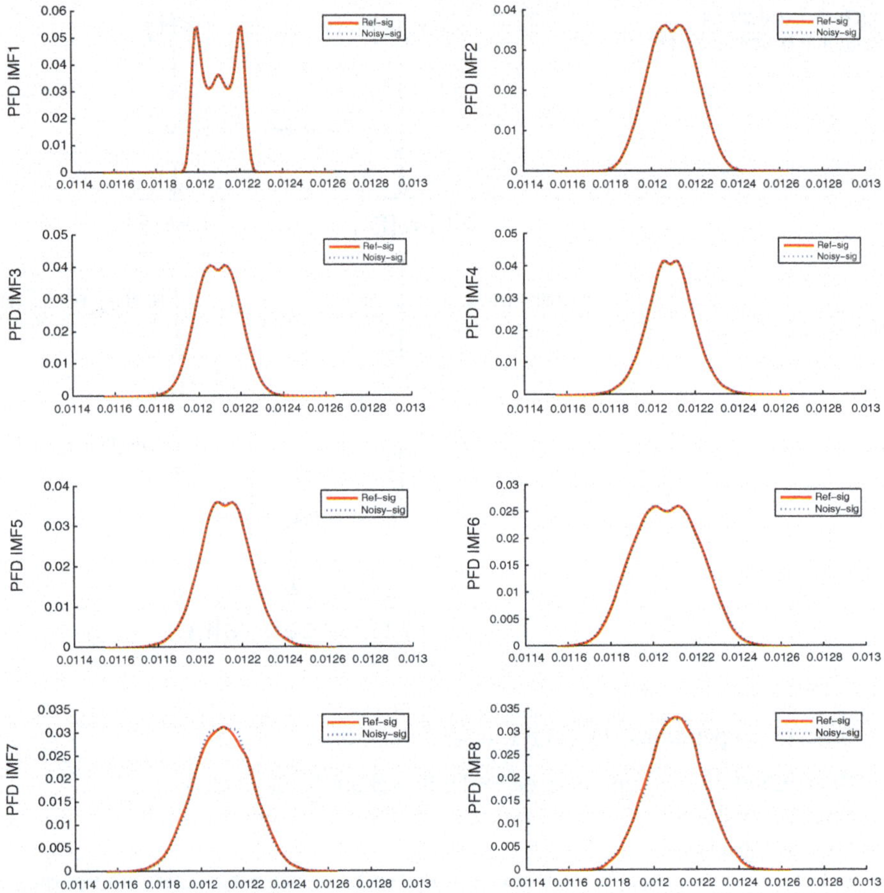

Fig. 6.6 PDF of the eight IMFs for L_0

Table 6.3 KLD mean values and maximum deviation of each IMF

	IMF_1	IMF_2	IMF_3	IMF_4	IMF_5	IMF_6	IMF_7	IMF_8
L_0	1.21×10^{-04}	2.83×10^{-04}	$6.77 \times e{-}^{04}$	0.0017	0.0056	0.0073	0.0195	0.0253
L_1	1.24×10^{-04}	3.88×10^{-05}	2.01×10^{-04}	5.39×10^{-04}	8.43×10^{-04}	0.0039	0.0199	0.0275
L_2	6.19×10^{-04}	6.27×10^{-05}	1.96×10^{-04}	4.49×10^{-04}	7.67×10^{-04}	0.0015	0.0087	0.0131
L_3	2.97×10^{-04}	4.26×10^{-05}	6.63×10^{-05}	1.61×10^{-04}	5.55×10^{-04}	0.0015	0.0062	0.01468
Δ_R	4.1	0.86	0.9	0.9	0.9	0.79	0.68	0.48

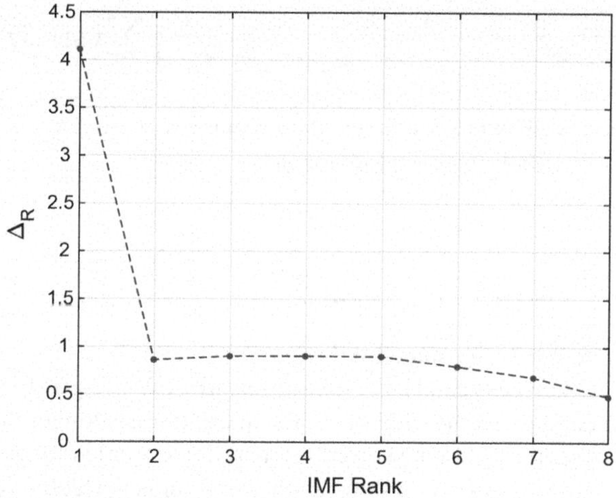

Fig. 6.7 Maximum deviation of the relative KLD mean value

6.5 Feature Analysis

In this section, we will consider bearing ball fault detection and analysis based on the seven selected $IMFs$ as previously described.

6.5.1 *Statistical Moments*

Feature analysis corresponds to the third step of the FDD and is based hereafter on the fact that each data distribution is commonly identified with its level of dispersion, asymmetry, and evolution around a mean value. Therefore, to detect the bearing fault presence, time-domain vibration analysis based on statistical moments (mean, variance, skewness, and kurtosis) will be used (Mezni et al. 2018; Press et al. 1992).

For a given dataset $X\{x_1, \ldots\ldots, x_N\}$ where N is the number of samples, the definition of statistical moments is given by the following expressions.

- The mean defined as the arithmetic average:

$$\overline{x} = \frac{1}{N}\sum_{i=1}^{N} x_i \tag{6.5}$$

- The variance is a measure characterizing the "width" or "variability" around the distribution's central value:

$$\sigma_{xi}^2 = \frac{1}{N-1}\sum_{i=1}^{N}(x_i - \overline{x})^2 \tag{6.6}$$

- The skewness is the third moment and characterizes the degree of asymmetry of a distribution around its mean:

$$S_{kxi} = \frac{1}{N}\sum_{i=1}^{N}\left[\frac{x_i - \overline{x}}{\sigma}\right]^3 \tag{6.7}$$

where σ is the standard deviation.

Skewness can be positive or negative. With a positive value, data are skewed right, meaning that the right tail of the distribution is longer than the left and with a negative value, the data are skewed left and the left tail is longer.

If $S_{kxi} = 0$, then the data distribution is perfectly symmetrical.

- The kurtosis is the fourth moment. As well as the skewness, the kurtosis is a nondimensional quantity. It reflects the "peakedness" or "flatness" of a distribution. The conventional definition of the kurtosis is

$$Kurt_{x_i} = \frac{1}{N}\sum_{i=1}^{N}\left[\frac{x_i - \overline{x}}{\sigma}\right]^4 \tag{6.8}$$

6.5.2 Application to the Retained IMFs

In the following, the four statistical moments (mean, variance, skewness, and kurtosis) will be evaluated on each retained IMF according to the following method:

1. Compute the EMD on the original signal (healthy or faulty);
2. Compute the four statistical moments for each of the seven selected IMFs;
3. Repeat operations (1) and (2) 200 times;
4. Add a weak AWGN signal with $SNR = 60\,\text{dB}$ to the original signal;
5. Compute the EMD on the resulting noisy signal;
6. Compute the four statistical moments for each of the seven selected IMFs;
7. Repeat operations (4)–(6) 200 times.

For each IMF, each statistical moment is calculated with 200 realizations of the healthy signals and 200 for the faulty ones.

Figures 6.8a, b and 6.9a, b display the results in the case of no load (L_0) and the smallest ball fault diameter (0.007 in.).

From these results, we can conclude that when a fault occurs, the variations of the variance and the kurtosis are more significant than those of the mean and the skewness. To evaluate quantitatively the relevance of these two criteria, their sensitivity will be calculated in the next section.

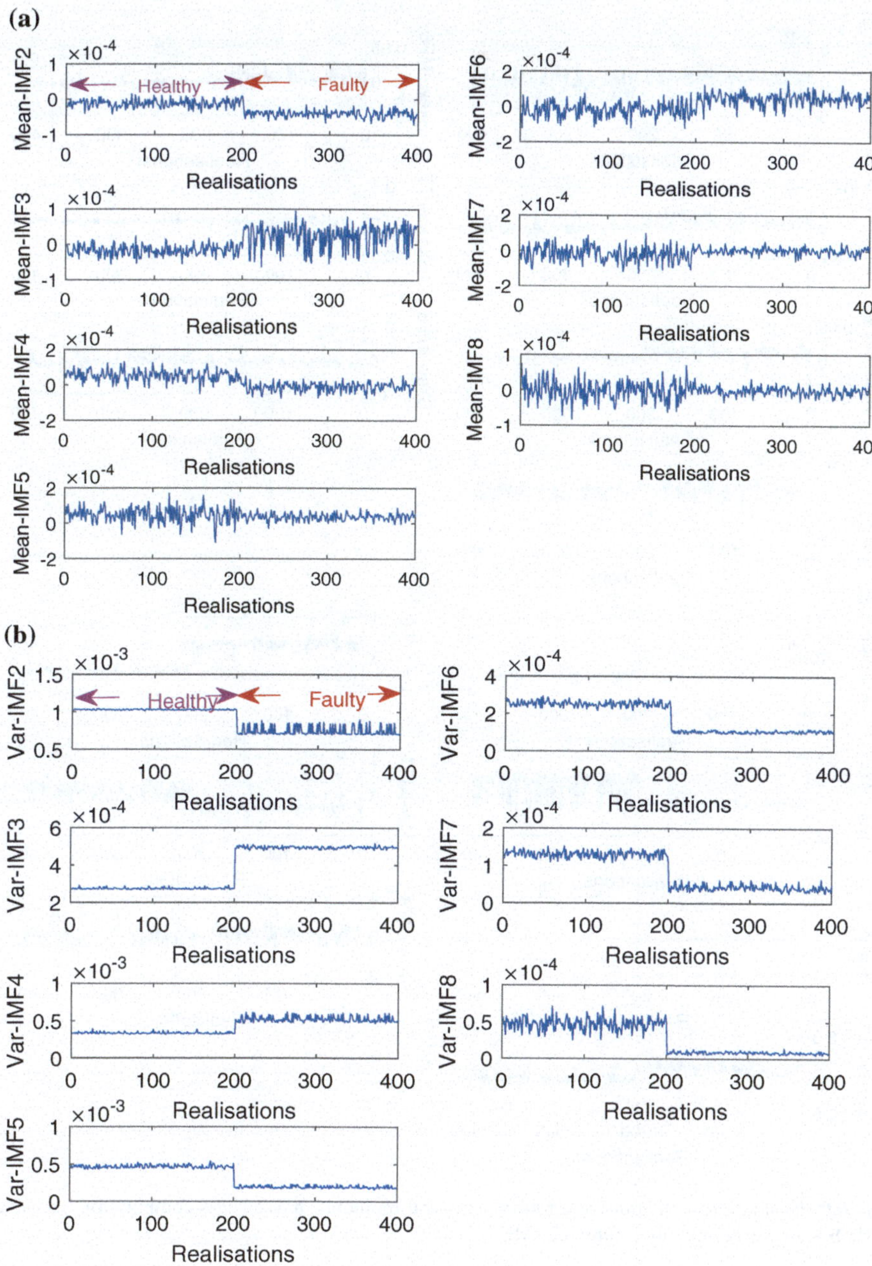

Fig. 6.8 Evaluation of the first and second statistical moments. **a** Mean computation for each IMF. **b** Variance computation for each IMF

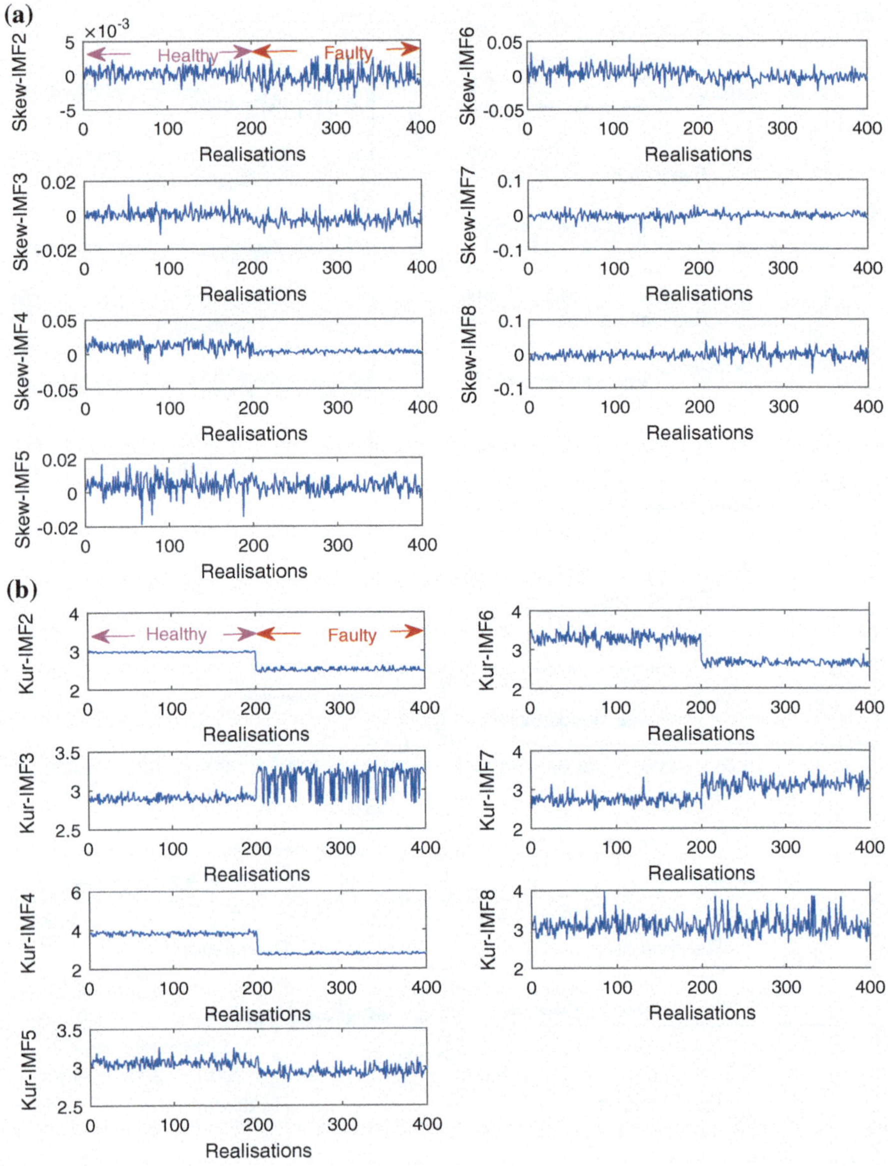

Fig. 6.9 Evaluation of the third and fourth statistical moments. **a** Skewness computation for each IMF. **b** Kurtosis computation for each IMF

6.5.3 Sensitivity Evaluation

In this section, we are interested in the evaluation of the sensitivity of the variance and the kurtosis for ball fault detection with respect to the fault depth as well as the load evolution.

The methodology is described in the flowchart displayed in Fig. 6.10.

1. Decomposition of the healthy signal using EMD;
2. Computation of each statistical moment (variance and kurtosis) for each selected IMF (Rank 2–8);
3. Repeat operations (1) and (2) 200 times;
4. Decomposition of the faulty signal using EMD;
5. Computation of each statistical moment for each selected IMF (Rank 2–8);
6. Repeat operations (4) and (5) 200 times;
7. Sensitivity evaluation for each criterion (variance and kurtosis).

The sensitivity criterion is defined as follows:

$$S(Cr) = \left| \frac{\overline{Cr_f} - \overline{Cr_h}}{Max(Cr_h - \overline{Cr_h})} \right| \tag{6.9}$$

where
Cr : criteria (i.e., variance or kurtosis).
$\overline{Cr_f}$: Mean of criteria under the faulty signal.
$\overline{Cr_h}$: Mean of criteria under the healthy signal.
$Max(Cr_h)$: The maximum of the criteria under the healthy signal.
$X_h(t)$ and $X_f(t)$ are, respectively, the original healthy signal and the faulty signal.

From the definition of the sensitivity criterion ($S(Cr)$), one can retrieve the following conclusions:

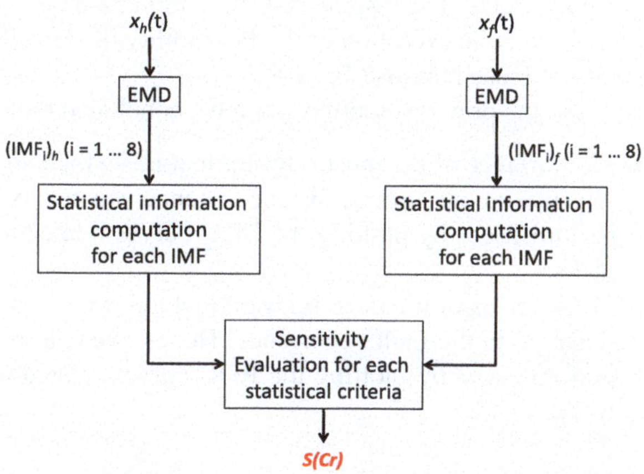

Fig. 6.10 Flowchart for the sensitivity computation

- if $0.5 < S(Cr) < 1$, then $0.5 < P_{MD} < 1$;
- if $S(Cr) = 1$, then $P_{MD} = 0.5$;
- if $1 < S(Cr) < 2$, then $0 < P_{MD} < 0.5$;
- if $S(Cr) \geq 2$, then $P_{MD} = 0$;

where P_{MD} is the probability of miss detection.

In the following only the incipient and the biggest ball faults will be considered:

- Fault F_1: 0.007 in.;
- Fault (F_3): 0.021 in.

For each IMF, the sensitivity (of the variance and the kurtosis) is normalized as follows:

$$NS = \frac{S(Cr)}{\max\left[S(Cr)\right]} \tag{6.10}$$

The results are displayed in Fig. 6.11 for the four operating points (L_0, L_1, L_2, and L_3) and the two fault levels (F_1 and F_3).

From Fig. 6.11, we can draw the following conclusions:

- For the variance, IMF_2, IMF_3, and IMF_4 are the most sensitive to the fault occurrence;
- For the kurtosis, IMF_2, IMF_4, and IMF_6 are the most sensitive to the fault occurrence.

In the next section, we can only focus on these most relevant IMFs to evaluate the fault detection performances.

6.5.4 *Fault Detection and Diagnosis Performances*

Receiver operating characteristic (ROC) curves are frequently used as tool for diagnosis performance evaluation. It is a graphical plot that illustrates the performances of a binary classifier system as its discrimination threshold is varied (Fawcett 2006). It represents the evolution of the Probability of Detection (P_D) along with the Probability of False Alarms (P_{FA}).

From the previous section, we have concluded that

- The variance of the most relevant features which are IMF_2, IMF_3, and IMF_4 is sensitive to the fault occurrence. Hence, we can evaluate the ball fault detection performances by plotting the ROC curves related to the variance for these three IMFs.
- The kurtosis of the most relevant features which are IMF_2, IMF_4, and IMF_6 is sensitive to the fault occurrence. Hence, we can evaluate the ball fault detection performances by plotting the ROC curves related to the kurtosis for these three IMFs.

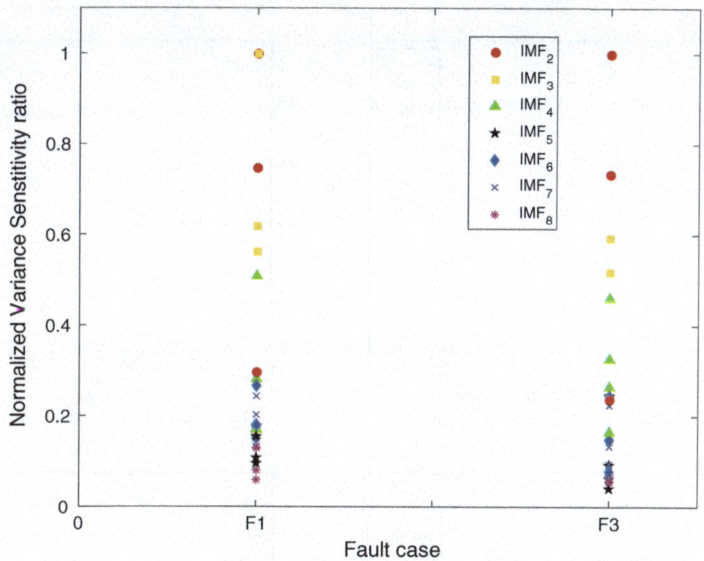

(a) Case of the Variance

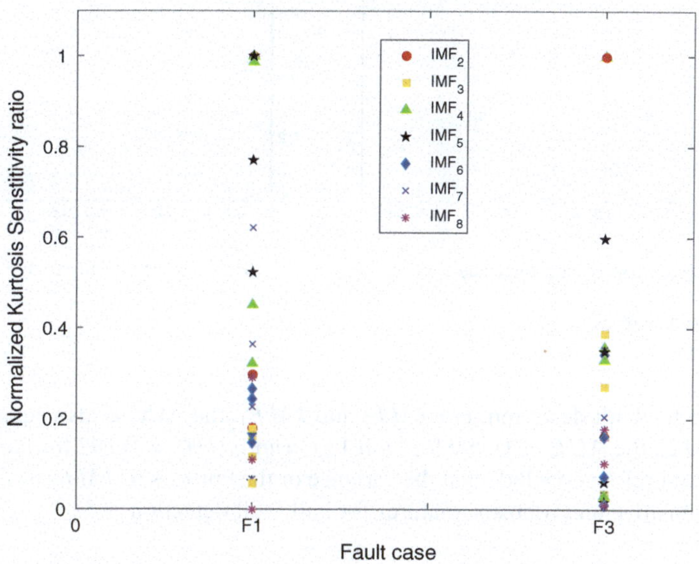

(b) Case of the kurtosis

Fig. 6.11 Normalized sensitivity for each IMF

Figure 6.12a, b displays the ROC curve if the variance is used as the feature for fault detection. The Area Under Curve (AUC) is equal to one, its maximum value for all the three IMFs. Figure 6.12c, d displays the ROC curve if the kurtosis is used as

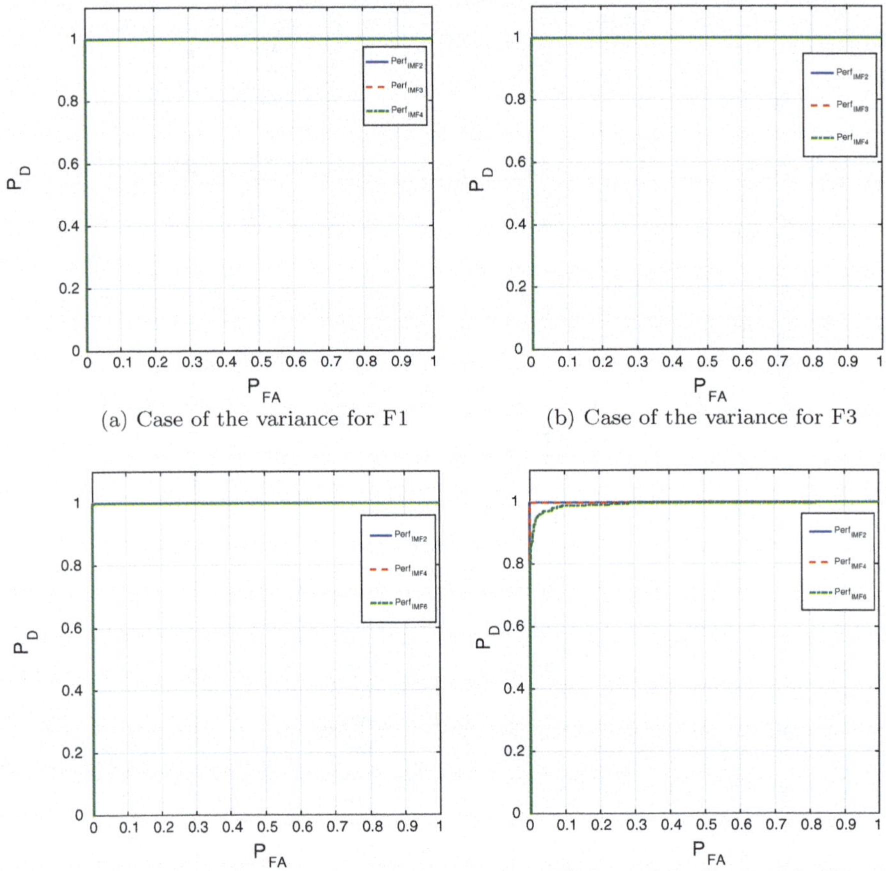

(a) Case of the variance for F1

(b) Case of the variance for F3

(c) Case of the kurtosis for F1

(d) Case of the kurtosis for F3

Fig. 6.12 ROC curves

the feature for fault detection. For IMF_2 and IMF_4, the AUC is also equal to one. But for IMF_6 the $AUC = 0.999$ for fault 1 (F_1) and $AUC = 0.992$ for fault 3 (F_3).

We can therefore conclude that the variance or the kurtosis of IMF_2 or IMF_4 are the most sensitive and robust signatures for ball fault detection.

6.6 Conclusion

In this case study, we have proposed and evaluated with experimental data a methodology to select the most relevant (sensitive to fault occurrence and robust to nuisances) Intrinsic Mode Functions (IMFs) for detecting bearing ball faults. The methodology is based on the analysis of the spectral content of vibration signals.

In the first step, the original signal is decomposed into 18 IMFs using Empirical Mode Decomposition (EMD). Only the first eight ones are retained as they have higher Signal-to-Noise Ratio (SNR), meaning higher energy. The robustness to load variations is evaluated with the Kullback–Leibler Divergence (KLD) known as a sensitive criterion to small changes. The results have shown that the first IMF is too sensitive to the load variation, and its use may lead to false alarms. Therefore, only the remaining ones (i.e., IMF_2–IMF_8) are retained for fault detection. From the analysis of the first four statistical moments, the variance and the kurtosis have appeared to be the most sensitive to the fault occurrence for both fault levels (the incipient and the most severe) and for all the operating points of the dataset.

Finally, the fault detection performances are evaluated through the Receiver Operating Curves (ROCs) of the variance (for IMF_2, IMF_3, and IMF_4) and the kurtosis (for IMF_2, IMF_4, and IMF_6). The results show that the ball fault is detected ($P_D = 1$) with no false alarms when using the variance of IMF_2, IMF_3, or IMF_4.

In conclusion, the ball fault information can always be retrieved either from IMF_2, IMF_4, or a combination of both when analyzing their variance or kurtosis.

References

Abdelkader, R., Derouiche, Z., Kaddour, A., & Zergoug, M. (2016). Rolling bearing faults diagnosis based on empirical mode decomposition: Optimized threshold de-noising method. In *2016 8th International Conference on Modelling, Identification and Control (ICMIC)* (pp. 186–191).

Abid, F. B., Zgarni, S., & Braham, A. (2016). Bearing fault detection of induction motor using SWPT and DAG support vector machines. In *IECON 2016-42nd Annual Conference of the IEEE Industrial Electronics Society* (pp. 1476–1481). IEEE.

Afgani, M., Sinanovic, S., & Haas, H. (2008). Anomaly detection using the Kullback-Leibler divergence metric. In *2008 1st International Symposium on Applied Sciences on Biomedical and Communication Technologies* (pp. 1–5).

Ali, J. B., Fnaiech, N., Saidi, L., Chebel-Morello, B., & Fnaiech, F. (2015). Application of empirical mode decomposition and artificial neural network for automatic bearing fault diagnosis based on vibration signals. *Applied Acoustics*, *89*, 16–27.

Amirat, Y., Choqueuse, V., & Benbouzid, M. (2013). EEMD-based wind turbine bearing failure detection using the generator stator current homopolar component. *Mechanical Systems and Signal Processing*, *41*(1), 667–678.

Anderson, A., & Haas, H. (2011). Kullback-Leibler divergence (KLD) based anomaly detection and monotonic sequence analysis. In *2011 IEEE Vehicular Technology Conference (VTC Fall)* (pp. 1–5).

Basseville, M. (1989). Distances measures for signal processing and pattern recognition. *Elsevier Signal Processing*, *18*(4), 349–369.

Bensaad, D., Guillet, F., Soualhi, A., & El Badaoui, M. (2017). Kurtosis analysis as a cycle-ratio function in gear and bearing fault detection. *International Journal of Condition Monitoring*, *7*(2), 36–40.

Bittencourt, A. C., Saarinen, K., Sander-Tavallaey, S., Gunnarsson, S., & Norrlof, M. (2014). A data-driven approach to diagnostics of repetitive processes in the distribution domain - applications to gearbox diagnostics in industrial robots and rotating machines. *Mechatronics*, *24*(8), 1032–1041.

Bouchikhi, E. H., Choqueuse, V., Benbouzid, M., Charpentier, J.-F., & Barakat, G. (2011). A comparative study of time-frequency representations for fault detection in wind turbine. In *IECON 2011-37th Annual Conference on IEEE Industrial Electronics Society* (pp. 3584–3589). IEEE.

Cai, Y.-P., Li, H.-H., Shi, L.-S., Bai, X.-F., & Shen, J.-W. (2011). Roller bearing fault detection using improved envelope spectrum analysis based on EMD and spectrum kurtosis. *Journal of Vibration and Shock*, *2*, 35.

Cheng, J., Yu, D., Tang, J., & Yang, Y. (2008). Application of frequency family separation method based upon EMD and local hilbert energy spectrum method to gear fault diagnosis. *Mechanism and Machine Theory*, *43*(6), 712–723.

Collacott, R. (1979). *Vibration monitoring and diagnosis: Techniques for cost-effective plant maintenance*. New York: Wiley.

Cover, T., & Thomas, J. (2005). *Elements of information theory*. New Jersey: Wiley.

Cusidó, J., Romeral, L., Ortega, J. A., Rosero, J. A., & Espinosa, A. G. (2008). Fault detection in induction machines using power spectral density in wavelet decomposition. *IEEE Transactions on Industrial Electronics*, *55*(2), 633–643.

CWRU, U. (2016). Bearing data centre. Case Western Reserve University. Retrieved May 2016.

Delpha, C., Diallo, D., Wang, T., Liu, J., & Li, Z. (2017a). Multisensor fault detection and isolation using Kullback Leibler divergence: Application to data vibration signals. In *International Conference on Sensing, Diagnostics, Prognostics, and Control (SDPC), Shanghai, China* (pp. 305–310). IEEE.

Delpha, C., Diallo, D., & Youssef, A. (2017b). Kullback-Leibler divergence for fault estimation and isolation: Application to gamma distributed data. *Mechanical Systems and Signal Processing Journal, Elsevier*, *93*(C), 118–135. https://doi.org/10.1016/j.ymssp.2017.01.045.

Duan, F., Corsar, M., & Mba, D. (2016). Using empirical mode decomposition scheme for helicopter main gearbox bearing defect identification. In *2016 Prognostics and System Health Management Conference (* PHM-*Chengdu)* (pp. 1–4).

Elbouchikhi, E., Choqueuse, V., Amirat, Y., Benbouzid, M. E. H., & Turri, S. (2017). An efficient Hilbert-Huang transform-based bearing faults detection in induction machines. *IEEE Transactions on Energy Conversion*, *32*(2), 401–413.

Fawcett, T. (2006). An introduction to ROC analysis. *Pattern Recognition Letters*, *27*(8), 861–874.

Feng, Z., Fan, Y., Liang, M., & Chu, F. (2013a). A nonstationary vibration signal analysis method for fault diagnosis of planetary gearboxes. *Proceeding of the CSEE*, *33*(17), 105.

Feng, Z., Liang, M., & Chu, F. (2013b). Recent advances in time-frequency analysis methods for machinery fault diagnosis: A review with application examples. *Mechanical Systems and Signal Processing*, *38*(1), 165–205.

Flandrin, P. (1989). Some aspects of non-stationary signal processing with emphasis on time-frequency and time-scale methods. *Wavelets*, 68–98.

Harmouche, J., Delpha, C., & Diallo, D. (2012). Faults diagnosis and detection using principal component analysis and Kullback-Leibler divergence. In *IEEE Industrial Electronics Society Conference (IECON 2012), Montreal, Canada* (pp. 3907–3912). IEEE.

Harmouche, J., Delpha, C., Diallo, D. (2013). A theoretical approach for incipient fault severity assessment using the Kullback-Leibler divergence. In *European Signal Processing Conference (EUSIPCO 2013), Marrakech, Morocco* (pp. 1–5). IEEE.

Harmouche, J., Delpha, C., & Diallo, D. (2014). Incipient fault detection and diagnosis based on Kullback-Leibler divergence using principal component analysis: Part I. *Elsevier Journal on Signal Processing*, *94*, 278–287.

Harmouche, J., Delpha, C., & Diallo, D. (2015a). Improved fault diagnosis of ball bearings based on the global spectrum of vibration signals. *IEEE Transactions on Energy Conversion*, *30*(1), 376–383.

Harmouche, J., Delpha, C., & Diallo, D. (2015b). Incipient fault detection and diagnosis based on Kullback-Leibler divergence using principal component analysis: Part II. *Elsevier Journal on Signal Processing*, *109*, 334–344.

Harmouche, J., Delpha, C., & Diallo, D. (2016). Incipient fault amplitude estimation using KL divergence with a probabilistic approach. *Elsevier Journal on Signal Processing*, *120C*, 1–7. https://doi.org/10.1016/j.sigpro.2015.08.008.

Hawley, S. D., Atlas, L. E., & Chizeck, H. J. (2010). Some properties of an empirical mode type signal decomposition algorithm. *IEEE Signal Processing Letters*, *17*(1), 24–27.

Huang, N. E., Shen, Z., Long, S. R., Wu, M. C., Shih, H. H., Zheng, Q., et al. (1998). The empirical mode decomposition and the Hilbert spectrum for nonlinear and non-stationary time series analysis. *Proceedings of the Royal Society of London A: Mathematical, Physical and Engineering Sciences*, *454*(1971), 903–995.

Huang, W., Shen, Z., Huang, N. E., & Fung, Y. C. (1999). Nonlinear indicial response of complex nonstationary oscillations as pulmonary hypertension responding to step hypoxia. *Proceedings of the National Academy of Sciences of the United States of America*, *96*(5), 1834–1839.

Hyvarinen, A. (1999). Survey on independent component analysis. *Neural Computing Surveys*, *2*, 94–128.

Kopsinis, Y., & McLaughlin, S. (2008). Improved EMD using doubly-iterative sifting and high order spline interpolation. *EURASIP Journal on Advances in Signal Processing*, *2008*(120), 120:1–120:8.

Kullback, S., & Leibler, R. (1951). On information and sufficiency. *The Annals of Mathematical Statistics*, *22*(1), 79–86.

Laha, S. (2017). Enhancement of fault diagnosis of rolling element bearing using maximum kurtosis fast nonlocal means denoising. *Measurement*, *100*, 157–163.

Lei, Y., Lin, J., He, Z., & Zuo, M. J. (2013). A review on empirical mode decomposition in fault diagnosis of rotating machinery. *Mechanical Systems and Signal Processing*, *35*(1), 108–126.

Li, Y., Xu, M., Liang, X., & Huang, W. (2017). Application of bandwidth EMD and adaptive multiscale morphology analysis for incipient fault diagnosis of rolling bearings. *IEEE Transactions on Industrial Electronics*, *64*(8), 6506–6517.

Liu, S., Yamada, M., Collier, N., & Sugiyama, M. (2013). Change-point detection in time-series data by relative density-ratio estimation. *Neural Networks*, *43*, 72–83.

Luo, L., Yan, Y., Xie, P., Sun, J., Xu, Y., & Yuan, J. (2012). Hilbert-Huang transform, Hurst and chaotic analysis based flow regime identification methods for an airlift reactor. *Chemical Engineering Journal*, *181–182*(Supplement C), 570–580.

Ma, J., & Jiang, J. (2011). Applications of fault detection and diagnosis methods in nuclear power plants: A review. *Progress in Nuclear Energy*, *53*(3), 255–266.

Mezni, Z., Delpha, C., Diallo, D., & Braham, A. (2018). Bearing fault detection using intrinsic mode functions statistical information. In *15th IEEE International Multi-Conference on Systems, Signals and Devices (SSD) (SSD'18)*.

Mohanty, S., Gupta, K., & Raju, K. (2017). Adaptive fault identification of bearing using empirical mode decomposition-principal component analysis-based average kurtosis technique. *IET Science, Measurement Technology*, *11*(1), 30–40.

Montesinos, M., Muñoz Cobo, J., & Pèrez, C. (2003). Hilbert-Huang analysis of BWR neutron detector signals: Application to DR calculation and to corrupted signal analysis. *Annals of Nuclear Energy*, *30*(6), 715–727.

Peng, Z., & Chu, F. (2004). Application of the wavelet transform in machine condition monitoring and fault diagnostics: A review with bibliography. *Mechanical Systems and Signal Processing*, *18*(2), 199–221.

Peng, Z., Tse, P. W., & Chu, F. (2005). A comparison study of improved Hilbert-Huang transform and wavelet transform: Application to fault diagnosis for rolling bearing. *Mechanical Systems and Signal Processing*, *19*(5), 974–988.

Press, W. H., Teukolsky, S. A., Vetterling, W. T., & Flannery, B. P. (1992). *Numerical recipes in C: The art of scientific computing*. New York: Cambridge Press University.

Qian, S. (2001). *Introduction to time-frequency and wavelet transforms*. Oxford: Prentice Hall.

Rai, A., & Upadhyay, S. (2017). Bearing performance degradation assessment based on a combination of empirical mode decomposition and k-medoids clustering. *Mechanical Systems and Signal Processing*, *93*(Supplement C), 16–29.
Riera-Guasp, M., Antonino-Daviu, J. A., & Capolino, G.-A. (2015). Advances in electrical machine, power electronic, and drive condition monitoring and fault detection: State of the art. *IEEE Transactions on Industrial Electronics*, *62*(3), 1746–1759.
Satish, L. (1998). Short-time Fourier and wavelet transforms for fault detection in power transformers during impulse tests. *IEE Proceedings - Science, Measurement and Technology*, *145*(2), 77–84.
Scheffer, C., & Girdhar, P. (2004). *Practical machinery vibration analysis and predictive maintenance*. Amsterdam: Elsevier.
Silva, J., & Narayanan, S. (2006). Average divergence distance as a statistical discrimination measure for hidden Markov models. *IEEE Transactions on Audio, Speech, and Language Processing*, *14*(3), 890–906.
Singh, J., Darpe, A., & Singh, S. (2017). Bearing damage assessment using Jensen-Rényi divergence based on EEMD. *Mechanical Systems and Signal Processing*, *87*, 307–339.
Srinivasan, R., Rengaswamy, R., & Miller, R. (2007). A modified empirical mode decomposition (EMD) process for oscillation characterization in control loops. *Control Engineering Practice*, *15*(9), 1135–1148.
Staszewski, W., & Tomlinson, G. (1994). Application of the wavelet transform to fault detection in a spur gear. *Mechanical Systems and Signal Processing*, *8*(3), 289–307.
Taylor, J. I. (1994). *The vibration analysis handbook*. Tampa: Vibration Consultants.
Tsao, W.-C., Li, Y.-F., & Pan, M.-C. (2010). Resonant-frequency band choice for bearing fault diagnosis based on EMD and envelope analysis. In *2010 8th World Congress on Intelligent Control and Automation* (pp. 1289–1294).
Venkatasubramanian, V., Rengaswamy, R., Kavuri, S. N., & Yin, K. (2003a). A review of process fault detection and diagnosis: Part III: Process history based methods. *Computers and Chemical Engineering*, *27*(3), 327–346.
Venkatasubramanian, V., Rengaswamy, R., Yin, K., & Kavuri, S. N. (2003b). A review of process fault detection and diagnosis: Part I: Quantitative model-based methods. *Computers and Chemical Engineering*, *27*(3), 293–311.
Wu, Z., & Huang, N. E. (2009). Ensemble empirical mode decomposition: A noise-assisted data analysis method. *Advances in Adaptive Data Analysis*, *01*(01), 1–41.
Xie, L., Zeng, J., Kruger, U., Wang, X., & Geluk, J. (2015). Fault detection in dynamic systems using the Kullback-Leibler divergence. *Control Engineering Practice*, *43*, 39–48.
Xu, D., Xu, Y., Chen, X., Zha, W., & Li, X. (2009). Life cycle vibration analysis based on EMD of rolling element bearing under ALT by constant stress. In *2009 8th International Conference on Reliability, Maintainability and Safety* (pp. 1177–1182).
Xuan, B., Xie, Q., & Peng, S. (2007). EMD sifting based on bandwidth. *IEEE Signal Processing Letters*, *14*(8), 537–540.
Xue, X., Zhou, J., Xu, Y., Zhu, W., & Li, C. (2015). An adaptively fast ensemble empirical mode decomposition method and its applications to rolling element bearing fault diagnosis. *Mechanical Systems and Signal Processing*, *62–63*, 444–459.
Yan, R., Gao, R. X., & Chen, X. (2014). Wavelets for fault diagnosis of rotary machines: A review with applications. *Signal Processing*, *96*, 1–15.
Yang, Z., Yang, L., Qing, C., & Huang, D. (2008). A method to eliminate riding waves appearing in the empirical AM/FM demodulation. *Digital Signal Processing*, *18*(4), 488–504.
Yesilyurt, I., and of Manchester. School of Engineering, U. (1997). Gearbox fault detection and severity assessment using vibration analysis. University of Manchester, UK.
Youssef, A., Delpha, C., Diallo, D. (2016a). Enhancement of incipient fault detection and estimation using the multivariate Kullback-Leibler divergence. In *European Signal Processing Conference (EUSIPCO 2016), Budapest, Hungary* (pp. 1408–1412). IEEE.

Youssef, A., Delpha, C., & Diallo, D. (2016b). An optimal fault detection threshold for early detection using Kullback-Leibler divergence for unknown distribution data. *Elsevier Journal on Signal Processing*, *120C*, 266–279. https://doi.org/10.1016/j.sigpro.2015.09.008.

Zeng, J., Kruger, U., Geluk, J., Wang, X., & Xie, L. (2014). Detecting abnormal situations using the Kullback-Leibler divergence. *Automatica*, *50*(11), 2777–2786.

Zhang, C., Chen, J.-J., & Xu, Y.-L. (2011). A bearing fault diagnosis method based on EMD and difference spectrum theory of singular value. *Journal of Vibration Engineering*, *24*(5), 539–545.

Zhang, Y., Bingham, C., Yang, Z., Ling, B.W.-K., & Gallimore, M. (2014). Machine fault detection by signal denoising-with application to industrial gas turbines. *Measurement*, *58*(Supplement C), 230–240.

Zhou, C., Liu, C., Wu, X., & Chi, Y. (2011). A demodulation method based on empirical mode decomposition and adaptive morphological filtering. *Mechanical Science and Technology for Aerospace Engineering*, *8*, 35.

Chapter 7
Fault Detection and Localization of Centrifugal Gas Compressor System Using Fuzzy Logic and Hybrid Kernel-SVM Methods

Bachir Nail, Nadji Hadroug, Ahmed Hafaifa and Abdellah Kouzou

Abstract In many industrial sectors, especially the petroleum sector, surveillance activity in rotating machinery is a very complex task and requires a great deal of information and essential data concerning the operation of these processes. The goal of this chapter is to realize a fault detection and localization strategy applied to a centrifugal gas compressor system, where this system is installed in a gas compressor station in Hassi R'mel in the south of Algeria. The proposed strategy is based on the method of hybrid Kernel-SVM method, where this method is used in many industrial sectors for detecting and diagnosing faults system; in our knowledge, this study is applied for the first time in centrifugal gas compressor system using experimental data. According to this proposed method, we can detect and monitor the state of thermodynamics of the studied system and obtain the important information for automatically take different situations for the intervention such as fault-tolerant control.

Keywords Centrifugal gas compressor · Thermodynamics behavior
Faults detection, diagnosis · Kernel-SVM method

B. Nail (✉)
Faculty of Science and Technology, University of Khemis Miliana, Road of Theniet El Had, 44225 Khemis Miliana, Algeria
e-mail: b.nail@univ-dbkm.dz

N. Hadroug · A. Hafaifa · A. Kouzou
Applied Automation and Industrial Diagnostics Laboratory, Faculty of Sciences and Technology, University of Djelfa, Djelfa, Algeria
e-mail: n.hadroug@univ-djelfa.dz

A. Hafaifa
e-mail: hafaifa.ahmed.dz@ieee.org

A. Kouzou
e-mail: kouzouabdellah@ieee.org

N. Hadroug · A. Hafaifa · A. Kouzou
Gas Turbine Joint Research Team, University of Djelfa, Djelfa, Algeria

N. Derbel et al. (eds.), *Diagnosis, Fault Detection & Tolerant Control*,
Studies in Systems, Decision and Control 269,
https://doi.org/10.1007/978-981-15-1746-4_7

7.1 Introduction

In the last few years, the maintenance of industrial systems during their operating modes is one of the main strategic problems facing the industry, from the design of a machine until its exploitation. Therefore, the diagnostic system is essential for ensuring the smooth and continuous operation of dynamic systems and for increasing their performances by guaranteeing better reliability. Indeed, the diagnostic system is used to provide the control system by the required real data of the dynamic system operating status in unfaulty (healthy) mode and in faulty mode. On the other side, the diagnostic system has to fulfill the requirement of robustness to avoid the practical cases of non-detection and false alarms, which means avoiding the eventual accidental and catastrophic situations. The diagnostic system is based mainly on comparing the actual behavior of the system with a reference behavior representing the healthy operation. This comparison allows to detect the behavioral changes that are due to the appearance of the faults.

The FDI diagnosis approaches implemented in the industries are generally divided into two main classes such as the diagnostic approaches based on mathematical model and the diagnostic approaches based on data analysis. The FDI mathematical model approaches include the observer's approach (Dai et al. 2009), the parity-space approach (Hafaifa et al. 2015), and system identification-based approaches (Bahareh et al. 2015). The FDI data analysis approaches are called the measurements approaches, and these approaches insure the detection of fault under certain conditions, including the artificial intelligence-based model (fuzzy logic, neural networks, etc.) (Hadroug et al. 2016), statistical approaches, and FDI approaches based on signal processing (Bassily et al. 2009; Isermann 2006).

The proposed diagnostic approach presented in this chapter is applied on the centrifugal gas compressor system which is used in many sectors and covers a very wide range of industrial applications. Indeed, this system is at the heart of many industrial sectors, such as the petroleum industry, the thermal and the nuclear power generation, the aeronautic and the space propulsion, the automobile industry, and the transport of gases (pipelines) (Saavedra et al. 2010; Maleki et al. 2016; Tabkhi et al. 2010; Mohamadi et al. 2014). A good understanding of this system operation is essential issue for increasing their performance and reducing their operating costs. In this case, one of the limits of using this system is determined by the stability limits, where it is well known that beyond these limits the operation system stability cannot be ensured. In order to take into account the characteristics of the studied gas compressor system and its operating conditions, it is necessary to establish a dynamic model that covers all the dynamic behaviors. This dynamic model presents an important element in the proposed diagnosis and detection approach, where the main objective is to obtain the best dynamical model based on previous works. Indeed, several recent researches have been dealt with the gas compressor systems modeling such as the industrial centrifugal compressor modeling based on fuzzy logic approach (Hafaifa et al. 2009), the centrifugal gas compressor parametric modeling based on system identification (Nail et al. 2018), the two shaft gas turbine modeling based

Table 7.1 General performance of centrifugal gas compressor BCL 505

Stages	1–5
Maximum discharge pressure	123 kg/cm^2
Maximum discharge temperature	121 °C
Efficiently—speed 3000 to 20000	rpm
Compressed gas	LNG

Fig. 7.1 Centrifugal gas compressor BCL 505 body, with real view in vertical joint plane

on linearized model (Hadroug et al. 2017), and the system simulation for dynamic centrifugal compressor model (Jiang et al. 2006).

The novelty of this work consists in proposing a new approach (hybrid Kernel-SVM method) for faults detection in a centrifugal gas compressor to ensure its safe operation, its availability for production, and to minimize its maintenance cost. This approach is applied to a BCL 505 centrifugal gas compressor used by Sonatrach in the HassiR'Mel gas compressor station in the south of Algeria.

7.2 Centrifugal Gas Compressor (BCL 505 System)

Centrifugal compressors are used in many industrial sectors, such as the oil industry, the production of thermal and nuclear energy, aerospace propulsion, automotive, water distribution, etc. Indeed, a good understanding of the operation of these devices is essential to increase their performance and reduce their operating cost. In this case, one of the limits of use of these systems is determined by its stability limits, limits beyond which stable operation of system is no longer ensured. The centrifugal gas compressor studied in this chapter is the BCL 505 type which is shown in Fig. 7.1, and its characteristics are presented in Table 7.1.

This compressor is constructed by Nuovo Pignone company and it is used in a gas compression heavy application such as in gas field production and gas network transportation, and it is equipped with a computer-based control room where the DCS is a part of it which allows to take directly input/output measurements from the installed sensors. The main function of the studied centrifugal gas compressor in this chapter is to ensure the pressure rise of the continuous flow of gas passing through it based on kinetic energy, where the increase of the gas pressure by a compressor is used to

- Reach a level of gas pressure.
- Compensate the pressure losses related to the circulation of the gas flow in a gas network.

The compressors can be classified according to their characteristics depending on the type of gas to be compressed such as air compressors and gas compressors, and/or depending on the movement of the moving parts such as linear or rotary motion, and/or depending on the operating principle such as volumetric compressors and dynamic compressors which are the application areas of the present chapter.

7.3 The Power and the Efficiency of the Studied System

The performance evaluation of a compressor is generally not limited to a single operating point. The compressor field shows the compression ratio as a function of the corrected flow at a constant corrected rotation speed to whom the iso-efficiency contours are often superimposed.

The limit at low flow rates is the pumping, which is characterized by the flow instability (sometimes until flow reversal), accompanied by pressure oscillations of great amplitude, which can eventually damage the machine. At high flows, the limit is the blockage, which corresponds to the appearance of a vibration phenomenon as sonic section in the floor.

Figures 7.2 and 7.3 represent the produced power by the centrifugal gas compressor and its efficiency under the healthy operation state, respectively, where the variation in the produced power (increase/decrease) depends on the market demand.

7.4 Preliminary Concepts About Fuzzy Logic

Initially, the concept of the fuzzy type-1 set has been proposed by Zadeh, the founder of the fuzzy logic (Zadeh 1965; Madau et al. 1996). A fuzzy set is characterized by a fuzzy membership function (the degree of belonging to the fuzzy set in [0,1]). Such sets are advisable in the case when there is an uncertainty at the level of the value of the membership itself, i.e., uncertainty can be either in the form of the membership function or in one of its parameters. That is, transition from an ordinary set to a

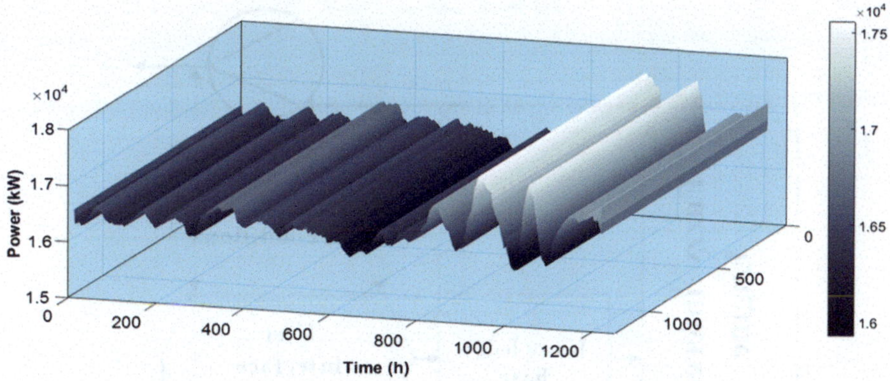

Fig. 7.2 The hourly evolution of the power of the centrifugal gas compressor system

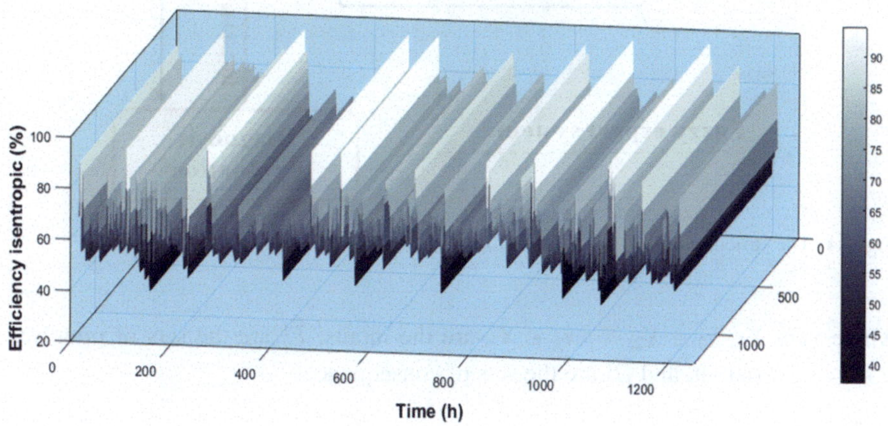

Fig. 7.3 The hourly evolution of the efficiency of centrifugal gas compressor system

fuzzy set is the direct consequence of the indeterminism of the value belonging to an element by 0 or 1. Similarly, when the functions of an element belonging to fuzzy numbers cannot be determined in real numbers within [0,1], then the fuzzy set type-1 is used, as shown in Fig. 7.4.

Depending on the form of primary membership, there are many types of fuzzy sets, among them, triangular, interval, Gaussian, and the Gaussian 2 are used in the fuzzy system fault detection in this chapter.

The fuzzy rules reside only in the nature of the membership functions. Therefore, the jth rule of a fuzzy system has the following form (Zadeh 1965):

$$\text{If } x_1 \text{ is } \widetilde{F}_1^i \text{ and } x_2 \text{ is } \widetilde{F}_2^i \text{ and } \cdots \text{ and } x_p \text{ is } \widetilde{F}_p^i \text{ Then } y \text{ is } \widetilde{G}^i \tag{7.1}$$

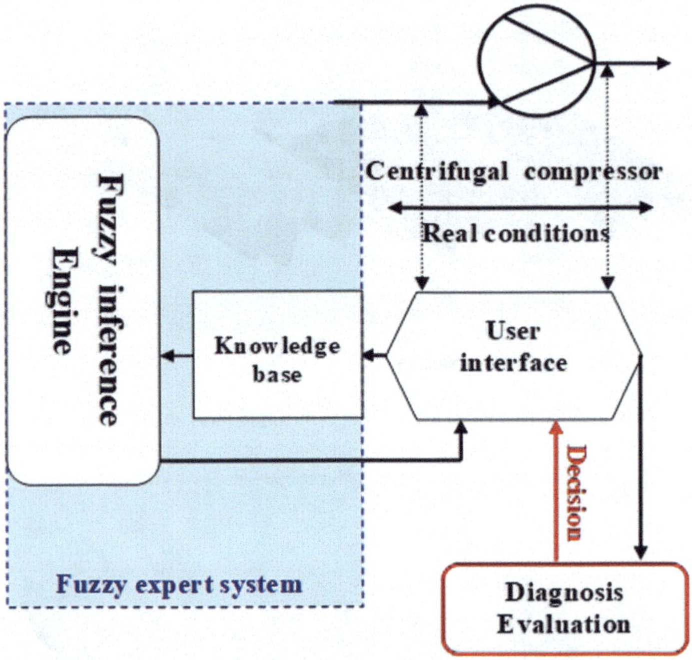

Fig. 7.4 The structure of the type-1 fuzzy logic system

where $x_1 \in X_1$, $x_2 \in X_2 \cdots x_p \in X_p$ are the inputs, $\widetilde{F}_j^i$ are the sets of premises, $y \in Y$ is the output, and $\widetilde{G}^i$ are the sets of consequences.

7.5 Resolution of the Optimization Problem of Fault Detection

In this section we have interested to use the binary classification "0" and "1" that is to say the approximation of decision functions to distinguish between two classes, and this by hyperplanes $h : x \longrightarrow y$ of equation $f(x) = \langle W, X \rangle + b = \sum_i^d w_i x_i + b = 0$, as shown in Fig. 7.5 assuming that X be of dimension d. We will then use the decision function $g(x) = \big(\langle W, X \rangle + b\big)$ to predict the class $(+/-)$ of the input X. The search for the optimal hyperplane, therefore, amounts to solving the following optimization problem which concerns the parameters W and b:

$$\min \frac{1}{2}||W||^2 \tag{7.2}$$
$$\text{under the constraints } y_i(W^T x_i + b) \geq 1\ ,\quad i = 1, 2 \cdots m$$

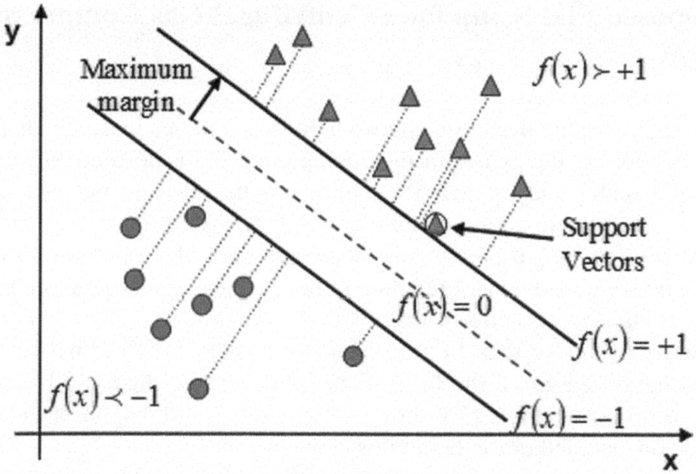

Fig. 7.5 The optimal hyperplane is perpendicular to the shortest straight segment joining an example of hyperplane learning

where x_i are the coordinates of a point X, which means that W is a linear combination of learning points.

According to the theory of optimization, an optimization problem has a dual form in the case where the objective function and the constraints are strictly convex. These convexity criteria are realized in Problem (7.2). To solve these types of problems, a so-called Lagrangian function is used which incorporates information on objective function and constraints and whose stationary character can be used to detect solutions. The SVM kernel is obtained by the direct application of the kernel trick (Hadroug et al. 2016). Just use the following changes:

$$L(\alpha) = \sum_{i=1}^{m} \alpha_i - \frac{1}{2} \sum_{i=1}^{m} \sum_{j=1}^{m} \alpha_i \alpha_j y_i y_j K\langle x_i, x_j \rangle \tag{7.3}$$

$$\text{subject to } f(x) = \langle W, X \rangle + b = \sum_{i=1}^{m} \alpha_i y_i K\langle x_i, x_j \rangle + b$$

where the Gaussian kernel is

$$K(x_i, x_j) = \exp\left(-\frac{||x_i - x_j||^2}{2\sigma^2}\right)$$

with the normalization constant, and this Gaussian kernel is a normalized kernel, that is, its integer over its complete domain is the unit for each "σ".

7.6 Proposed FDI Setup for a Centrifugal Gas Compressor Plant

The faults diagnosis and detection approach proposed in this research chapter takes into consideration all the phases of the operating cycle of the centrifugal gas compressor system under study, such as the evolution of the pressure and the temperature of the gas in the centrifugal gas compressor and the evolution of mass and volume flows as function of the gas pressure and temperature. In the same time, it takes into account the power absorbed as a function of the gas, and other operating conditions that allow to obtain precise information about the dynamic behavior and to maximize the availability of this system. In this diagnostic system, a hybrid between the real data of estimation based on the fuzzy logic observer and the hybrid kernel-SVM algorithms is proposed. Figure 7.6 shows the proposed fault diagnosis and detection approach details studied presented in this chapter.

On the other side, it is well known that the industrial systems have complex behaviors, and they are characterized by uncertain variables or parameters as a function of time, and this constraint complicates their control task and implies many difficulties in achieving the good performances of such systems. For this purpose, this work proposes a real-time fault diagnosis and detection approach, where the main aim is to detect and to localize the defective components in the studied centrifugal gas compressor system.

This proposed approach is based on the calculation of the residues $r(k)$ following Eq. (7.2), which presents the errors between the optimal Y_{op} and the observed $\widehat{Y}$ outputs, respectively. On the other side, the residues are the inputs for the fuzzy type-2 system, when the system is under healthy operating state, and these residues have generally a null average and a determined variance.

In practice, the residues do not have exactly zero value in the absence of faults, because the obtained model of the studied system does not take into account all the internal and external parameters, which means that only the preponderant parameters are taken into account and that certain simplification has been considered. On the other side, the measurements performed on the system are often affected by measurement noise.

The residues are expressed as follows:

$$r(k) = Y_{op}(k) - \widehat{Y}(k) \tag{7.4}$$

where $r(k) \in [r_p(k), r_T(k)]^T \in \mathbb{R}^{2\times 1}$.

In this context, an elementary detection method consists in comparing the value of the obtained residues with a predefined threshold (modeling error function). An alarm is triggered each time this threshold is crossed:

$$\begin{cases} r(k) \leq \varepsilon \Longleftrightarrow d(k) = 0 \\ r(k) > \varepsilon \Longleftrightarrow d(k) \neq 0 \end{cases} \tag{7.5}$$

where $d(.)$ presents the vector of the faults.

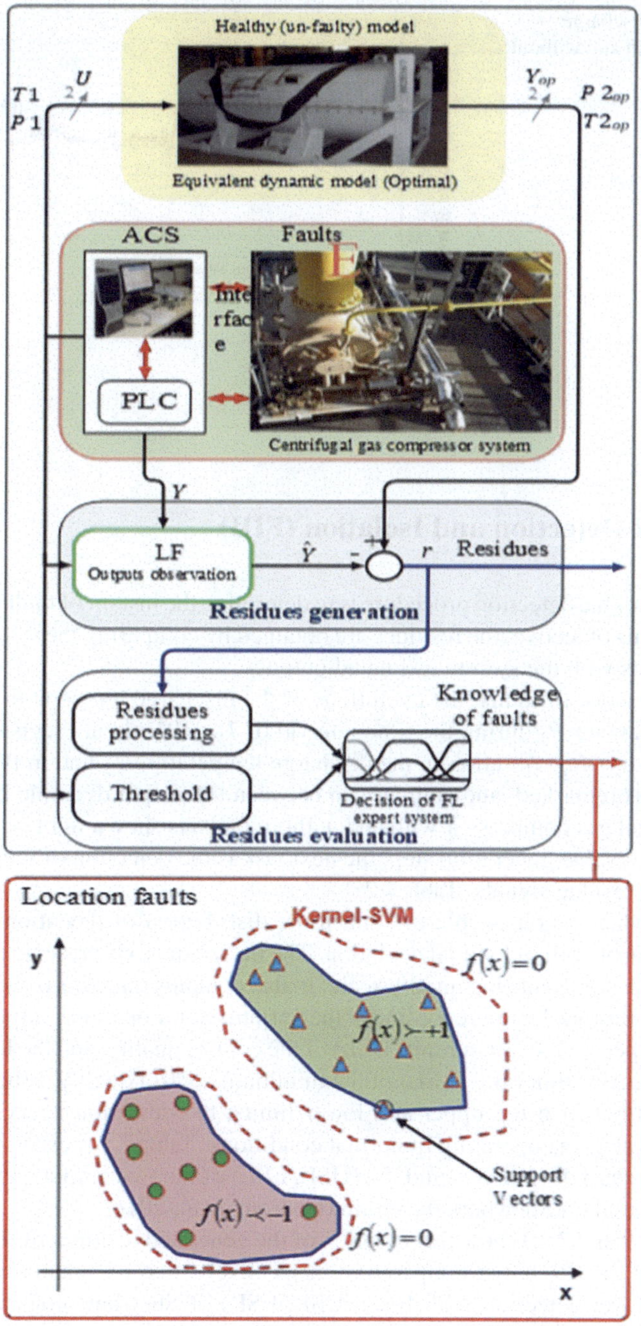

Fig. 7.6 Real-time fault detection diagnosis configuration for centrifugal gas compressor

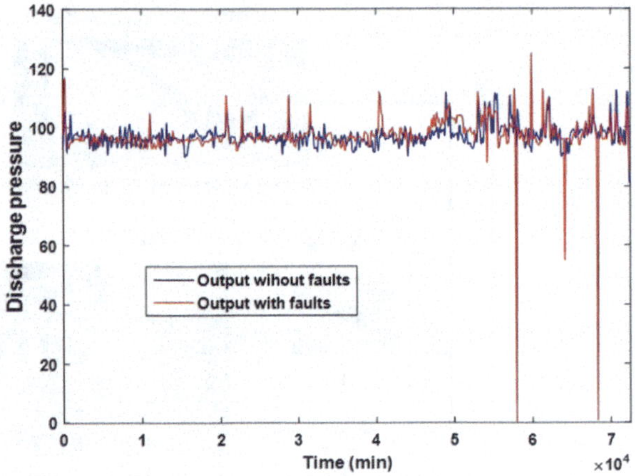

Fig. 7.7 The discharge pressure P2 with and without faults

7.7 Fault Detection and Isolation (FDI)

The purpose of the detection procedure is to determine the instant of fault occurrence. To achieve this objective, the residues are obtained by comparing the system optimal model outputs with the system estimated outputs.

In the presence of faults, the evolutions of the discharge temperature T_2 and the discharge pressure P_2 during the time interval of 7×10^4 min are registered via the DCS, where a rising vibration in the discharge temperature T_2 and in the discharge pressure P_2 is remarked, and comparison between the output dynamic behaviors of the centrifugal gas compressor with and without faults is shown in Figs. 7.7 and 7.8.

After the residue generation step, the next task is the detection of faults based on the obtained residue signals, Table 7.2.

The Shewhart graph is able to control the distribution of deviations instead of trying to control each individual deviation. The horizontal axis presents the time and the vertical axis presents the quality scale. It also contains three horizontal lines: the middle line presents the reference line of the normal operation mode output, the upper line is the upper specification limit (USL) of the control quality, and the lowest line is the lower specification limit (LSL) of the minimum control quality. When decisions are confined between the upper and lower limits, the deviation is acceptable and the centrifugal gas is operating in normal conditions. Table 7.1 presents the average value m, standard deviation s, and the (USL, LSL) of the two output signals that are calculated based on Shewhart algorithm (Hadroug et al. 2016).

Figures 7.9 and 7.10 show the variation of the generated residues of the discharge pressure and the discharge temperature, respectively, that are included within the upper and lower detection fault lines (USL, LSL) of Shewhart graph for the two outputs of the studied centrifugal gas compressor system.

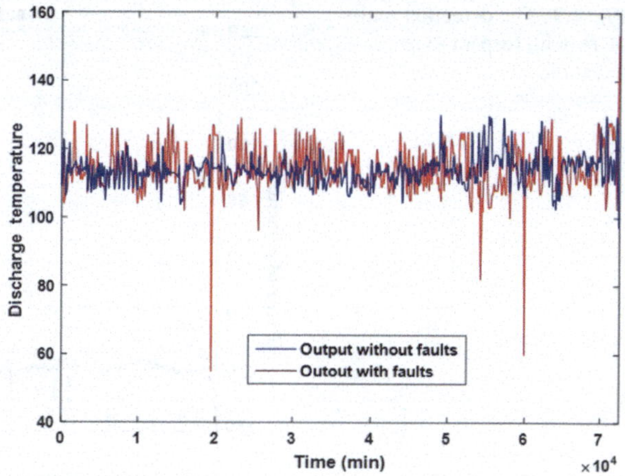

Fig. 7.8 The discharge temperature T2 with and without faults

Table 7.2 The threshold fault detection

Error	m	s	(USL, LSL)
E_{T2}	–0.0015	3.3063×10^{-8}	±7.2755
E_{P2}	0.0014	3.7597×10^{-9}	±21.631

When the residues exceed the upper and lower allowed limits (threshold), the system operates with faults that may not appear externally at the beginning. Therefore, the problem which needs to be solved is to find an efficient way which allows to indicate immediately the faults when they occur. To solve this problem, a type-1 fuzzy logic system is suggested as an expert model in order to detect and identify the type of faults occurring. In this case, many tests have been done to select the best fuzzy set type-1 that gives a robust performance and good results; in this work, the Gaussian 2 membership function was chosen as the best one.

Twenty faults are detected related to the discharge pressure output P_2, 12 faults were accurate during the first 12 h, and 8 faults during the time interval from 968 to 1142 h, regarding the discharge temperature output T_2; around 400 faults are detected, and these faults have occurred during almost the operating time along the interval time from 1 to 1208 h.

The detection faults of the discharge pressure and the discharge temperature, respectively, are shown in Figs. 7.11 and 7.13 based on 2D thermal card, where the red color symbolizes the detected fault and each fault is represented by stem.

Figures 7.12 and 7.14 present the diagnostic technique (Kernel-SVM) that showed and located clearly the point with defects. In this study, tracking techniques that signaled T_2 and P_2 and detection of defects in the centrifugal compressor to provide key information for system development intervention and fault reparation have been proposed.

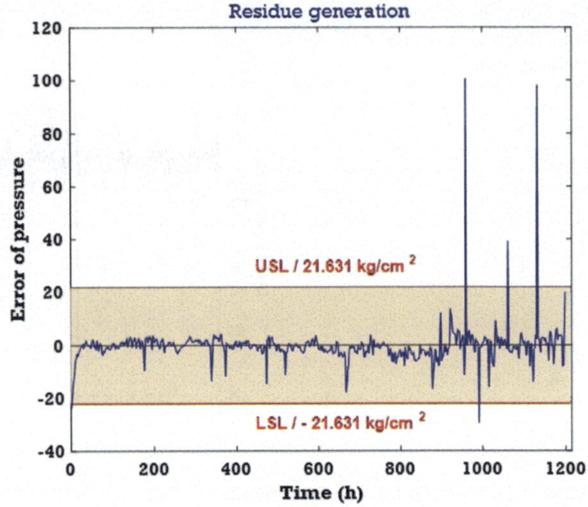

Fig. 7.9 The detection faults of P_2 with respect to a threshold

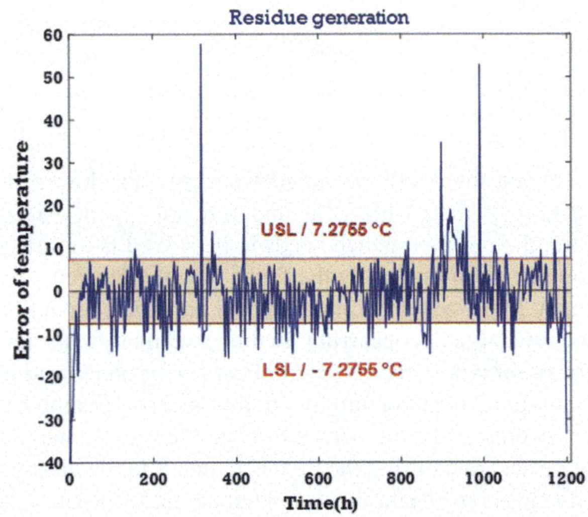

Fig. 7.10 The detection faults of T_2 with respect to a threshold

After the fault detection based on the output signals of the centrifugal gas compressor T_2 and P_2 through the proposed fault detection setup, a maintenance schedule is required for the system to identify the nature of the faults, the resulting damages, and the affected components, where the main aim is to ensure the required change and reparation in the system to restart the system operating mode again.

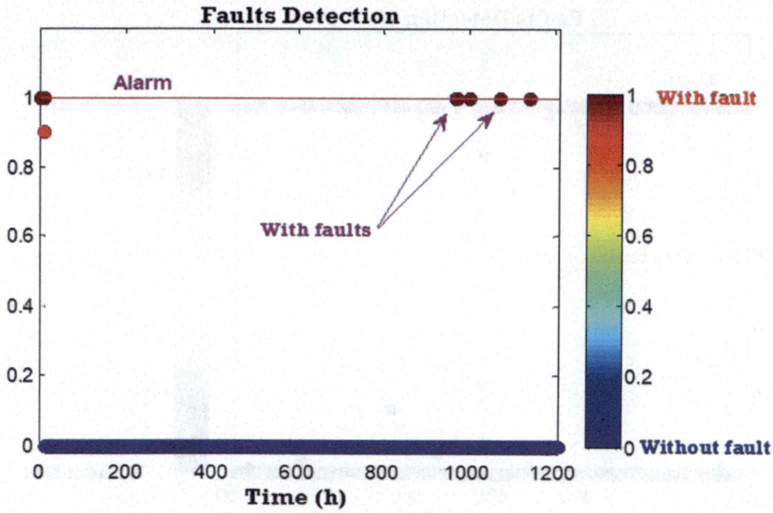

Fig. 7.11 The fault evaluation of P_2 gas compressor system examined

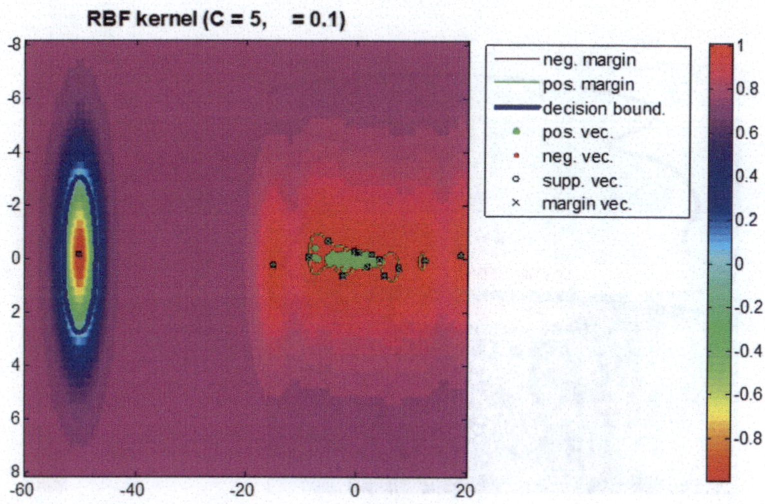

Fig. 7.12 The fault detection using the hybrid method K-SVM

During the experimental study, the system was completely opened for performing the maintenance, checking the inside body of the studied system, and for the validation of the proposed fault diagnosis and detection presented in this chapter and its accuracy for the determination and assessment of the fault levels. Indeed, after careful examination, the damage, observed and realized in the components of the BCL 505 gas compressor, is presented in the following figures:

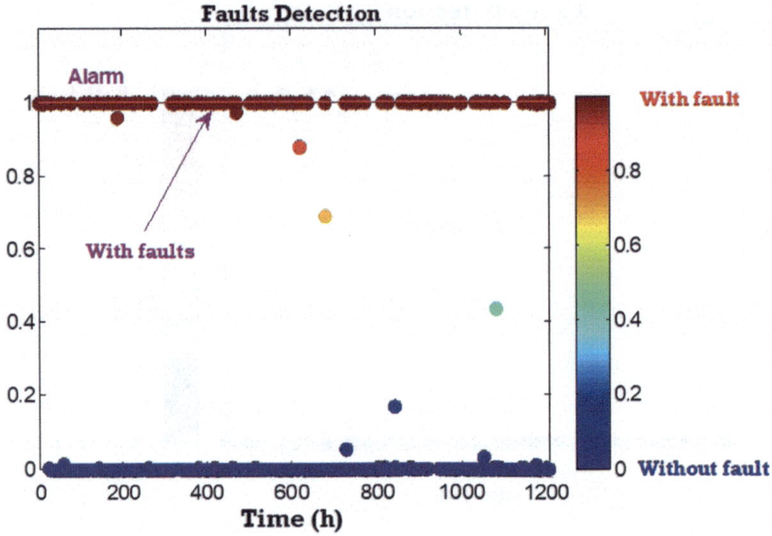

Fig. 7.13 The fault evaluation of T_2 gas compressor system examined

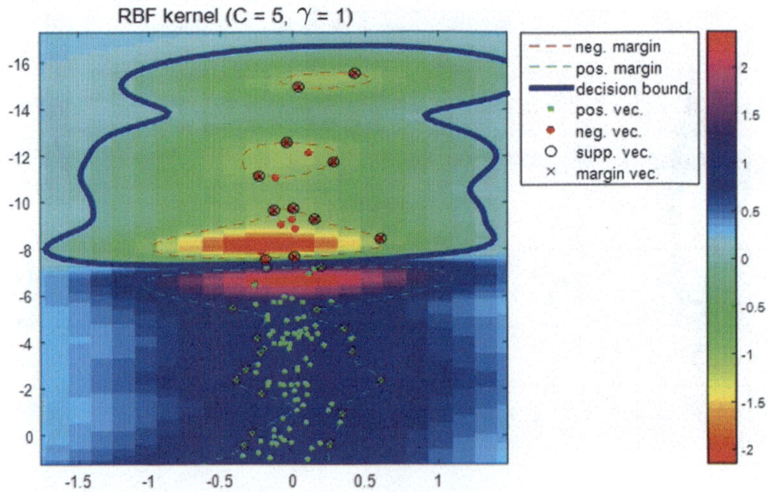

Fig. 7.14 The fault detection using the hybrid method K-SVM

- Scratch on the blades of the impeller 1, as shown in Fig. 7.15.
- O-ring joint defects, as shown in Fig. 7.16.

After viewing and identifying the affected parts in the centrifugal gas compressor system, Table 7.3 confirms the results obtained from the proposed setup for the fault detection. In comparison with the constructor operating system documents, it can be said that the obtained results are within the norms given by the constructor.

Fig. 7.15 The status of scratch on the blades of the impeller 1

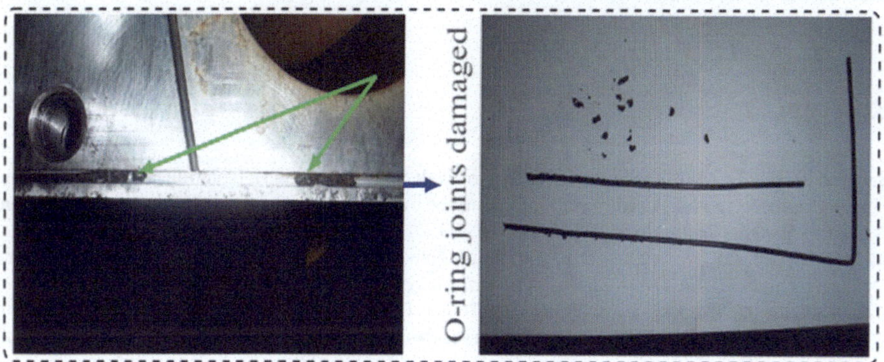

Fig. 7.16 The status of O-ring joints defective

Table 7.3 Fault identification and classification. ↗ Rise. ↙ Fall

Faults	Outputs	Signal	Fault identification
F_1	P_2	↗	· Distortions and incision in the level of blades
	T_2	↗	
F_2	P_2	↙	· O-ring joints defective
	T_2	↗	· Friction between the blades and wheel-space
$^{r}F_3$	P_2	↙	· Looseness in screws
	T_2	↙	· O-ring joints defective

7.8 Conclusion

The faults diagnosis and localization approach is proposed in this chapter and applied on the centrifugal gas compressor BCL 505 based on experimental data obtained on site from the measurement of the real-time acquisition control system (ACS). The main purpose of the proposed approach is to improve the energy efficiency by improving the operating mode and monitoring the performance of the BCL 505 centrifugal gas compressor used in gas transportation station and studied in this chapter.

This proposed fault diagnosis and detection approach is a combination of the two fault detection and isolation (FDI) approaches that are mainly based on the optimal identified healthy parametric equivalent model, fuzzy system, and the hybrid Kernel-SVM method.

The proposed fault diagnosis and detection approach investigated in this chapter is a promising approach which can be applied for different heavy industrial systems to improve their efficiency, especially in the area of petrol and oil industrial applications such as the gas turbine, the turbo-alternator, the turbo gas compressor, etc.

References

Bahareh, P., Nader, M., Khashayar, K., & Sensor fault detection. (2015). Isolation, and identification using multiple-model-based hybrid kalman filter for gas turbine engines. *IEEE Transactions on Control Systems Technology*, *24*(4), 1184–1200.

Bassily, H., Lund, R., & Wagner, J. (2009). Fault detection in multivariate signals with applications to gas turbines. *IEEE Transactions on Signal Processing*, *57*(3), 835–842.

Dai, X., Gao, Z., Breikin, T., & Wang, H. (2009). Disturbance attenuation in fault detection of gas turbine engines: A discrete robust observer design. *IEEE Transactions on Systems Man and Cybernetics Part C (Applications and Reviews)*, *39*(2), 234–239.

Hadroug, N., Hafaifa, A., Kouzou, A., & Chaibet, A. (2016). Faults detection in gas turbine using hybrid adaptive network based fuzzy inference systems. *Diagnostyka*, *17*(4), 3–17.

Hadroug, N., Hafaifa, A., Kouzou, A., & Chaibet, A. (2017). Dynamic model linearization of two shafts gas turbine via their input/output data around the equilibrium points. *Energy*, *120*(2), 488–497.

Hafaifa, A., Laaouad, F., & Laroussi, K. (2009). Fuzzy modelling and control for detection and isolation of surge in industrial centrifugal compressors. *Automatic Control Journal of the University of Belgrade*, *19*(1), 19–26.

Hafaifa, A., Guemana, M., & Daoudi, A. (2015). Vibrations supervision in gas turbine based on parity space approach to increasing efficiency. *Journal of Vibration and Control*, *21*(8), 1622–1632.

Isermann, R., & Fault-Diagnosis Systems. (2006). *An Introduction from Fault Detection to Fault Tolerance*. Berlin Heidelberg: Springer.

Jiang, W., Khan, J., & Dougal, R. A. (2006). Dynamic centrifugal compressor model for system simulation. *Journal of Power Sources*, *158*(2), 1333–1343.

Madau, D., & Feldkamp, L. A. (1996) Influence value defuzzification method. In *5th IEEE International Conference on Fuzzy Systems* (Vol. 3, pp. 1819–1824).

Maleki, S., Bingham, C., & Zhang, Y. (2016). Development and realization of changepoint analysis for the detection of emerging faults on industrial systems. *IEEE Transactions on Industrial Informatics*, *12*(3), 1180–1187.

Mohamadi, B., Mohamad, F., & Tabkhi, J. S. (2014). Exergetic approach to investigate the arrangement of compressors of a pipeline boosting station. *Energy Technology*, *2*(8), 732–741.

Nail, B., Kouzou, A., Hafaifa, A., & Chaibet, A. (2018). Parametric identification and stabilization of turbo-compressor plant based on matrix fraction description using experimental data. *Journal of Engineering Science and Technology*, *13*(6), 1850–1868.

Saavedra, I., Bruno, J. C., & Coronas, A. (2010). Thermodynamic optimization of organic rankine cycles at several condensing temperatures: Case study of waste heat recovery in a natural gas compressor station. *Proceedings of the Institution of Mechanical Engineers, Part A, Journal of Power and Energy*, *224*(7), 917–930.

Tabkhi, F., Pibouleau, L., Hernandez-Rodriguez, G., Azzaro-Pantel, C., & Domenech, S. (2010). Improving the performance of natural gas pipeline networks fuel consumption minimization problems. *AIChE Journal*, *56*(4), 946–964.

Zadeh, L. (1965). Fuzzy sets. *Information and Control*, *8*(3), 338–353.

Chapter 8
Performance Investigation of an Improved Diagnostic Method for Open IGBT Faults in VSI-Fed IM Drives

M. A. Zdiri, B. Bouzidi and H. Hadj Abdallah

Abstract The design of power conversion systems has become an important topic. In particular, the voltage source inverter fed induction motor drives are given an increasing interest especially in medium and high power industrial applications such as variable speed electric motor drives, electric and hybrid propulsion systems, aeronautics, and robotics. In such applications, the drive reliability could be affected by faulty operations which could be caused either by the system topology or by the control strategy or both. Consequently, a special attention should be paid to the fast fault detection and identification and to the drive reconfiguration, which allied to fault-tolerant control implementation. Doing so, the shutdown of the drive system is so avoided. Accordingly, this chapter proposes an improved diagnosis method dedicated to VSI-fed IM drives considering the implementation of the direct rotor flux oriented control. This proposed diagnostic method is focused on faulty operations essentially related to open IGBT's faults in the voltage source inverters, especially that these latter are prone to suffer critical faults linked to several stress and to the complexity of their topologies. The improved diagnostic method is fast, quite simple, and just requires only the combination of the information of the slope of the current and the normalized current without the use of another additional sensor and hardware. Simulation results have proved the high performance of the proposed diagnostic method in terms of fast fault detection associated with a high robustness against the issue of false alarms, against load torque and speed fast variations.

Keywords Diagnostic method · Open IGBT faults · VSI · IM · False alarms

M. A. Zdiri (✉) · H. H. Abdallah
Control and Energy Management Laboratory, National Engineering School of Sfax, Sfax, Tunisia
e-mail: zdirimohamedali@gmail.com

H. H. Abdallah
e-mail: hadjabdallahhsan@gmail.com

B. Bouzidi
Research Laboratory on Renewable Energies and Electric Vehicles, National Engineering School of Sfax, Sfax, Tunisia
e-mail: badii.bouzidi@enis.tn

N. Derbel et al. (eds.), *Diagnosis, Fault Detection & Tolerant Control*, Studies in Systems, Decision and Control 269,
https://doi.org/10.1007/978-981-15-1746-4_8

8.1 Introduction

Voltage source inverters (VSIs) already occupy a certain level of maturity and an important place in several industrial applications such as variable speed electric motor drives, electric and hybrid propulsion systems, aeronautics, and robotics. In such applications, the drive reliability could be affected by faulty operations which could be caused either by the system topology or by the control strategy or both. Referring to statistical studies, it has been proven that 60% of the failures of the conversion system result from failures of the circuit boards, solder boards, and semiconductors in the device modules and 40% from failures of control strategies (Yang et al. 2010).

Due to the fact that the VSIs are often exposed to several stress and are characterized by complex topologies, these latter are prone to suffer from critical faults. Referring to literature, the VSI can be affected by faulty operations which can generally be caused by three types of fault: (i) IGBT's faults (faults appeared in the IGBTs), (ii) controller faults (faults linked to the drivers), and (iii) current and voltage sensor faults (faults linked to sensors) (Diallo et al. 2005). Through other studies, based on survey regarding the reliability in power electronic converters, it has been confirmed that the power switch failures have a high percentage in the VSIs faulty operations.

Generally, most of the VSIs use IGBTs as power switches because of their high switching frequency, high efficiency, and their performance and ability to handle short circuit for more than 10 ms (Ubale et al. 2013; Trabelsi et al. 2017). However, these IGBTs are prone to suffer critical faults due to electrical and thermal effects. The gate-misfiring faults, the short-circuit faults, and the open-circuit faults are considered as the major IGBT faults of the VSIs. Concerning the third class of IGBT's faults, the open-circuit faults can be caused by thermal cycling, by very high collector current, or by other types of faults essentially related to the control circuit (Trabelsi et al. 2017; Campos-Delgado and Espinoza-Trejo 2011; Jung et al. 2013). Such VSI faults can consequently affect the drive reliability of the applications, and can even cause the shutdown of the drive system. In this trend, in attempt to bypass these harmful influences and to improve both system reliability and availability, a special attention should be paid to the development of fault diagnostic methods and to the drive reconfiguration, which allied to fault-tolerant control implementation (Rodriguez et al. 2007).

Considering the open-switch fault in VSI-fed electric motor drive, several approaches which investigated the diagnostic method have been reported in the literature. In what follows, an analysis of the most recent works is presented. In Mendes et al. (1998), the Park's vector approach was introduced for the first time by *Mendes* which has been recognized as an effective diagnostic tool to localize and detect the faulty IGBT. However, this diagnostic method required more complex pattern recognition algorithms and was not suitable for the integration into the drive controller. Thus, *Mendes*, in Mendes and Cardoso (1999), introduced the average value of the Park's vector current approach, in which the faulty IGBTs have been detected when the vector modulus wasn't null and have been located by a simple identification of the vector phase. In Caseiro et al. (2009), a diagnostic method has been reported considering other developments including the derivative of the vector phase. Other diagnostic method for open-circuit faults in VSI is based on Park's vector approach which was presented in Zidani et al. (2008). This latter used probabilistic and fuzzy approaches to define different model boundaries under faulty and healthy operations, in order to avoid false alarms and increase the certainty of the diagnostic method. These approaches have some drawbacks such as the sensitivity to transients and load dependence. In Rothenhagen and Fuchs (2005), *Rothenhagen* presented a diagnostic method based on the use of the normalized average currents to detect and identify the faulty IGBTs, in an attempt to make the fault detection independent of load torque. Similar work is presented in Sleszynski et al. (2009), where the normalized average current approach has been used to detect multiple IGBT open-circuit faults. These normalization current methods have been penalized by large detection times and high complexity of the implementation schemes. A comparative study for several diagnostic methods was presented in Lu and Sharma (2009) in order to analyze the performance of these methods. In Trabelsi et al. (2017), confirmed that the performance and robustness of the diagnostic methods depend essentially on the estimation of the system parameters and on the model accuracy. Indeed, and according to this reviewer, the estimation of the current signals involved more computational times and mathematical operations which make the real-time implementations of these methods very difficult.

New algorithms concerning multiple faults for a VSI-fed three-phase electric motor are introduced in Estima and Cardoso (2011), Arafa et al. (2013), using the normalized current average values and the errors of the normalized current average absolute values. In Estima and Cardoso (2013), an improved diagnostic method for open IGBT faults considering the information of the motor phase currents and their corresponding reference signals was proposed. Furthermore, Trabelsi et al. (2017) presented a new approach based on the information of the slope vector and the measured phase currents. These algorithms have been characterized by their robustness against false alarms during transient operation.

In this trend, this chapter is aimed to investigate the performance of an improved diagnosis method for open IGBT faults in VSI-fed IM controlled by the direct rotor flux oriented control (RFOC) in an attempt to bypass the drawbacks of the existing diagnostic methods. The proposed diagnostic method is based on the information of the normalized currents combined with a new diagnosis variable which can be provided using the information of the slope of the current vector in (α-β) frame. Simulation works highlight the high performance of the advanced diagnostic method in terms of simple implementation scheme, fast diagnosis, low cost, and robustness against false alarms.

8.2 Proposed Fault Diagnostic Method

The block diagram scheme of the proposed fault diagnostic method is shown in Fig. 8.1. This diagnosis method results from an enhanced version of the method introduced by Estima and Cardoso (2011). This improved diagnostic method is

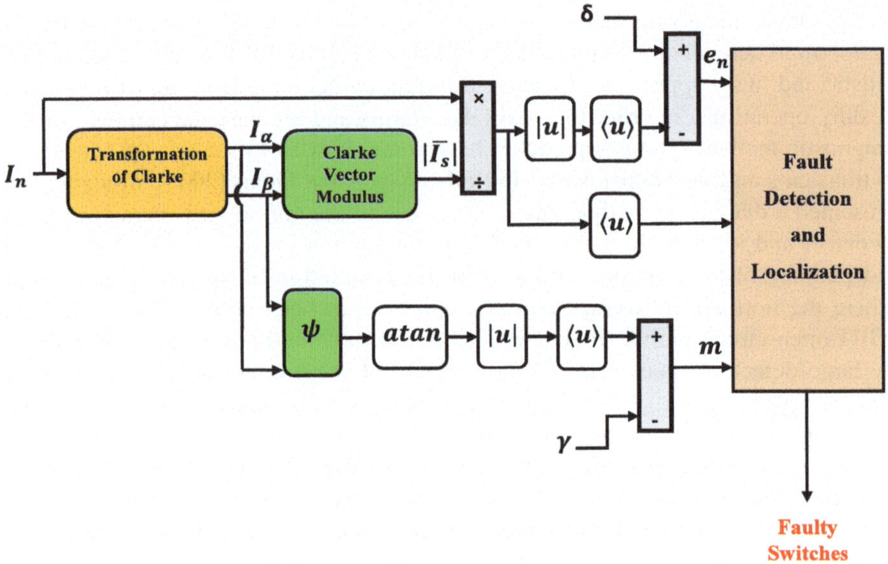

Fig. 8.1 Block diagram of the improved diagnostic method

advantageous of other diagnostic methods because it is based only on the combination of the information of the slope of currents and the normalized currents and it avoids the use of other extra sensors or hardware. Doing so, the compactness and the cost-effectiveness of the implementation scheme of this method are gained.

The improved diagnostic method is based on *Clarke*'s transformation of the motor phase current which transforms the three-phase system (I_{as}, I_{bs}, I_{cs}) into a two-phase system (I_α, I_β) as follows:

$$I_\alpha = \sqrt{\frac{3}{2}}\, I_{as}, \tag{8.1}$$

$$I_\beta = \frac{1}{\sqrt{2}}\, I_{as} + \sqrt{2}\, I_{bs}. \tag{8.2}$$

Accordingly, *Clarke*'s vector modulus is expressed as follows:

$$|\overline{I}_s| = \sqrt{{I_\alpha}^2 + {I_\beta}^2}. \tag{8.3}$$

To be robust against the motor operating condition variations, the proposed method is based on the use of the normalized currents instead of the measured ones. Therefore, the normalized currents are performed by dividing each three-phase currents by *Clarke*'s vector modulus. The obtained normalized currents are given by the following equation:

$$I_{nN} = \frac{I_n}{|\overline{I}_s|}, \tag{8.4}$$

where $n = as$, bs, and cs, and I_n is the balanced three-phase sinusoidal current system given by

$$I_n = \begin{cases} I_{as} = I_{max}\ \sin(\omega_s t) \\ I_{bs} = I_{max}\ \sin\left(\omega_s t - \frac{2\pi}{3}\right) \\ I_{cs} = I_{max}\ \sin\left(\omega_s t + \frac{2\pi}{3}\right), \end{cases} \tag{8.5}$$

where I_{max} is the current's maximum amplitude and ω_s is the stator current frequency.

Through Eqs. 8.3 and 8.5, it can be proven that *Clarke*'s vector modulus can be given by the following expression:

$$|\overline{I}_s| = \sqrt{\frac{3}{2}}\, I_{max}. \tag{8.6}$$

The normalized currents can be rewritten as follows:

$$\begin{cases} I_{asN} = \sqrt{\frac{2}{3}}\ \sin(\omega_s t) \\ I_{bsN} = \sqrt{\frac{2}{3}}\ \sin\left(\omega_s t - \frac{2\pi}{3}\right) \\ I_{csN} = \sqrt{\frac{2}{3}}\ \sin\left(\omega_s t + \frac{2\pi}{3}\right). \end{cases} \tag{8.7}$$

Referring to Eq. 8.7, one can clearly notice that the amplitude of the normalized currents is equal to $\sqrt{\frac{2}{3}}$, and so independent of the amplitude of the three-phase motor currents "I_{max}".

The average absolute values of the three normalized currents $\langle |I_{nN}| \rangle$ are given as follows:

$$\langle |I_{nN}| \rangle = \omega_s \int_0^{\frac{1}{\omega_s}} |I_{nN}|\ d\omega_s t. \tag{8.8}$$

Accordingly, the average absolute value of the normalized currents in healthy case "δ" is given by the following equation:

$$\delta = \frac{1}{\pi}\sqrt{\frac{8}{3}} \approx 0.5198. \tag{8.9}$$

Finally, the diagnosis variables "e_n" are obtained considering the errors between the normalized average absolute values of currents in healthy cases and the ones in faulty cases as follows:

$$e_n = \delta - \langle |I_{nN}| \rangle. \tag{8.10}$$

From Eq. 8.10, one can confirm that under healthy operating conditions, the diagnosis variables e_n will take values near to zero.

It is to be noted that, only in the case of an open IGBT fault, the three diagnosis variables e_n are used to detect and locate just the affected motor phase and are enabled to carry information about the faulty IGBT. Indeed, and considering the case of an open IGBT fault, the diagnosis variable linked to the affected phase will assume a positive value. However, for the two other phases, the corresponding diagnosis variables "e_n" will have negative values. So, in order to perform a complete inverter diagnosis and locate the faulty IGBT, one can combine the information extracted from variables e_n together with the ones from normalized current average values $\langle I_{nN} \rangle$.

To bypass the diagnostic method problems and to achieve a robust performance of the diagnostic algorithm against false alarms, high speed, and load variations, we propose in this chapter a new diagnosis variable "m". This latter is based on the information of the slope of the current vector defined as follows:

$$\psi = \frac{I_{\alpha k}}{I_{\beta k}}, \tag{8.11}$$

where $I_{\alpha k}$ and $I_{\beta k}$ are the *Clarke* component currents at instant kT_s and T_s is the sampling period.

The angle of Clarke trajectory deviation "ϕ" is given by the following expression:

$$\phi = \arctan(\psi). \tag{8.12}$$

Therefore, the average absolute value of the Clarke trajectory angle deviation "$\langle |\phi| \rangle$" is expressed as follows:

$$\langle |\phi| \rangle = \omega_s \int_0^{\frac{1}{\omega_s}} |\phi| \, d\omega_s t. \tag{8.13}$$

Finally, the new diagnosis variable "m" is obtained from the errors between the Clarke trajectory angle deviation average absolute value in faulty operation case $\langle |\phi| \rangle$ and the one in healthy operation case "γ" as follows:

$$m = \langle |\phi| \rangle - \gamma, \tag{8.14}$$

where $\gamma \approx 0.785$.

The values taken by the three normalized current average values $< I_{nN} >$ and the diagnosis variables e_n and m allow the generation of distinct fault signatures corresponding to the occurring default type in the VSI if it is single or phase. So, in order to make an accurate and complete diagnostic method, the information given by these diagnosis variables are combined together. Doing so, these three diagnosis variables are formulated according to the following expressions:

$$E_n = \begin{cases} N \text{ for} & e_n < 0 \\ 0 \text{ for} & 0 \le e_n < k_f \\ P \text{ for} & k_f \le e_n < k_d \\ D \text{ for} & e_n \ge k_d, \end{cases} \tag{8.15}$$

$$S_n = \begin{cases} L \text{ for} & \langle I_{nN} \rangle < 0 \\ H \text{ for} & 0 < \langle I_{nN} \rangle, \end{cases} \tag{8.16}$$

$$M = \begin{cases} SS \text{ for} & m \le k_p \\ S \text{ for} & k_p < m < 0 \\ B \text{ for} & 0 < m < k_g \\ BB \text{ for} & m \ge k_g. \end{cases} \tag{8.17}$$

The threshold values (k_f, k_d, k_p, and k_g) can be empirically established by analyzing the three diagnosis variables under different faulty and healthy operating conditions. It is to be underlined that, since, the proposed diagnosis method is based on the use of normalized currents, and the thresholds values (k_f, k_d, k_p, and k_g) are independent from the motor operating condition variations.

The considered diagnostic method enables the detection and the identification of nine distinct signatures of IGBT faults which can be divided into a single IGBT open-circuit fault and a single-phase open-circuit fault as summarized in Table 8.1.

To highlight the performance of the proposed diagnostic method, the direct rotor flux oriented control (RFOC) strategy was applied to control the speed of an induction motor feeding by a voltage source inverter. The implementation scheme of the control strategy is illustrated in Fig. 8.2.

8.3 Simulation Results

In order to evaluate the performance and the robustness of the diagnostic method against false alarms and the variation of both speed and load, simulation work has been carried out using the MATLAB/Simulink environment and considering two

Table 8.1 IGBT's open-circuit fault detection and identification

Faulty IGBTs	E_{as}	E_{bs}	E_{cs}	S_{as}	S_{bs}	S_{cs}	M
T_1	P	N	N	L			S
T_4	P	N	N	H			S
T_2	N	P	N		L		B
T_5	N	P	N		H		B
T_3	N	N	P			L	B
T_6	N	N	P			H	B
T_1, T_4	D						SS
T_2, T_5		D					BB
T_3, T_6			D				BB

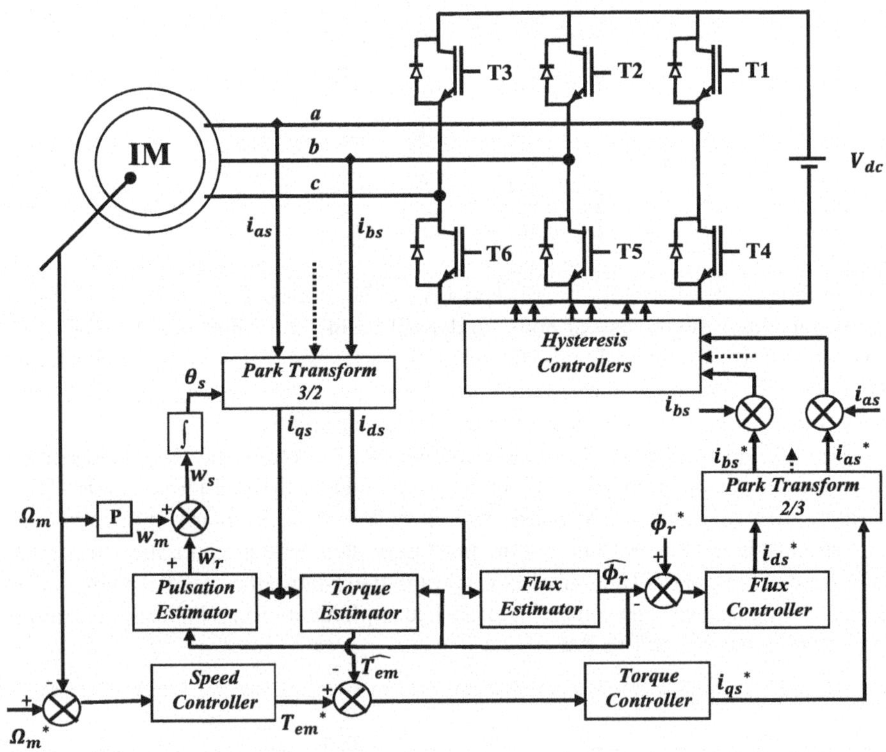

Fig. 8.2 Block diagram of the direct RFOC strategy for voltage source inverter fed induction motor drive

Table 8.2 IM parameters

Rated power (kw)	1.1
Rated line voltage (volt)	600
Supply frequency (HZ)	50
Rated speed (rpm)	2820
Rated load torque (Nm)	3.5
Number of pole pairs	1
Stator resistance (Ω)	6.863
Rotor resistance (Ω)	7.67
Stator inductance (H)	0.708
Rotor inductance (H)	0.708
Mutual inductance (H)	0.684
Moment of inertia (kg.m^2)	0.0033
Friction factor (Nm.s/rad)	0.0035

distinct failure conditions: (i) a single IGBT open-circuit fault and (ii) a single-phase open-circuit fault.

It is to be noted that the open IGBT fault is performed by removing the gate signal of the corresponding IGBT and keeping the respective antiparallel diode connected.

The parameters of the used IM are illustrated in Table 8.2.

Considering all operating conditions, the load torque is kept equal to 1.75 Nm which represents 50% of the rated torque and the reference speed is chosen equal to 200 rad/s which represents 68.21% of the motor nominal speed. The threshold values k_f, k_d, k_p, and k_g of the diagnosis variables are chosen equal to 0.06, 0.275, −0.408, and 0.161, respectively. A more detailed explanation of the selection of these thresholds is illustrated in Sect. 8.4.1.

8.3.1 *Healthy Operation Case*

Figure 8.3 presents some features of the induction motor drive at steady-state healthy operation.

Referring to Fig. 8.3a which illustrates the current vector trajectory in (α, β) plane, one can confirm that the current vector describes a circular and smooth trajectory.

Referring to Fig. 8.3b which illustrates the angle of *Clarke* trajectory deviation ϕ versus time, one can notice that ϕ takes value in the range of ± 1.57 rad.

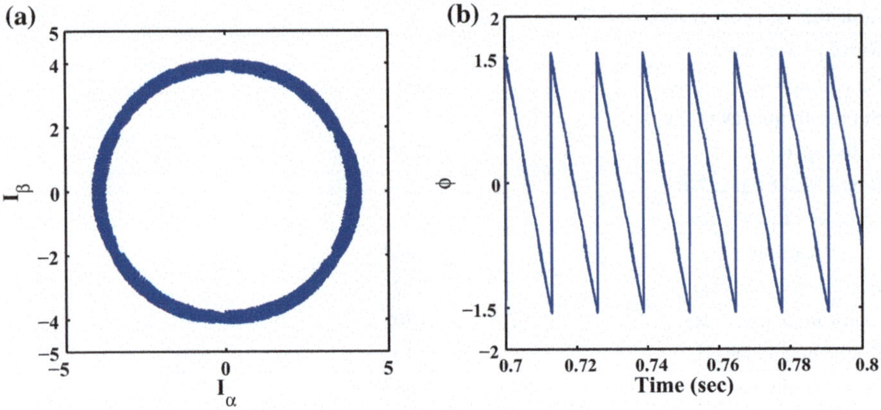

Fig. 8.3 Simulation results of VSI-fed IM drive at steady-state healthy operation. **Legend: a** current vector trajectory in (α, β) plane, **b** ϕ

8.3.2 *Single IGBT Open-Circuit Fault*

Figure 8.4 presents the simulation results of the motor phase currents, the diagnosis variables E_n, the normalized current average values S_n, and the diagnosis variable M considering an open-circuit fault appearing in the T_1 IGBT at $T = 0.71$ s.

Referring to Fig. 8.4, it is to be noticed that when the IGBT T_1 open-circuit fault occurs, the diagnosis variable E_{as} immediately increases and converges to a value of about 0.255. However, the two other diagnosis variables E_{bs} and E_{cs} decrease to reach values of about -0.098 and -0.089, respectively.

Concerning the normalized current average values, and due to the fact that the flow of the "a" phase current can be made by only the bottom IGBT T_4 (since that the open-circuit fault appears in upper IGBT T_1 of the first phase), a large negative average value of this affected phase is obtained.

Regarding the diagnosis variable M, it decreases from zero in the healthy case to a negative value equal to -0.388 in the faulty case.

As conclusion, one can confirm that the open-circuit fault is detected and identified at the instant $T = 0.7124$ s, when the diagnosis variable E_{as} each reaches a threshold value k_f.

8.3.3 *Single-Phase Open-Circuit Fault*

Figure 8.5 shows the simulation results of the motor phase currents, the diagnosis variables E_n, the normalized currents average values S_n, and the diagnosis variable M considering a single-phase open-circuit fault appearing in IGBTs T_1 and T_4 of the "a" phase at $T = 0.8$ s.

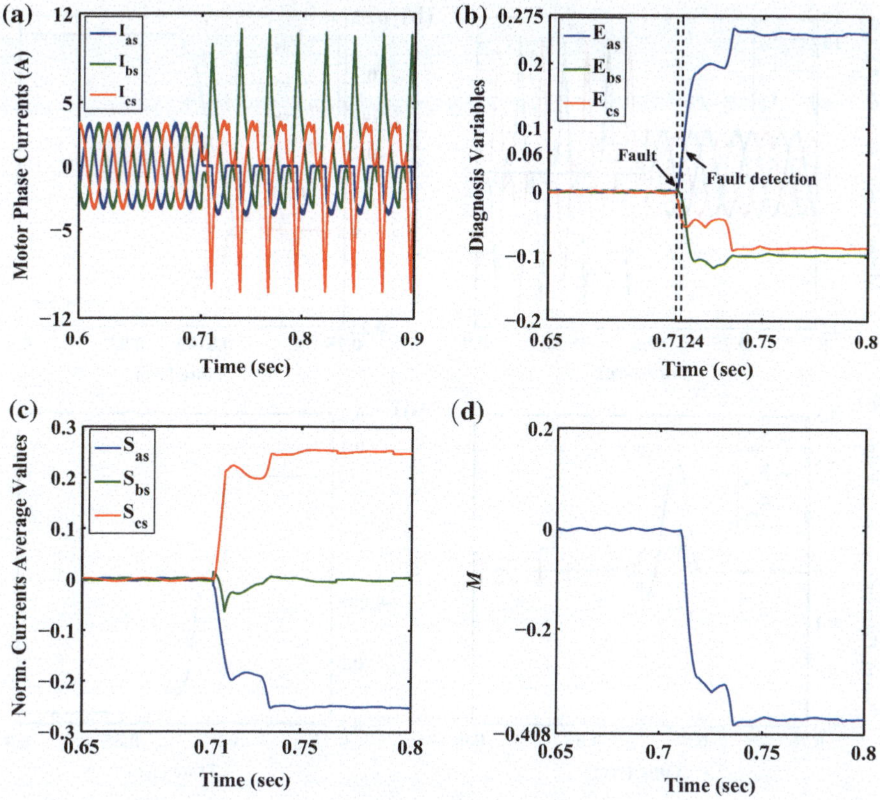

Fig. 8.4 Simulation results of VSI-fed IM drive in the case of single IGBT T_1 open-circuit fault appears at the instant T = 0.71 s. **Legend: a** motor phase currents, **b** diagnosis variables E_n, **c** normalized currents average values S_n, and (d) diagnosis variable M

Referring to Fig. 8.5b, it can be seen, when the IGBTs T_1 and T_4 open-circuit faults occur, that the diagnosis variable E_{as} immediately increases to reach a final value of 0.508 and the two other remaining diagnosis variables E_{bs} and E_{cs} decrease to converge to a negative value of -0.183 each.

Referring to Fig. 8.5c, it is to be underlined that the information provided by the normalized current average values S_n is not relevant for the considered fault type.

Referring to Fig. 8.5d, it is to be noticed that when the IGBTs', T_1 and T_4, open-circuit faults occur, the diagnosis variable M decreases from a null value corresponding to the healthy case and converges immediately to a negative value of about -0.767.

As considered in the simulation work, the single-phase open-circuit fault appeared in IGBTs T_1 and T_4 at $T = 0.8$ s. Therefore, the diagnosis variables E_{as} and M reached the defined thresholds k_d and k_p, respectively at the time $T = 0.8142$ s. So,

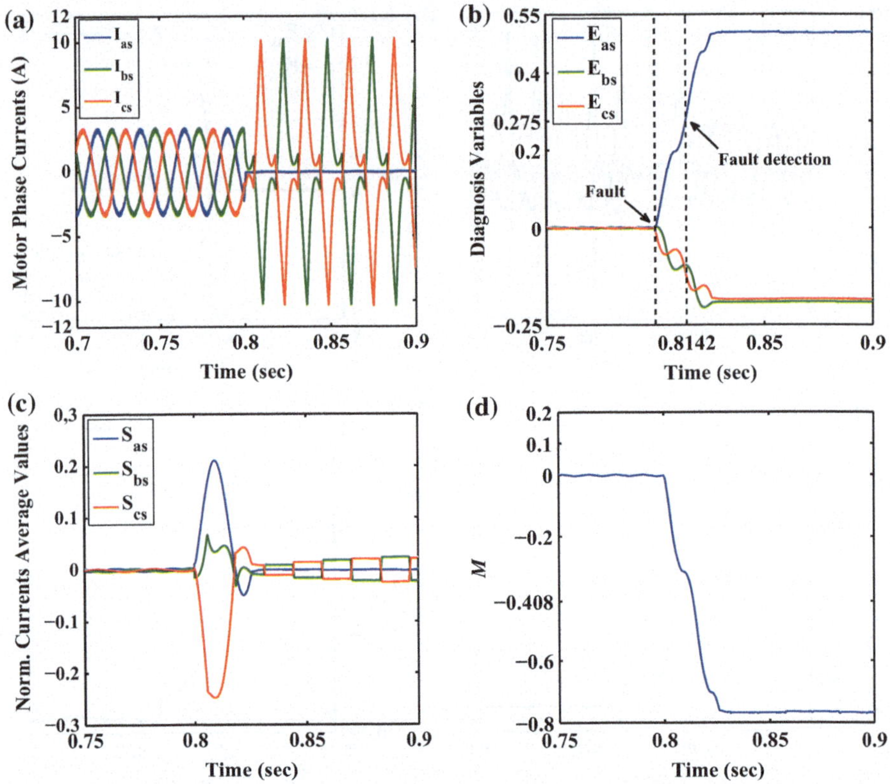

Fig. 8.5 Simulation results of VSI-fed IM drive in the case of a single-phase open-circuit fault appearing in IGBTs T_1 and T_4 at $T = 0.8$ s. **Legend: a** motor phase currents, **b** diagnosis variables E_n, **c** normalized current average values S_n, and **d** diagnosis variable M

one can conclude that the “a” phase open-circuit fault is detected and identified by this proposed diagnostic method after 0.0142 s.

8.3.4 Algorithm Performance During Load and Speed Variations

This paragraph is aimed to prove the robustness of the proposed diagnostic method against the issue of false alarms caused by load and speed fast variations. To do so, simulation work has been carried out considering

- a load torque variation versus time as follows:
  - from 0.2 to 0.4 s, the load torque is taken constant equal to 2.6 Nm which represents 74.29% of the rated one,

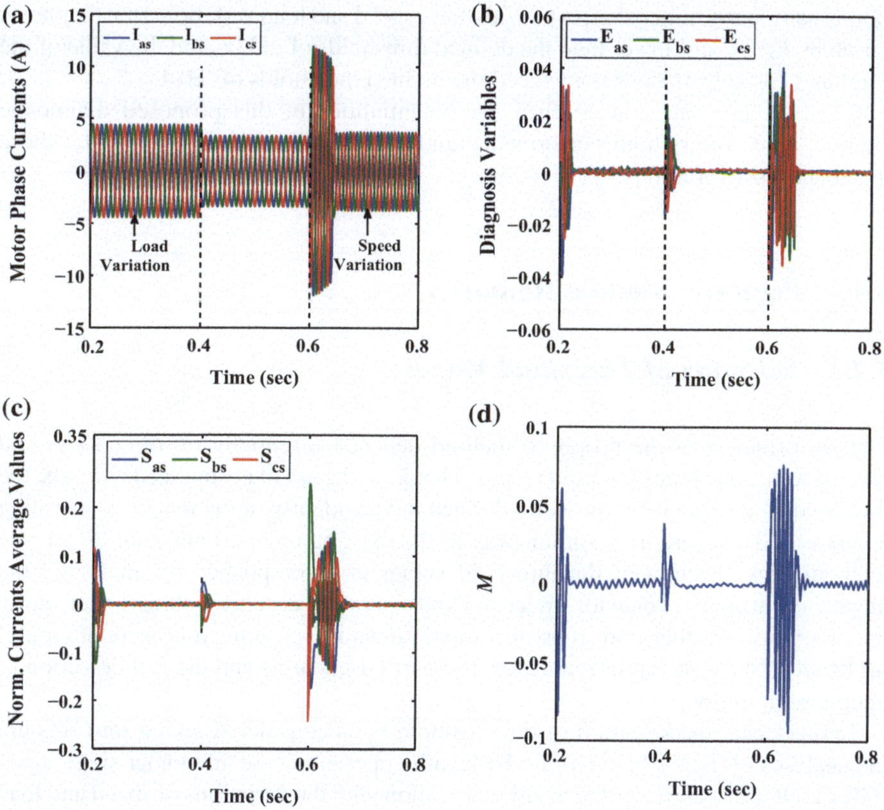

Fig. 8.6 Simulation results of VSI-fed IM drive under load and speed fast variations. **Legend: a** motor phase currents, **b** diagnosis variables E_n, **c** normalized currents average values S_n, and **d** diagnosis variable M

  - at 0.4 s, a load torque step from 2.6 to 1.75 Nm is applied,
  - from 0.4 to 0.8 s, the load torque is taken constant equal to 1.75 Nm which represents 50% of the rated one.

- a speed variation versus time as follows:
  - from 0.2 to 0.6 s, the speed is taken constant equal to 200 rad/s which represents 67.73% of the nominal one,
  - at 0.6 s, a speed step from 200 to 280 rad/s is applied,
  - from 0.6 to 0.8 s, the speed is taken constant equal to 280 rad/s which represents 94.82% of the nominal one.

The obtained results are illustrated in Fig. 8.6.

Referring to Fig. 8.6 which illustrates the motor phase currents, the diagnosis-variables E_n, the normalized current average values S_n, and the diagnosis variable

M, one can clearly notice that, even with these speed and load variations, the diagnosis variables E_n are still lower than the defined thresholds of ± 0.06 and the value of the diagnosis variable M does not exceed the defined thresholds of ± 0.1.

Consequently, one can confirm the potentialities of this proposed diagnostic method which offers high performance and avoids false alarms, even under these speed and load variations.

8.4 Diagnostic Method Remarks

8.4.1 Selection of Threshold Values

The performance of the proposed method depends essentially on the correct and precise selection of the threshold values. Thanks to the use of normalized currents, the threshold values can be empirically defined independently of the machine operating conditions. Therefore, in a similar way to the existing open-circuit fault diagnostic methods, the selection of the threshold values is accomplished by analyzing the diagnosis variables' behavior under healthy operation case as well as under faulty operation case. Furthermore, these threshold values are set taking into account a trade-off between the robustness against the issue of false alarms and the fast detection of open-circuit faults.

In this trend, the selection of the threshold k_f is considered taking into account the analyses of the E_n behavior under healthy operation case and under single open IGBT fault. Let's consider the healthy operation with the variations of speed and load as illustrated in Fig. 8.6, in which one can clearly notice that the E_n values are null in steady-state operation and the E_n oscillations increase but their amplitudes are always below 0.04 in transient operation. However, in the case of faulty operation considering a single IGBT open-circuit fault, the diagnosis variable E_n of the corresponding affected phase will assume a distinct positive value ($0.04 < E_n \leq 0.255$), while the E_n of the other phases will have negative values. So, theoretically, the threshold k_f can have a value of about 0.04. However, it is to be noted that the choice of a low k_f value means fast fault detection but worse robustness against false alarms. On the other side, the choice of a large k_f value increases the robustness but decreases the speed detection. For this reason, it is always better to consider a safety margin of 0.02, and therefore the threshold k_f value is kept equal to 0.06.

The threshold value k_d performs an important role in case of single-phase open-circuit fault. Indeed, considering this type of fault, it can be verified that E_n corresponding to the faulty leg reach values greater than 0.255. This means that, theoretically, the threshold k_d can have a value slightly higher than 0.255. Assuring also a safety margin of 0.02, k_d is chosen equal to 0.275. It is to be underlined that the condition $k_f < k_d$ is verified which proves the coherence of the proposed diagnostic method.

The selection of k_p and k_g is done by analyzing the behavior of the diagnosis variable M under: (i) a healthy case, (ii) a single IGBT open-circuit fault case, and (iii) a single-phase open-circuit fault case. So, one can notice that

- under healthy operation case, M is null,
- under a single IGBT open-circuit fault case, considering the first leg, M reaches a minimum value of −0.388. However, for the other legs, M reaches a maximum value of 0.141,
- under a single-phase open-circuit fault case, considering the first leg, the variable M converges to a minimum value of about −0.767. However, for the other legs, M converges to a maximum value of about 0.265.

Assuming a safety margin of 0.02, the values of the thresholds k_p and k_g can be taken equal to −0.408 and to 0.161, respectively.

8.4.2 *Fast Fault Algorithm Detection*

Let's consider

- the case when the fault occurs in the upper (bottom) IGBT of the inverter leg. One can confirm that during the positive (negative) current half period of the corresponding faulty leg, the phase current tends immediately to zero. So, the detection and the localization of this IGBT fault are relatively fast,
- the case when the fault occurs in the bottom (the upper) IGBT of the affected phase and during the positive (negative) current half period. One can confirm that this fault will be located and detected at the next half period of the current. So, a low detection speed is performed.

As conclusion, it is to be noticed that the detection speed and the localization of the open IGBT fault depend on (i) the instant of the fault occurrence and (ii) the position of the faulty IGBT of the affected inverter leg (upper or bottom IGBT).

8.5 Conclusion

In this chapter, an improved diagnostic method for open-circuit faults of a voltage source inverter fed induction motor drive controlled by a direct RFOC strategy has been proposed. The proposed diagnostic method was based on the use of information provided from the three motor phase currents, their normalized average values, and the new diagnostic variable M under healthy and faulty operations. This method was achieved considering a simple implementation scheme without the need of an extra sensor.

The performance investigation of this advanced diagnostic method has been carried out considering a single IGBT open-circuit fault case and a single-phase open-circuit fault case.

Simulation results highlighted the attractive performance of the proposed method in terms of low cost, fast fault detection associated with a high robustness against false alarms and against speed and load torque fast variations. Furthermore, this work can be extended to other topologies of converters and other fault types.

References

Arafa, S. M., Haitham, Z. A., & Din, A. S. Z. E. (2013). Open-circuit fault diagnosis of three-phase induction motor drive systems. *Journal of Electrical Engineering*, *13*, 60–68.

Campos-Delgado, D. U., & Espinoza-Trejo, D. R. (2011). An observer-based diagnosis scheme for single and simultaneous open-switch faults in induction motor drives. *IEEE Transactions on Industrial Electronics*, *58*, 671–679.

Caseiro, J. A., Mendes, A. S., & Cardoso, A. M. (2009). Fault diagnosis on a pwm rectifier ac drive system with fault tolerance using the average current park's vector approach. In *IEEE International Electric Machines and Drives Conference* (pp. 695–701). Miami

Diallo, D., Benbouzid, M. E. H., Hamad, D., & Pierre, X. (2005). Detection and diagnosis in an induction machine drive: A pattern recognition approach based on concordia stator mean current vector. *IEEE Transactions on Energy Conversion*, *20*, 512–519.

Estima, J. O., & Cardoso, A. J. M. (2011). A new approach for real-time multiple open-circuit fault diagnosis in voltage source inverters. *IEEE Transactions on Industry Applications*, *47*, 2487–2494.

Estima, J. O., & Cardoso, A. J. M. (2013). New algorithm for real-time multiple open-circuit fault diagnosis in voltage-fed pwm motor drives by the reference current errors. *IEEE Transactions on Industrial Electronics*, *60*, 3496–3505.

Jung, S. M., Park, J. K., Kim, H. W., Cho, K. Y., & Youn, M. J. (2013). An mras-based diagnosis of open-circuit fault in pwm voltage-source inverters for pm synchronous motor drive systems. *IEEE Transactions on Power Electronics*, *28*, 2514–2526.

Lu, B., & Sharma, S. (2009). A literature review of igbt fault diagnosis and protection methods for power inverters. *IEEE Transactions on Industry Applications*, *45*, 1770–1777.

Mendes, A. M. S., & Cardoso, A. J. M. (1999). Voltage source inverter fault diagnosis in variable speed ac drives, by the average current park's vector approach. In *IEEE International Electric Machines and Drives Conference* (pp. 704–706). London

Mendes, A. M. S., Cardoso, A. J. M., & Saraiva, E. S. (1998). Voltage source inverter fault diagnosis in variable speed ac drives, by park's vector approach. In *Seventh International Conference on Power Electronics and Variable Speed Drives* (pp. 538–543). Orlando

Rodriguez, M. A., Claudio, A., Theilliol, D., & Vela, L. G. (2007). A new fault detection technique for igbt based on gate voltage monitoring. In *Proceedings of Power Electronics Society Conference* (pp. 1001–1001). Orlando

Rothenhagen, K. & Fuchs, F. W. (2005). Performance of diagnosis methods for igbt open circuit faults in three phase voltage source inverters for ac variable speed drives. In *Annual Power Electronics Specialists Conference* (pp. 1–10). Dresden

Sleszynski, W., Nieznansk, J., & Cichowski, A. (2009). Open transistor fault diagnostics in voltage-source inverters by analyzing the load currents. *IEEE Transactions on Industrial Electronics*, *56*, 4681–4688.

Trabelsi, M., Boussak, M., & Benbouzid, M. (2017). Multiple criteria for high performance real-time diagnostic of single and multiple open-switch faults in ac-motor drives: Application to igbt-based voltage source inverter. *Electric Power Systems Research, 144*, 136–149.

Ubale, M. R., Dhumale, R., & Lokhande, S. D. (2013). Open switch fault diagnosis in three-phase inverter using diagnostic variable method. *International Journal of Research in Engineering and technology 2*, 636–640.

Yang, S., Xiang, D., Bryant, A., Mawby, P., Ran, L., & Tavner, P. (2010). Condition monitoring for device reliability in power electronic converters. *IEEE Transactions on Power Electronics, 25*, 2734–2752.

Zidani, F., Diallo, D., Benbouzid, M. H., & Naït-Saïd, R. (2008). A fuzzy-based approach for the diagnosis of fault modes in a voltage-fed pwm inverter induction motor drive. *IEEE Transactions on Industrial Electronics, 55*, 586–593.

Chapter 9
Study of an HVDC Link in Dynamic State Following AC Faults and Commutation Failures Based on Modeling and Real-Time Simulation

Mohamed Mankour, Leila Ghomri and Mourad Bessalah

Abstract A 12-pulse high-voltage direct current (HVDC) system based on line-commutated converter (LCC) is modeled and studied in this work. This study is performed and implemented in a real-time simulator using RT-LAB platform HYPER-SIM OP-5600. Both internal and external faults create perturbations, interruption in the transit of power, and valve stress in the dynamic state. However, AC fault represents the most common unsymmetrical fault that can occur in the power systems, and also the misfiring control leads to this instability of the whole system and provides an overvoltage across the converter valve. Therefore, understanding the HVDC system fault behavior is very important for modeling, design, and validation of the control system during and after the different kinds of faults. In this study, the HVDC model chosen is based on the first CIGRÉ HVDC benchmark. Besides, the inverter is connected to a weak AC system. Commutation failure represents also the most common faults that can occur in the inverter valves during the conversion process. This kind of malfunction can be triggered after both the internal and external faults. Therefore, a single phase to ground AC fault of the inverter bus and a misfiring control is applied at one of the inverter valves. Those kinds of faults are applied in the goal to test the control of the system, their influence on the DC recovery, and observing the rising of the commutation failure on the inverter valves in front of two different values of time rising of the mechanism addressed to maintain stability, to ensure a good recovery from the fault and also inhibiting the commutation failures at the converter thyristors. The results are validated by mean of digital real-time simulator HYPER-

M. Mankour (✉) · L. Ghomri · M. Bessalah
National Polytechnic School Maurice Audin, ENPO-MA, Oran, Algeria
e-mail: med_mank@yahoo.com

L. Ghomri
Abdelhamid Ibn Badis University, Mostaghanem, Algeria
e-mail: lilaghomri@yahoo.fr

M. Bessalah
e-mail: mb.enpo@yahoo.com

N. Derbel et al. (eds.), *Diagnosis, Fault Detection & Tolerant Control*,
Studies in Systems, Decision and Control 269,
https://doi.org/10.1007/978-981-15-1746-4_9

SIM (OP-5600) using the simulation in the loop (SIL) test. The performance of the LCC-HVDC system is investigated where the obtained results show the DC control influence to maintain a good and a safe stability of the system during and after the perturbation.

Keywords HVDC system · Real-time simulation · HYPERSIM simulator DC control · AC fault

9.1 Introduction

Since the first application of the high-voltage direct current (HVDC) technology, it has made a wide utilization around the world for its high-capacity networking transmission and for application in the long distance or by means of submarine cable (Arrillaga 1998). Most of HVDC systems are based on line-commutated converter (LCC) utilizing thyristor semiconductor technology. However, it causes some problems in their application especially in case of AC system faults with energy loss to the AC system during the fault that is unavoidable (Padiyar 2013). During disturbance caused by reduction in voltage in one or all phases at AC system in the inverter side, a commutation failure rises at the valve group and leads to increased reactive power consumption which depress AC bus voltage at the inverter side (Lingxue et al. 2007; Jovcic 2007) particularly for weak AC system (CIGRE WG 14.05 1995).

Recovery from disturbance, system stability, and the probability of commutation failures depends on interaction AC–DC and on the strength of AC system where the converter terminal is connected (Arrillaga 1998; Mankour et al. 2017); the AC system strength is represented by the short-circuit ratio (SCR), where the SCR is the ratio of the AC system short-circuit capacity of the connected AC system to DC link power at the converter terminal. This ratio divides the power system strength into three main categories: strong, weak, and very weak system (Arrillaga 1998; CIGRE WG 14.05 1995).

In order to understand the behavior the HVDC system in dynamic condition, various simulations and comparisons with different software are used to achieve this study, such as PSCAD/EMTDC, PSB/Simulink, and PSCAD/Simulink in reference (Faruque et al. 2006) and with PSS/E in reference (Kwon et al. 2015). Today, the real-time simulation using the real-time simulators become an important tool for power systems and electrical networks, especially the HVDC links to test and validate their control in different states and also to validate the protection methods by means of the approaches available in real-time simulation domain. In this study, a full digital real-time simulator (DRTS) is used; this DRTS is HYPERSIM (OP-5600) simulator developed by RT-LAB (Barry et al. 2000; Guay et al. 2012); HYPERSIM offers an extensive software and hardware package designed to simulate online or offline complex power system (Guay et al. 2017). In domain of supercomputer and fast processors, the simulations of complex and dynamic systems are achievable in real time (Faruque et al. 2015; Sybille and Giroux 2002; Babazadeh et al. 2017) and its application is mainly classified into three categories as follows: Hardware in the loop (HIL), rapid control prototyping (RCP), and simulation in the loop (SIL). In this study, the third approach is used, the SIL. With a powerful enough simulator, both controller and plant can be simulated in real time in the same simulator. The benefit of the SIL over the two other subscribed approaches is that neither inputs nor outputs are used during the simulation. Hence, since both controller and plant models run on the same simulator, timing with the outside world is no longer critical with no impact on the validity of results. Therefore, SIL can be classified as an ideal for a class of simulation called accelerated simulation (Bélanger et al. 2010).

An SIL approach is used in this study to show, examine the dynamic performance of the HVDC inverter, and demonstrate relationship between the DC control, the stability, and the recovery from both the instability and the commutation failures. This study is implemented to analyze the electromagnetic transients of a 12-pulse HVDC system based on LCC converter feeding a weak AC system. The simulation is achieved after two kinds of faults (the internal and the external faults), where the external fault is a single-phase AC fault to ground at the inverter side and the internal fault is a misfiring control on the inverter valve (a commutation failure).

9.2 Commutation and Commutation Failure Description

A typical three-phase 6-pulse inverter is illustrated in Fig. 9.1, where AC system voltage, leakage inductance transformer X_c, and both DC current and DC voltage are presented.

Due to impedance in the supply network and leakage inductance of the converter transformer, the valve current cannot change from one valve to the other suddenly

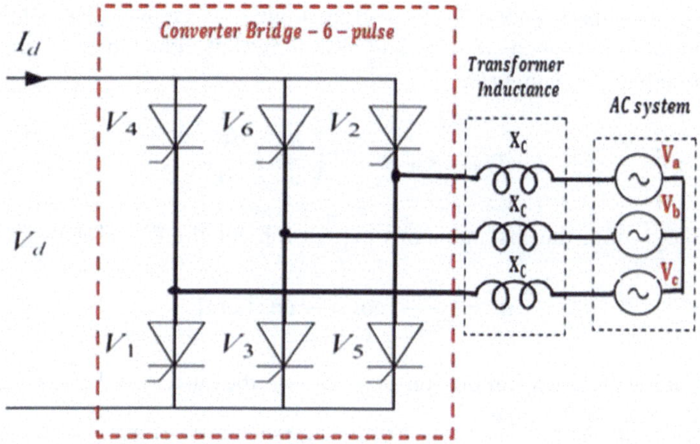

Fig. 9.1 Typical three-phase 6-pulse inverter

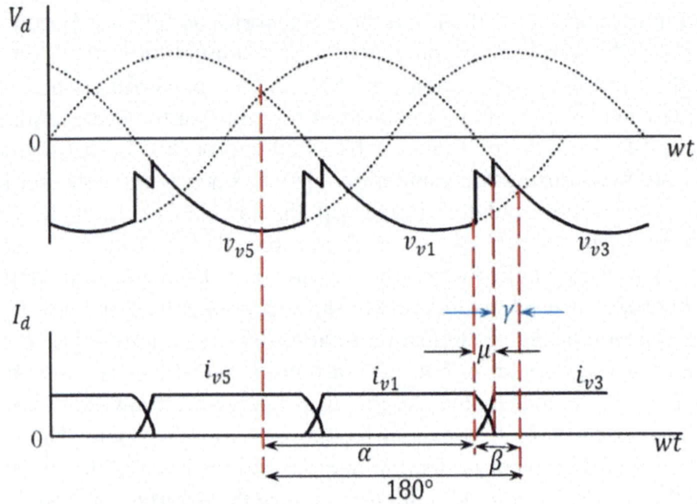

Fig. 9.2 Inverter operation showing commutation overlap

and thus commutation cannot be instantaneous. In fact, for each valve, commutation process takes a period of time (Arrillaga 1998)-(Jovcic 2007).

Commutation process between valve 1 and valve 3 in normal condition is presented in Fig. 9.2. In fact, during the commutation, a short-circuit current I_{cc} is created through the two reactance X_c by voltage V_{ba} during conducting between valves 1 and 3. Then, this short-circuit current will be decreased gradually and even the current goes out from the valve 1 to the valve 3 and that is explained by ($I_d = i_{v1} + i_{v3}$); the DC current rests constant during commutation.

The time of short-circuit current annulations is measured by commutation period between the two valves, known as the overlap angle μ given in Eq. 9.4. At the beginning of commutation when $\omega t = \alpha$, and assuming that inductances of transformer winding are equal, we can write

$$v_{ba} - v_{ac} = \sqrt{2}\left(\frac{X_c}{\omega}\right)\frac{di_v}{dt} \tag{9.1}$$

After integration, the instantaneous expression for the commutating current is thus

$$i_{cc} = \frac{V_{LL}}{2X_c}[\cos\alpha - \cos(\omega t)] \tag{9.2}$$

Therefore, at the end of commutation $\omega t = \alpha + \mu$, the current valve 3 is equal to the DC current, then Eq. 9.2 gives

$$i_{v3} = I_d = \frac{V_{LL}}{2X_c}[\cos\alpha - \cos(\mu + \alpha)] \tag{9.3}$$

Then, the overlap angle μ is expressed as

$$\mu = \arccos\left(\cos\alpha - \frac{X_c I_d \sqrt{2}}{V_{LL}}\right) \tag{9.4}$$

where α represents the delay angle, β is the advance angle ($\beta = 180 - \alpha$), X_c is the reactance of the transformer converter, I_d represents the DC current, and V_{LL} is the line-to-line RMS voltage dependent on AC system voltage.

The commutation failures represent an adverse dynamic event that can occur in converter valves; this phenomenon is signaled when either one or different converter valves that are supposed to turn-off continues to conduct (stay in conduction) (Oketch 2016). Hence, the current that passes through the converter valve is not transferred to next valve in the exact firing sequence. It can be seem clearly the difference between the successful and fail commutation process in Figs. 9.3 and 9.4.

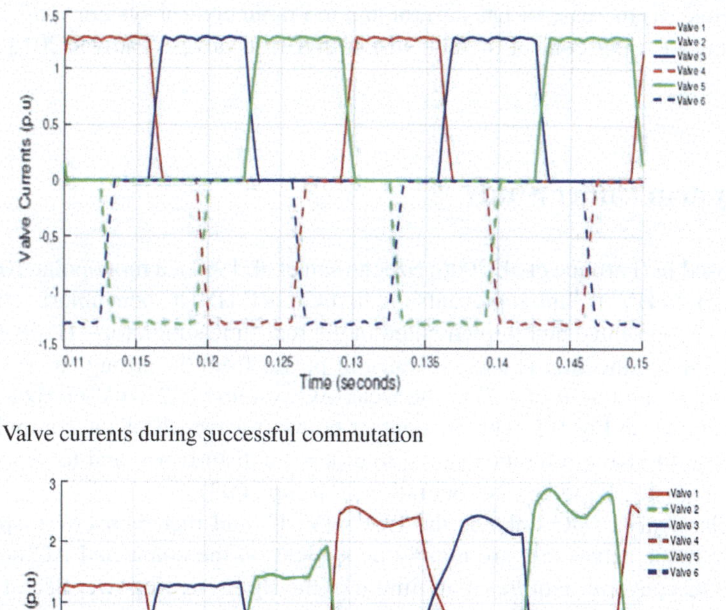

Fig. 9.3 Valve currents during successful commutation

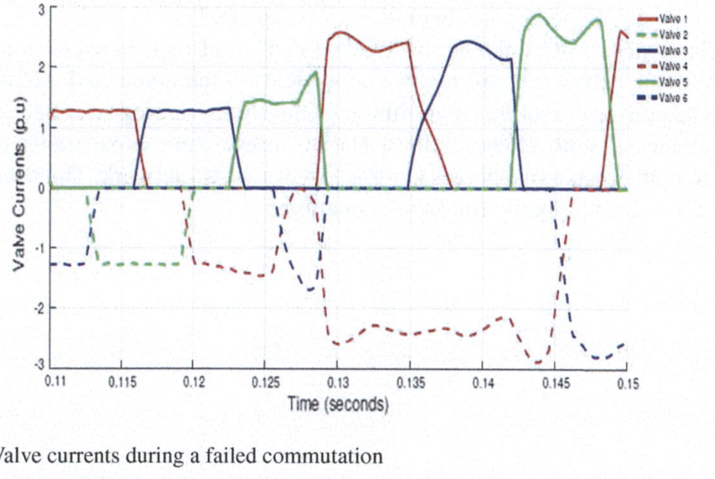

Fig. 9.4 Valve currents during a failed commutation

The difference between the rectifier and the inverter is that the inverter is more prone to commutation failure and it presents the commonest fault that can occur in its valves during the conversion process, because its firing angle is large ($90 < \alpha < 180$), unlike the rectifier which has delay angle ($\alpha < 90$); that is why rectifier rarely fails to commutation failure. For that, to make a successful inverter operation and to avoid any commutation failure and overshooting of the current valve, an extinction angle γ must be taken in consideration because thyristor requires a turn-off time for a successful commutation (9.1) and this angle needs to be maintained at the minimum value (Arrillaga 1998)–(Padiyar 2013). Therefore, a limit is imposed at the maximum value of the inverter firing angle:

$$\alpha_{max-inv} = 180 - (\mu + \gamma_{min}) \tag{9.5}$$

Increasing of the overlap angle μ or decreasing of extinction angle γ_{min} which causes reduction in voltage or increase in current or both can lead to commutation failures, and they basically cause by the following causes: (i) faults in connected AC systems, (ii) increase of DC current and too rapid control system, and (iii) malfunctioning in firing control or fault in the thyristor valves (Padiyar 2013; Group 2007).

9.3 System Under Study

As proposed in (Faruque et al. 2006; Szechtman et al. 1991), a monopolar 1000 MW (500 kV, 2 kA) HVDC link represents the first CIGRÉ HVDC benchmark. However, the model is presented and implemented in the real-time simulator—HYPERSIM.

A DC interconnection is used to transmit power from the strong AC system (1) (315 kV, 60 Hz and SCR of 4.7) to the weak AC system (2)(230 kV, 60 Hz and SCR of 2), as shown in Fig. 9.5. The element of point-to-wave (POW) is compulsory in the network to give a reference signal to clock the disturbance and to synchronize the data acquisition and breaker operations (Group 2007).

The classical HVDC links as the LCC-HVDC, and their converters are based on the thyristor valves and are mainly dependent on the connected AC networks. The AC/DC converter requires a minimum of the short-circuit power Scc of the AC network connected with. However, these HVDC links are not able to transit the active power into neither a passive network nor a very weak AC network. The short-circuit ratio (SCR) is defined by the following equations:

$$SCR = \frac{S_{cc}}{P_d} \tag{9.6}$$

$$S_{cc} = \frac{U^2}{Z_s} \tag{9.7}$$

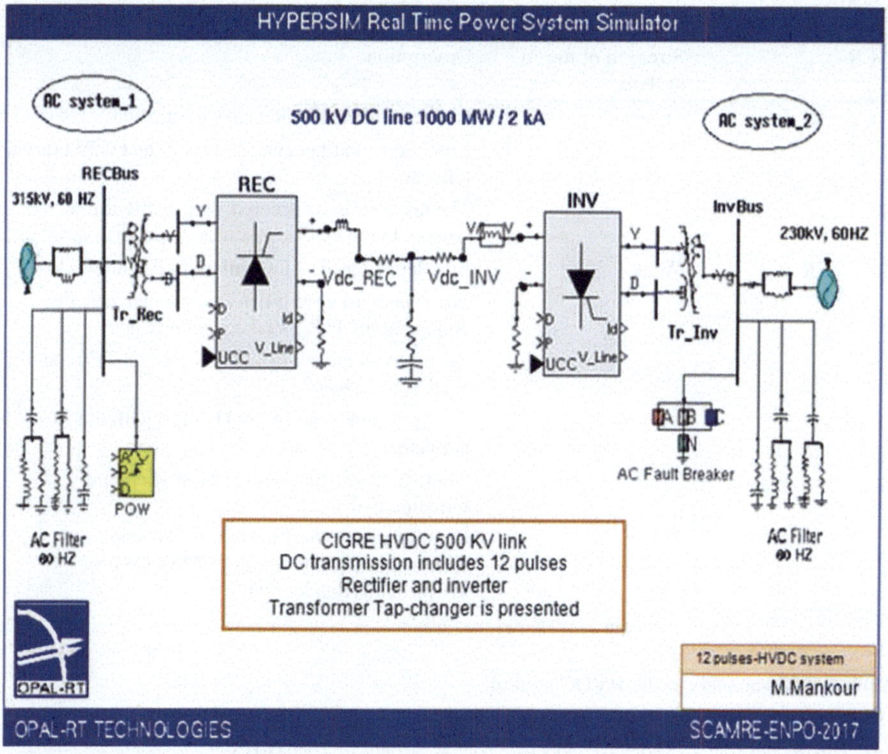

Fig. 9.5 HVDC system in HYPERSIM interface

where S_{cc} is the power of short circuit of the AC system between the PCC and the HVDC converter (MVA), P_d is the direct power of the converter (MW), U is line-to-line voltage, and Z is the equivalent impedance of the AC network at the fundamental frequency.

The control of HVDC plays a vital role in different interactions of the AC/DC phenomenon. However, the evaluation of the AC network strength must be considered. In the literature (CIGRE WG 14.05 1995)–(Transmission 1997), the AC networks are classified into three categories according to the SCR. Table 9.1 gives the different groups.

9.3.1 The AC System

The AC networks, both AC system (1) and AC system (2), are modeled to represent the AC supply networks of the rectifier and inverter sides. Each one has internal impedance and magnitudes of their impedances which are represented as $(R - R/L)$ networks having the same damping function of frequency such as the resonances at the fundamental and third harmonic (Arrillaga 1998).

Table 9.1 Classification of the AC network according to the SCR

SCR	Strength of the AC system	Description
$SCR \geq 3$	Strong	Limited interaction and easy controlled
		Little or no-fault occurrence such the commutation failures
		The network is considered very stable if the voltage in the PCC is taken as controlled variable
$2 \leq SCR < 3$	Weak	Interdependence of DC link with AC network These networks are taken as weak because the voltage at the PCC is taken as invariable
		Require voltage control ability like OLTC or static VAR compensator
$SCR < 2$	Very weak	Strong dependence of the DC link with the AC network Another specification must be studied and implemented
		Require strong VAR generator like static synchronous compensator (STATCOM) or synchronous condenser)

Table 9.2 Parameters of the HVDC system

Parameters	Rectifier	Inverter
AC voltage base	315 kV	230 kV
System frequency	60 Hz	60 Hz
Nominal DC voltage	500 kV	500 kV
Nominal DC current	2 kA	2 kA
Minimum angle	18	18
DC parameters	$R = 2.5\,\Omega$ $L = 0.589\,\text{H}$	$R = 2.5\,\Omega$ $L = 0.589\,\text{H}$ $C = 26.0\,\mu\text{F}$

Table 9.2 gives the detailed parameters of the HVDC link, and the diagram of AC system is shown in Fig. 9.6. The rest of the filter parameters is detailed in reference (Mankour et al. 2018).

9.3.2 Converter Transformer

Both converter transformers (rectifier and inverter) with tap changer control are modeled with three-phase transformer (three windings per phase): a Y primary with grounding impedance, a floating Y secondary, and delta tertiary. The leakage resistances and inductances of windings are considered.

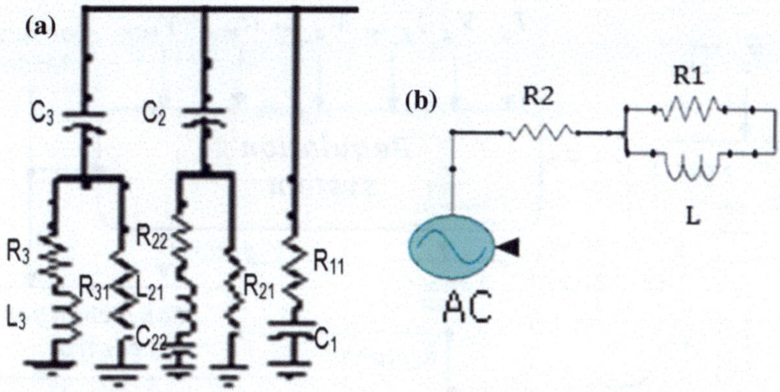

Fig. 9.6 **a** Diagram of the AC filter, **b** AC system representation

9.3.3 DC Line

The DC line is modeled using an equivalent-T network with smoothing reactor 0.5968 H connected to both rectifier and inverter sides. The reactor smoothing is represented by decoupling reactor to divide the task into two subtasks in order to decrease the calculation load (Group 2007). The benefit of the smoothing reactor is for prevention of intermittent currents, limitation of the DC fault currents, prevention of resonance in the DC currents, and reducing harmonic current in the DC side.

9.3.4 AC Filter and Reactive Support

The HVDC converter stations generate harmonic currents. For the twelve pulse converter, the characteristic harmonics are of the order of

$$n = 12k \mp 1\ (k = 1, 2, 3 \ldots etc) \tag{9.8}$$

Therefore, current harmonics of the order of 11, 13, and 25 and higher are generated on AC side of 12-pulse HVDC converters. For this reason, damped filters are installed to limit the amount of harmonics to the level required by network.

The reactive power consumed by the HVDC converter depends on many factors such as the active power, the transformer reactance, and the control angle. Therefore, the reactive power increases with increase of the active power. The capacitive banks are also installed to recompense the reactive consumed by converters in conversion process. Each filter and capacitor banks are installed on both HVDC sides (the rectifier and the inverter).

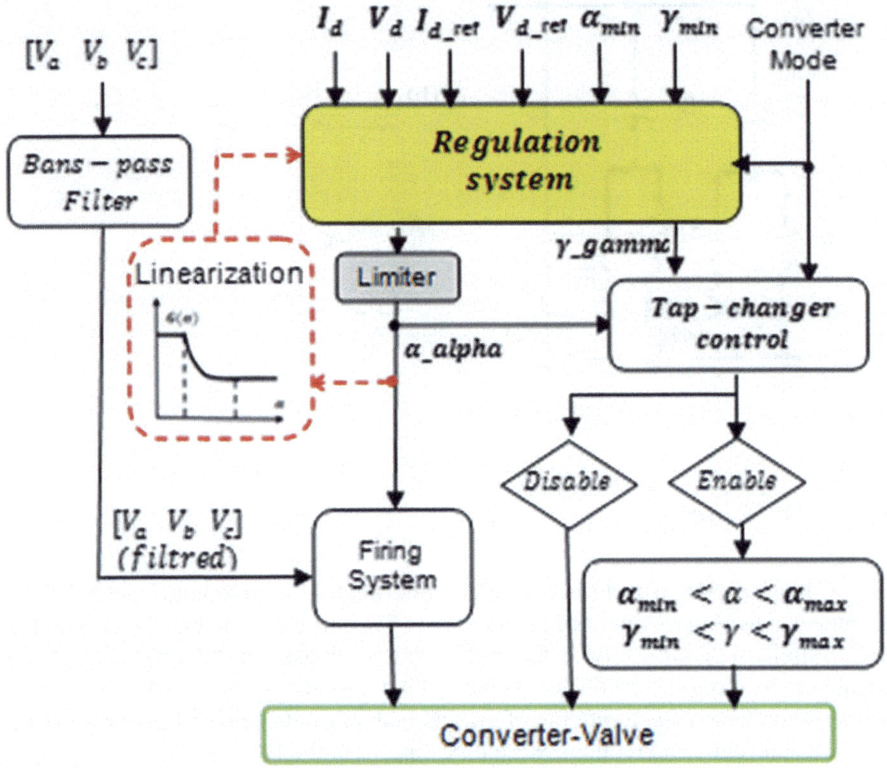

Fig. 9.7 Control system diagram of converter

9.3.5 Converter

Two 12-pulse converters are used for both station rectifier and inverter sides; each valve of converter is composed of many thyristors in series and has an RC parallel snubber.

9.3.6 Control System

The basic units of system control are tap changer control, synchronization regulation of firing angles (delay angle (α) and firing angle (γ)), voltage-dependent current order limiter (VDCOL), and the protection against commutation failures, as shown in Fig. 9.7. The function of the regulation system for both converter rectifier and inverter are two current and voltage regulators of the proportional and integral type (PI) operating in parallel to calculate (α_{rec}) and adjust the delay angle (γ_{inv}) of rectifier and maintain extinction angle of inverter at a minimum values; the voltage

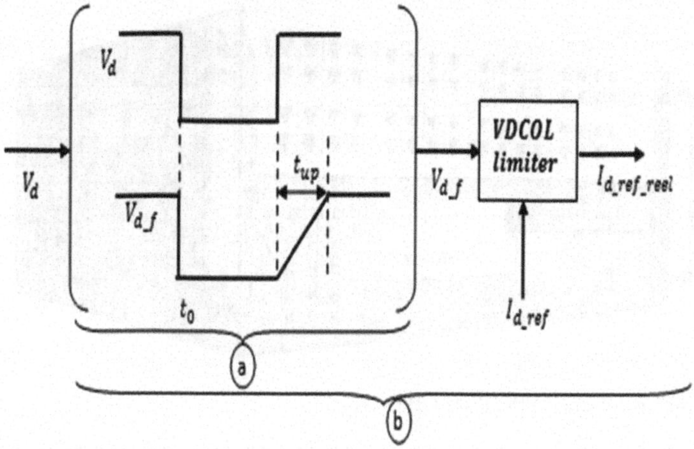

Fig. 9.8 **a** First-order dynamic filter, **b** Voltage-dependent current order limiter

is regulated with a slope determinate by current margin ($\Delta I_d = 0.1$ pu) and voltage margin ($\Delta V_d = 0.05$ pu).

9.3.6.1 VDCOL Function

VDCOL function, voltage-dependent current order limiter (VDCOL), is used to adjust the current reference I_{d-ref} set point at the correspondent value of the DC voltage V_{d-line}. Another benefit of this function is the good recovery from the DC power transit, and the minimization of the risk of commutation failure is also considered (CIGRE WG 14.05 1995)–(Arrillaga 1998). In disturbance state (e.g., Faults on AC side or DC line fault), VDCOL changes automatically the current reference I_{d-ref}; this change is done by reducing the I_{d-ref} set point when V_{d-line} drops under a threshold value determined by the function.

As shown in Fig. 9.8, the VDCOL limiter includes two functions: the dynamic filtering of the DC voltage and the calculation of the current reference $I_{d-ref-reel}$ based on the filtered voltage, when DC voltage drops, the current reference I_{d-ref} falls immediately but increases more slowly following the slope determined by a rising time t_{up}.

9.4 Implementation and Interpretation of Results

9.4.1 Real-Time Platform

The RT-LAB platform consists of HYPERSIM OP-5600, where the HYPERSIM is real-time digital electromagnetic transient simulator developed by Hydro-Quebec Research Institute (IREQ) (Sybille and Giroux 2002). HYPERSIM offers an exten-

Fig. 9.9 Hardware of the HYPERSIM OP-5600

sive software and hardware package designed to simulate online or offline those complex power systems.

HYPERSIM is power system simulator that provides a simulation in multiple CPUs, where the model splits on CPU (see Fig. 9.9). HYPERSIM can run in offline (as fast as possible) and in real-time simulation (for different topologies, SIL or in HIL) (Guay et al. 2017).

The features of the HYPERSIM shown in Fig. 9.9 are as follows: two CPU Intel Xeon Six-core 3.46 GHZ 12M cache, Four Memory 2 GB and X8DTL-I-O Supermicro Motherboard, processor 5600/5500 series. The HVDC system implemented in the simulation is modeled in single phase.

9.4.2 Simulation Results

A detailed digital simulation of the monopolar HVDC link is performed using real-time simulator (HYPERSIM OP-5600 simulator), 50 μs time step is used for the simulation, and real-time simulation is selected. Both converter rectifier and inverter are blocked at first and then deblocked when achieving steady state after the simulation is started.

As described in Sect. 9.2, faults in the AC side and internal faults in the converter can trigger the commutation failures. Also, the SCR influences into the interaction between AC/DC. Therefore, three different tests have been studied in this work to check, study, and investigate the influence of the control system on the recovery from the different kinds of faults and on the enhancement of the system stability.

In the first two simulations, a single phase to ground fault at the inverter bus is applied by changing the time rising (t_{up}) of the VDCOL function (85 and 9 ms) in order to show the influence of this function on the DC recovery, also to study the commutation failure rising or inhibiting. Second, an internal fault is applied at an inverter valve to follow the same steps in the first two tests (the rising time is taken as in the normal state).

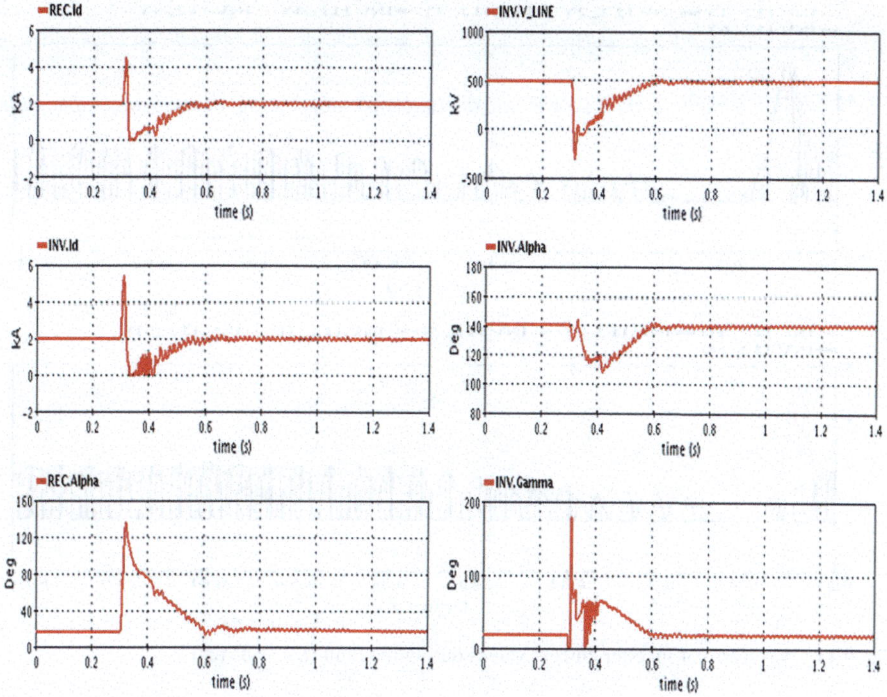

Fig. 9.10 Single phase to ground, 6-cycle fault at inverter side ($t_{up} = 85\,\text{ms}$)

9.4.2.1 Single Phase to Ground at Inverter Side with ($T_{up} = 85\,\text{ms}$)

A single phase to ground fault was applied to the phase A of the inverter AC bus, and the duration of the fault was six cycles at $t = 3\,\text{s}$ (see Fig. 9.10). First test is for the rise time ($t_{up} = 85\,\text{ms}$).

When the fault is applied, the DC voltage (V_{inv}) drops to zero due to reduction in the AC voltage at the inverter bus and an overshoot in the DC current is observed momentarily for both the rectifier and the inverter ($I_{d-inv} = 5.6\,\text{kA}$ and $I_{d-rec} = 4.4\,\text{kA}$). In this situation, the rectifier firing angle α_{rec} is increased by the current controller to reduce the increasing DC current, and is forced into inverter region and reaches 134 deg; this fault creates a commutation failure and the inverter extinction angle γ_{inv} falls to zero while α_{inv} decreases. The DC current reaches its minimum value ($I_{d-min} = 600\,\text{A}$) determined by the VDCOL limiter ($I_{d-min} = 0.3\,\text{pu}$).

As shown in Fig. 9.11, we can observe a number of commutation failures at the valves group and an increasing in DC current (valves 1 and 4) because they are conducting current at the same time. The system comes back to steady state gradually according to the function VDCOL limiter after clearing the fault at $t = 4\,\text{s}$, and no rising of the commutation failures in the valves' group.

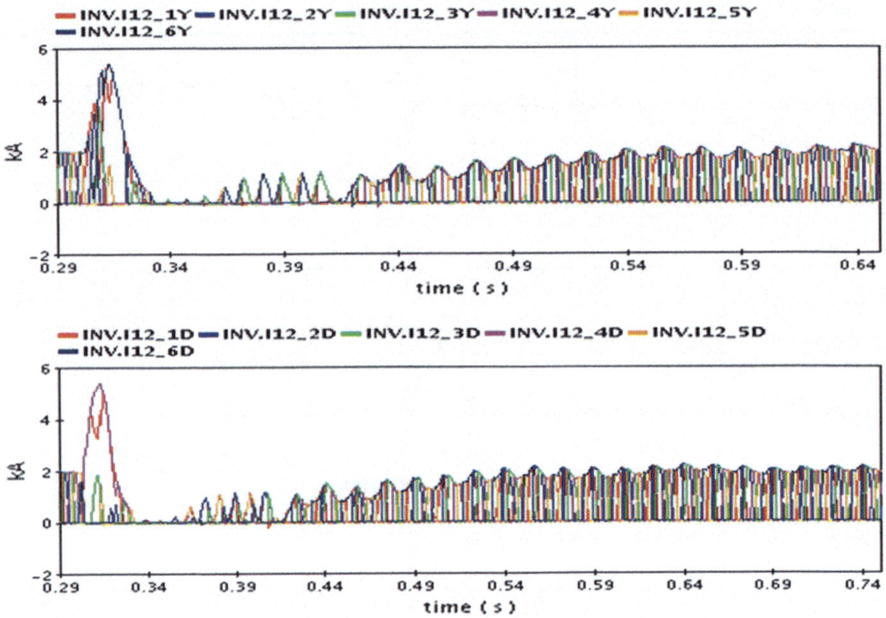

Fig. 9.11 Waveform of inverter current valves during the fault ($t_{up} = 85\,\text{ms}$)

9.4.2.2 Single Phase to Ground at Inverter Side with ($T_{up} = 9\,\text{ms}$)

The same fault with a same duration was applied at the phase A of the AC inverter side but the time rising of the VDCOL function ($t_{up} = 9\,\text{ms}$) at this case is less than 10 time if compared with the first case, the duration of the fault is six cycles (see Figs. 9.12 and 9.13).

Commutation failures are observed at the inverter valves. The fault causes an overshoot of the DC current and a collapse of the DC voltage. Compared to the waveforms of Figs. 9.10 and 9.11, repetitive commutation failures occurred in the inverter valves after clearing the fault at $t = 4\,\text{s}$, due to the short time of rise time of the VDCOL function which causes an overshoot of the current valve than the DC current. In addition, the delay angle α_{rec} is forced another time into inverter region to reduce the overshoot of the DC current created by the second commutation failures, but the system finally recovers to the steady state after reducing the DC current I_{d-min} by the VDCOL function after the fault is cleared.

9.4.2.3 Misfiring Fault

The simulator provides the possibility to applied misfiring fault at the inverter valves by trigging a commutation failure. For this, a commutation failure is applied at the valve 5 at $t = 4\,\text{s}$ in order to observe the behavior of the HVDC under this circum-

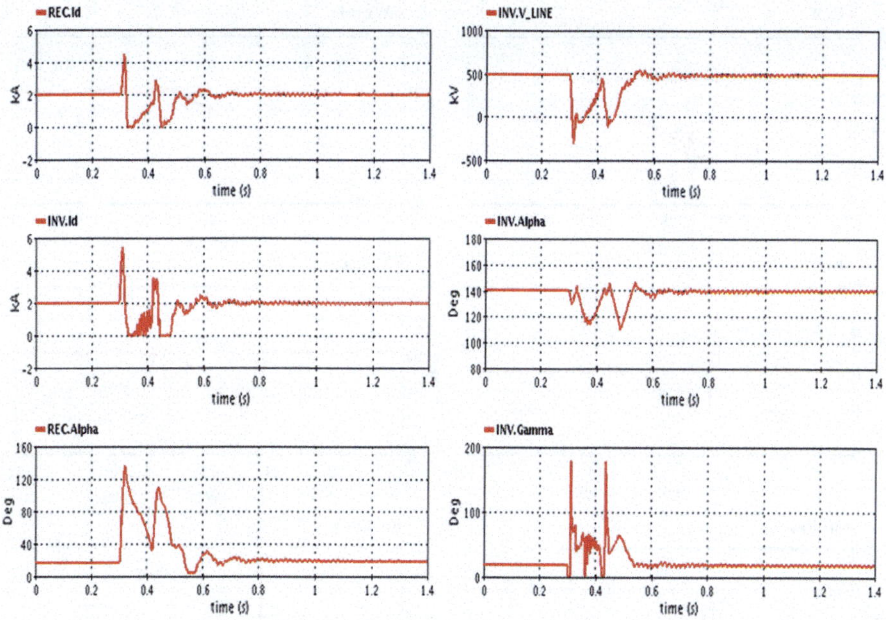

Fig. 9.12 Single phase to ground, 6-cycle fault at inverter side ($t_{up} = 9\,\text{ms}$)

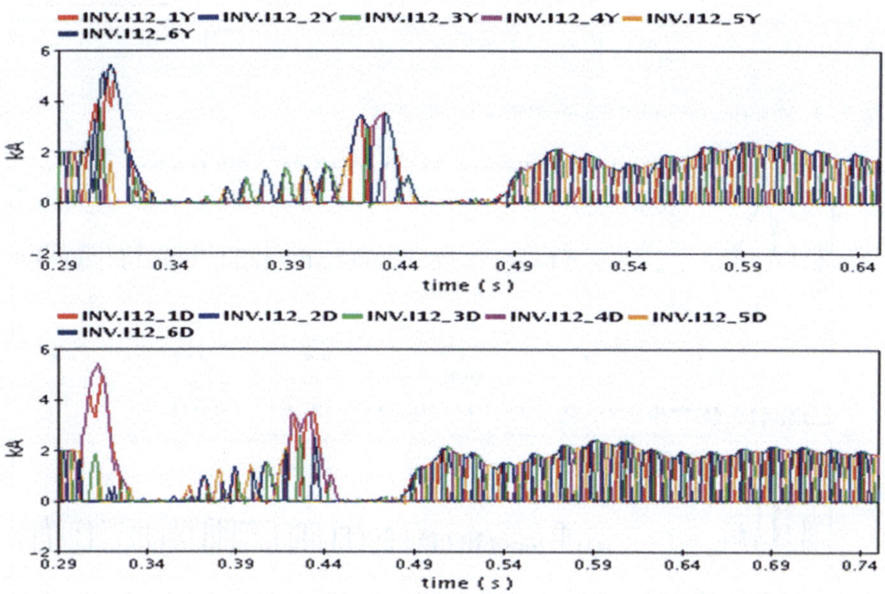

Fig. 9.13 Waveform of inverter current valves during the fault ($t_{up} = 9\,\text{ms}$)

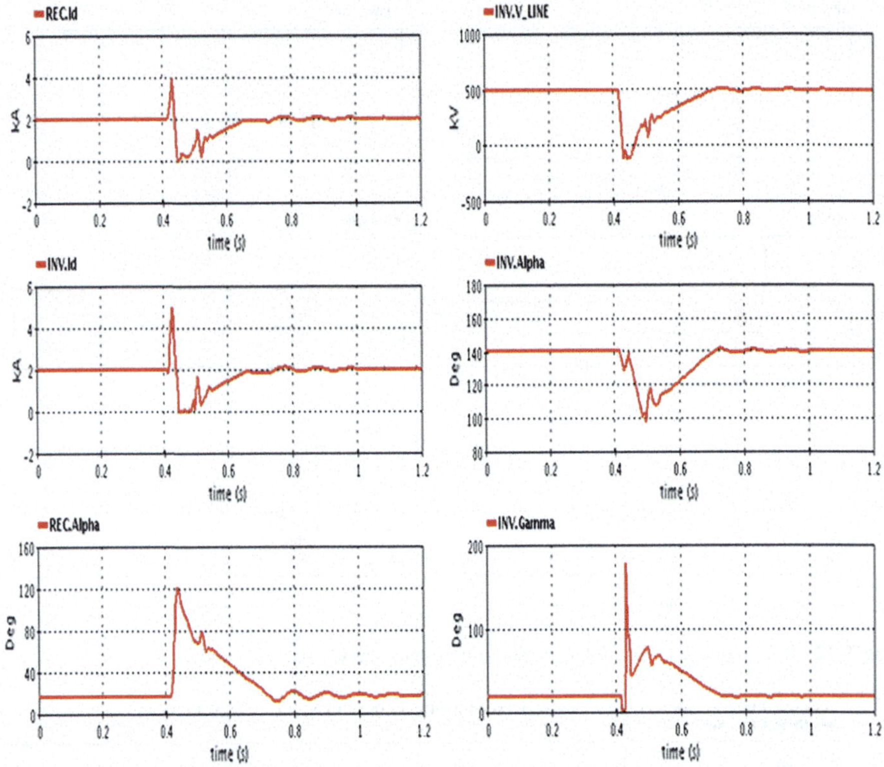

Fig. 9.14 Misfiring control at the inverter valve Nbr 5

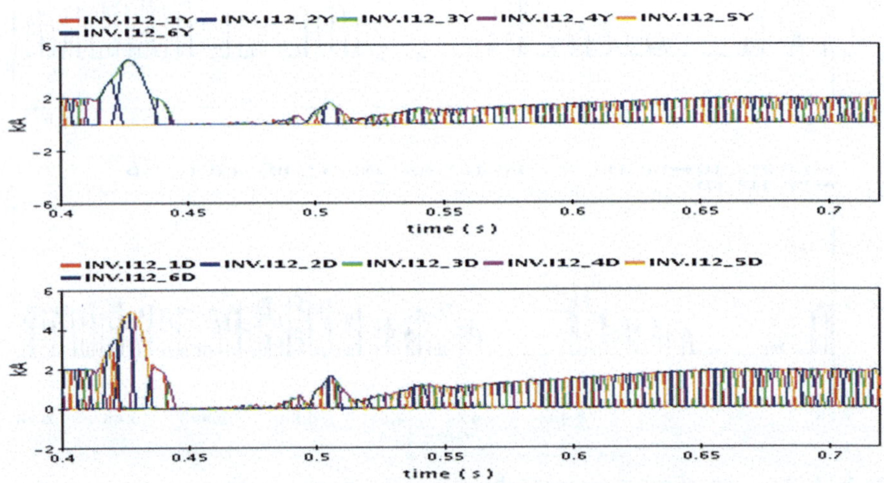

Fig. 9.15 Current waveform of inverter valves during the misfiring fault

stance. As shown in Fig. 9.14, the DC current increases quickly on both sides the rectifier and the inverter and the DC voltage becomes negative because of increasing the rectifier delay angle to $\alpha_{rec} = 120\,\text{deg}$ by the reaction of the controller. Commutation failures can be seen in Fig. 9.15, especially at the valves 1 and 5, because they are conducting in this same time. The inverter delay angle decreases and his extinction angle falls to zero. After the fault is cleared, the VDCOL function rises and readjusts the current to the real reference 1 pu gradually. The slope of the VDCOL is imposed to reduce the consumption of the reactive power by the converter during the faults, reduce the overshooting the current to protect the valves, and avoid the rising of the commutations failures. The rise time is $t_{up} = 85\,\text{ms}$. Finally, the system returns to the steady state in approximately 0.25 s.

9.5 Conclusion

In this study, we have presented the dynamic behavior of the HVDC system using the RT-LAB platform (Real-time power system simulator HYPERSIM OP-5600). Three tests were achieved to demonstrate the role and the importance of the control system based on the voltage-dependent current order limiter (VDCOL) function to detect the faults, recovery to steady state, and avoid the repetitive or the persistent commutation failures on the inverter station. Based on the obtained simulation, not only the external fault can perturb the DC link, but also the internal fault such as the misfiring fault at the inverter valve can also lead to interruption in the transit of the active power, valve stress, and temporary overvoltage in the converter valves.

The action of the VDCOL function is by reduction of the current reference to limit both the DC current and the DC line voltage, and also to reduce the consumption of the reactive power during and after the faults. Choosing the adequate rising time of this function is very important to control the DC recovery in the dynamic state and protects the converter valves from the different undesirable phenomena.

Based on this accurate simulation using the HYPERSIM simulator, as the first step, further study of another type of converters based on GTO and IGBT valves known as voltage source converters VSC-HVDC can be implemented, and the possibility of studying the HVDC system behavior following other fault scenarios such as DC fault which can be also obtained. The validation of these results on the real industrial application of an HVDC link will be made in the future work as the second step.

References

Arrillaga, J. (1998). *High voltage direct current transmission*. The Institution of Electrical Engineers, USA

Babazadeh, D., Muthukrishnan, A., Mitra, P., Larsson, T., & Nordström, L. (2017). Selection of DC voltage controlling station in an HVDC grid. *Elsevier-Electric Power Systems Research, 144*, 224–232.

Barry, A. O., Guay, F., Guerette, S., & Giroux, P. (2000). Digital Real-time Simulation for Distribution System. In *International Conference on Transmission and Distribution Construction, Operation and Live-Line Maintenance*.

Bélanger, J., Paquin, J. N., & Vene, P. (2010). The what where and why of real-time simulation. In *IEEE PES General Meeting*.

Faruque, M. O., Strasser, T., Lauss, G., Jalili-Marandi, V., Forsyth, P., Dufour, C., et al. (2015). Real-time simulation technologies for power systems design, testing and analysis. *IEEE Power and Energy Technology Systems Journal*, *2*, 63–73.

Faruque, M.O., Zhang, Y., & Dinavahi, V. (2006) Detailed modeling of CIGRE HVDC benchmark system using PSCAD/EMTDC and PSB/SIMULINK. *IEEE Transactions Power Delivery*, *21*, 378–387.

Group, H. H. Q. (2007). *HYPERSIM reference guide manual 9.2*. IEEE Std

CIGRE WG 14.05 (1995). *Commutation failures causes and consequences, interaction between AC and dc Systems*.

Guay, F., Cardinal, J., Lemiex, E., & Guerette, S. (2012). Digital real-time simulator using IEC 61850 communication for testing devices. In *CIGRE Canada*.

Guay, F., Chiasson, P., Verville, N., Tremblay, S. & Askvid, P. (2017) *New hydro-québec real-time simulation interface for HVDC commissioning studies*. In *International Conference on Power Systems Transients*.

Jovcic, D. (2007). Thyristor-based HVDC with forced commutation. *IEEE Transactions on Power Delivery*, *22*, 557–564.

Kwon, D. H., Moon, H., Kim, R., Chan, G. K., & Moon, S. (2015). Modeling of CIGRE benchmark HVDC system using PSS/E compared with PSCAD. In *5th IEEE International Youth Conference on Energy*.

Lingxue, L., Zhang, Y., Zhong, Q., & Fushuan (2007). Identification of commutation failures in HVDC systems based on wavelet transform. In *14th International Conference on Intelligent System Applications to Power Systems*.

Mankour, M., Khiat, M., Ghomri, L., & Bessalah, M. (2017). Dynamic performance of an HVDC link based on protection function against commutation failures. In *5th International Conference on Control & Signal Processing*.

Mankour, M., Khiat, M., Ghomri, L., Chaker, A., & Bessalah, M. (2018). Modeling and real time simulation of an HVDC inverter feeding a weak AC system based on commutation failure study. *Elsevier, ISA Transactions*, *77*, 222–230.

Oketch, I. (2016). *Commutation failure prevention for HVDC: Improvement in algorithm for commutation failure prevention in LCC HVDC*. Chalmers University of Technology, Gothenburg, Sweden.

Padiyar, K. (2013). *HVDC power transmission systems* (2nd ed.). India: New Academic Science Limited.

Sybille, G., & Giroux, P. (2002). Simulation of FACTS controllers using the MATLAB power system blockset and HYPERSIM real-time simulator. In *IEEE Power Engineering Society Winter Meeting*.

Szechtman, M., Wess, T., & Thio, C. V. (1991). First benchmark model for HVDC control studies. *Electra*, *135*, 54–67.

Transmission and Distribution Committee of the IEEE Power Engineering Society, (1997). *IEEE guide for planning DC links terminating at AC locations having low short-circuit capacities*.

Chapter 10
Influence of Design Variables on the Performance of Permanent Magnet Synchronous Motor with Demagnetization Fault

Manel Fitouri, Yemna Bensalem and Mohamed Naceur Abdelkrim

Abstract This chapter focuses on the study of the effects of design variables on the permanent magnet (PM) demagnetization of a permanent magnet synchronous motor (PMSM) using a two-dimensional (2D) finite element analysis (FEA) software package. The aim is to investigate the shape design of a permanent magnet considering the effect of the pole embrace and the magnet thickness on demagnetization by using finite element method (FEM). As a result, it is shown that pole embrace and magnet thickness are the most important geometrical dimensions of the PMSM in terms of demagnetization fault. The effects of demagnetization fault on the magnetic field distribution, the back electromotive force (EMF) induced, and the electromagnetic torque of PMSM are examined. Using the simulated model, a technical method, based on fast Fourier transform (FFT) analysis of induced voltage and electromagnetic torque, is exported to detect the demagnetization fault. Finally, the technique used and the obtained results show clearly the possibility of extracting signatures to detect and locate faults and verifying the effectiveness of the analysis results.

Keywords Finite element method (FEM) · Permanent magnet synchronous motor (PMSM) · Demagnetization fault · Permanent magnet · Fast Fourier transform (FFT) analysis

M. Fitouri (✉) · Y. Bensalem · M. N. Abdelkrim
Research Laboratory Modelling, Analysis and Control of Systems (MACS), National Engineering School of Gabes, Gabes, Tunisia
e-mail: manel.fitouri@gmail.com

Y. Bensalem
e-mail: bensalem.yemna@yahoo.fr

M. N. Abdelkrim
e-mail: naceur.abdelkrim@enig.rnu.tn

N. Derbel et al. (eds.), *Diagnosis, Fault Detection & Tolerant Control*, Studies in Systems, Decision and Control 269,
https://doi.org/10.1007/978-981-15-1746-4_10

10.1 Introduction

Permanent magnet synchronous motor has become the most advantageous power equipment in many industrial applications with high efficiency, high power density, and extended high achievable speed operation (Apostoaia 2014). Due to these advantages, PMSM is one of the most preferred motor types in fields such as weapon systems, industry, medicine, and spacecraft.

Permanent motors are susceptible to many types of faults in industrial applications and are classified into two groups: One is an electrical fault such as fault in connection of the stator windings and stator short-circuited turns (Fitouri et al. 2016), and a demagnetization in rotor magnets (Faiz and Mazaheri-Tehrani 2017). The other is a mechanical fault including damaged bearings faults (Pacas 2017) and static and dynamic eccentricity faults (Kang et al. 2017).

One of the problems with this type of permanent magnet motor is related to its demagnetization, which is caused by conditions such as strong magnetic field, high working temperature, self-demagnetization, high mechanical stress, or a combination of these factors.

Demagnetization fault is counted the most significant and frequently occurred type of faults in PM motor because it has always been a major concern that can seriously degrade the machines.

This type of fault is typically caused by the excessive field weakening, the high temperature stress, and some combination of electrical, mechanical, and the physical damage (Ruiz et al. 2009; Kim et al. 2017). This phenomenon affects the motor performance characteristics such as the electromotive force (EMF) (Kim et al. 2006; Farooq et al. 2008), the electromagnetic torque, and the distribution of flux density (Choi and Jahns 2015). For this reason, some authors have studied their proposed model on demagnetization fault using analytical methods based on Maxwell's equations (Liang et al. 2016; Farooq et al. 2008), magnetic equivalent circuits (Choi and Jahns 2015; Bianchi and Jahns 2004), and permeance network methods (Ahmed Farooq et al. 2009).

In this work, it is proposed the use of the finite element (FE) model that is the nonlinear model which analyzes the performance of the machine and simulates the demagnetization fault in each part of permanent magnet.

The key objective of this paper is to investigate the shape design of a permanent magnet considering the effect of the pole embrace and the magnet thickness on demagnetization by using FEM.

As well known, the one major reason for diagnosis of this fault at the initial stage of its occurrence is the ever-increasing cost of the permanent magnets. Many methods have been used to detect demagnetization fault in PMSM (Urresty et al. 2011, 2012; Diaz et al. 2018; Moosavi et al. 2015). Spectral analysis is used for realizing the electromagnetic torque and the induced voltage in terms of frequency. Two-dimensional fast Fourier transform (FFT) is applied on the space frequency obtained from FEM simulations using ANSYS Maxwell Software. The proposed model is denoted that all the simulations are realized by FEM, and the diagnosis of the fault is performed for both healthy and faulty machines.

This work is organized as follows. After presenting the introduction in Sect. 10.1, we derived an analytical model for the electromagnetic torque, the back EMF induced, and the flux density in a permanent magnet (PM) due to the demagnetization fault of PMSM in Sect. 10.2. Then, the proposed FE model is described in Sect. 10.3. Section 10.4 explains the model of the demagnetization fault of the machine, and simulation results of the proposed model are shown and analyzed, as well as the electromagnetic characteristics of the PM model of healthy and faulty conditions are illustrated. In Sect. 10.5, the proposed fault detection technique (fast Fourier transform) is analyzed. Finally, some concluding remarks are drawn in Sect. 10.6.

10.2 Analytical Model

In this section, an analytical model is developed to exploit the simulation results of FEM. Figure 10.1 shows the simplified 2D model. The aim is pursed by evaluating the effect of the demagnetization fault on the induced voltage, the electromagnetic torque, and the flux density (Shin et al. 2017; Kim et al. 2014).

The governing equation can be obtained using the Maxwell's equation (Vaseghi 2009). In the PM region,

$$B = \mu_0(H + M) \tag{10.1}$$

$$\nabla \times H = 0 \tag{10.2}$$

Consequently,

$$\nabla \times B = \mu_0(\nabla + M) \tag{10.3}$$

The magnetic vector potential A is defined as

$$\nabla \times A = B \tag{10.4}$$

$$\nabla^2 A = -\mu_0\mu_r \nabla \times H - \mu_0 \nabla \times M = -\mu_0\mu_r J - \mu_0 \nabla \times M \tag{10.5}$$

The governing equation in the regions of the rotor core (region I), air gap (region II), and stator core (region IV) can be expressed by Laplace's equation in (10.2). Besides, the governing equation of the PM (region III) is represented by Poisson's equation in (10.3).

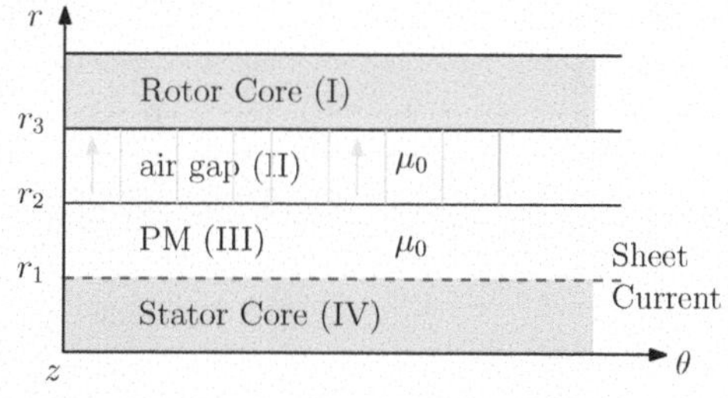

Fig. 10.1 Simplified 2D model

$$\nabla^2 A^{I,II,IV} = 0 \tag{10.6}$$

$$\nabla^2 A^{III} = -\mu_0(\nabla \times M) \tag{10.7}$$

where M is the magnetization of PMs, J is the current density, and μ_0 is the permeability of the air.

In the permanent magnet region, the vector potential only has z-components and its solution is derived by considering the solution of Poisson's equations (Shin et al. 2017):

$$\begin{aligned} A_{zn}{}^{III} &= A_0^{III} + B_0^{III} Ln\,(r) \\ &+ \sum_{n=1}^{\infty} \left[\left(A_n^{III} r^{-n} + B_n^{III} r^n + \frac{r\mu_0 n}{(n^2-1)} M_n \cos n\theta_0 \right) \sin n\theta \right. \\ &+ \left. \left(C_n^{III} r^{-n} + D_n^{III} r^n - \frac{r\mu_0 n}{(n^2-1)} M_n \sin n\theta_0 \right) \cos n\theta \right] i_z \end{aligned} \tag{10.8}$$

$$\begin{aligned} A_{zn}{}^{II} &= A_0^{II} + B_0^{II} Ln\,(r) \\ &+ \sum_{n=1}^{\infty} [(A_n^{II} r^{-n} + B_n^{II} r^n) \sin n\theta + (C_n^{II} r^{-n} + D_n^{II} r^n) \cos n\theta] i_z \end{aligned} \tag{10.9}$$

where M_n was defined in (10.10), and $M_r n$ and $M_\theta n$ are the radial and tangential components of the magnetization, respectively.

$$M_n = M_{rn} - \frac{1}{n} M_{\theta n} \tag{10.10}$$

The boundary conditions of the PMSM to be satisfied by the magnetic field components in regions III are

$$R = r_1 \;:\; B_{\theta n}^{III} = \mu_0 M_{\theta n} \tag{10.11}$$

$$R = r_2 \;:\; B_{\theta n}^{III} - \mu_0 M_{\theta n} = \mu_0 B_{\theta n}^{II} \tag{10.12}$$

By (10.3), the flux density of the normal and tangential components are deduced from the potential vector by

$$B_r = \frac{1}{r}\frac{\partial A}{\partial \theta} i_r \tag{10.13}$$

$$B_\theta = -\frac{\partial A}{\partial r} i_\theta \tag{10.14}$$

10.2.1 Electromagnetic Torque Expression

The electromagnetic torque expression is calculated using the Maxwell's stress tensor and it is expressed as follows:

$$C_e = \frac{L\,R^2}{\mu_0}\int_0^{2\pi} B_{rg}\,B_{\theta g}\,d\theta = \frac{L\,R^2\,\pi}{\mu_0}\sum_{n=1}^{\infty}(B_{rcn}\,B_{\theta cn} + B_{rsn}\,B_{\theta sn}) \quad (10.15)$$

where L is the length of the machine, R is the radius of the integration surface, and B_{rcn}, $B_{\theta cn}$, B_{rsn}, and $B_{\theta sn}$ are the radial and tangential components of the flux density at radius R.

$$\begin{aligned} B_{rcn} &= n\left(A_n^{II}\,r^{-n-1} + B_n^{II}\,r^{n-1}\right)\\ B_{\theta cn} &= n\left(C_n^{II}\,r^{-n-1} - D_n^{II}\,r^{n-1}\right)\\ B_{rsn} &= n\left(C_n^{II}\,r^{-n-1} + D_n^{II}\,r^{n-1}\right)\\ B_{\theta sn} &= n\left(A_n^{II}\,r^{-n-1} - B_n^{II}\,r^{n-1}\right) \end{aligned} \quad (10.16)$$

10.2.2 Back Electromotive Force (EMF) Expression

In order to compute the back EMF of a PMSM, we first determine the flux linking the stator coil of one-phase winding due to the PM:

$$\lambda_{PM} = N\,\phi = N\,R\,L\int_0^{2\pi} B_{rn}\,(\theta, t)\;d\theta \quad (10.17)$$

where N is the number of turns per coil of one-phase winding, and L and R are the coil pitch angle and the length of the motor and the radius of the integration surface, respectively. Therefore, the total fluxes ψ_{PM} linking all the coils of a phase are calculated as follows:

$$\psi_{PM} = N_c\,K_{dn}\,\lambda_{PM} \quad (10.18)$$

where N_c is the number of coils of one-phase winding and K_{dn} is a distribution factor. Using (10.17) and (10.18), the expression of the induced back EMF per phase can be derived as follows:

$$e = -\frac{d\psi_{PM}}{dt} = \omega_r\sum_{n=1}^{\infty} K_n\,\sin\,(n\,p\,\omega_r\,t) \quad (10.19)$$

where ω_r is the rotating speed of the rotor.

The analytical magnetic flux density distributions have been used to calculate the important quantities, such as electromagnetic torque and the back EMF induced. We derived the analytical solution for magnetic flux distribution in order to show the effects of demagnetization fault in permanent magnet on back EMF and electromagnetic torque waveforms. The preferred validation analytical model is provided by a FE model of a PM model employing to study the effect of the motor performance on the demagnetization fault. In this case, the analytical model is carried out based on FE solutions as described in Sect. 10.3.

10.3 Finite Element Model of PMSM

The finite element method is a computer-based numerical technique for calculating the parameters of electromagnetic devices. It can be used to calculate the flux density, the flux linkages, the inductance, the torque, the induced EMF, etc. A FEM, which addresses dynamic material properties of the magnets and yet can be implemented readily in the software, is presented for finding the parameters of the magnetization circuit. It offers unlimited flexibility in the geometrical shape, the material properties, and the boundary conditions in different regions of the machine (Ebrahimi et al. 2008). Also, it provides detailed information about the machine nonlinear effects (based on its geometry and material properties). This modeling approach is able to obtain an accurate and a complete description of an electrical machine. The magnetic circuit is modeled by a mesh of small elements. Their field values are assumed to be a simple function of position within these elements, enabling interpolation of results (Prasad and Ram 2013).

For the design of PMSM, the toolbox RMxprt can be used. This tool allows its user to design predefined motor patterns. As mentioned above, the simulation time is one of the core problems of the co-simulation. Via RMxprt and motor design (winding scheme/slot to pole ratio, etc.), the PMSM can be shared into its symmetry parts. This means, instead of doing the simulation of the complete motor, only one of its symmetric components is simulated and then the complete motor behavior can be extrapolated.

Figure 10.2 illustrates the RMxprt model of the PMSM model.

When the RMxprt model was analyzed, it was exported to the 2D model using the ANSYS Maxwell, and it only computes solution for only one of the symmetries. In this case, the machine has been divided into four equal parts and only one part is represented on 2D model as shown in Fig. 10.2.

The geometrical of 2-D model and the mesh division is depicted in Fig. 10.3.

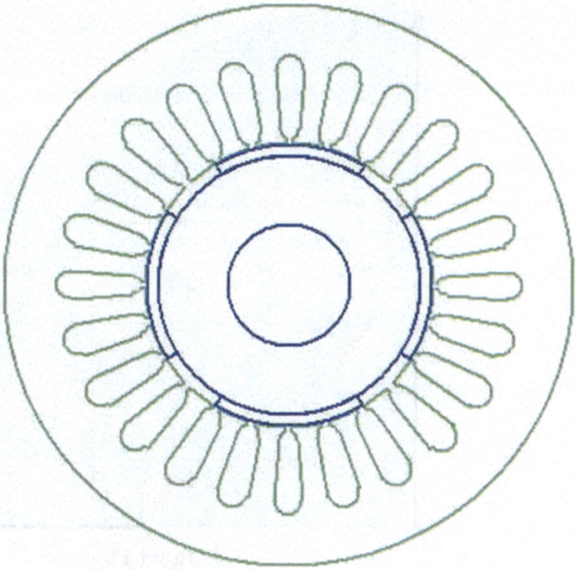

Fig. 10.2 RMxprt model of the PMSM machine

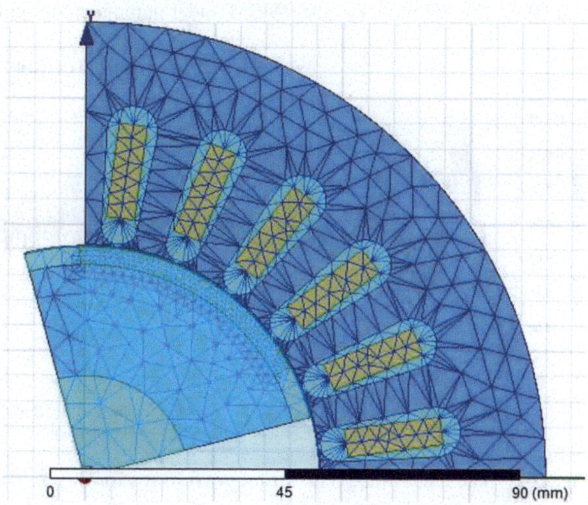

Fig. 10.3 Finite Element Model of PMSM and the mesh division

10.4 Demagnetization Fault Modeling of PMSM

For the study of demagnetization fault related to shape design, the effect of PM dimension on motor performance is studied. The effect of the pole embrace and the magnet thickness is investigated by the FEM. By changing the values of these parameters, the geometry of the rotor of the studied machine can be varied (Kim et al. 2005, 2006; Kudrjavtsev et al. 2017). After validation of PM model, the analysis of the imposed parameters on motor performance has been conducted.

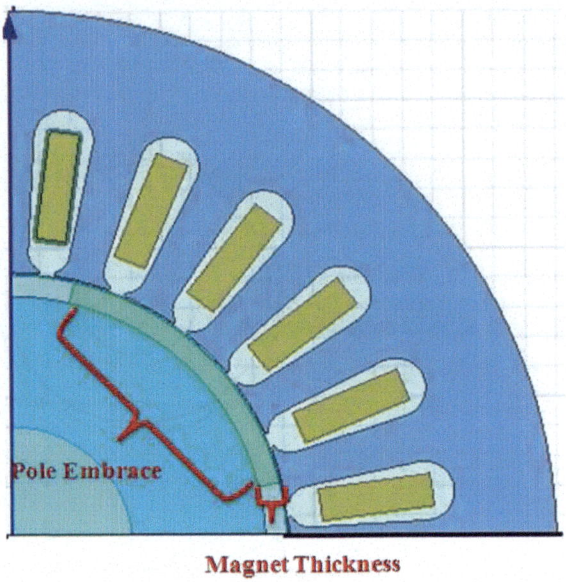

Fig. 10.4 The 2D model of the PMSM under demagnetization fault

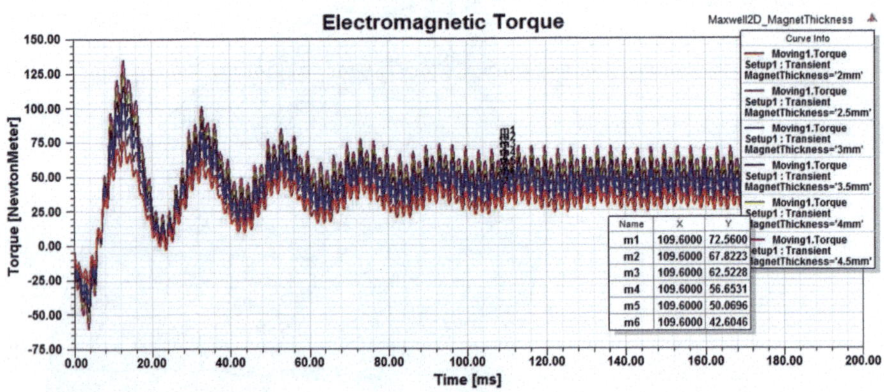

Fig. 10.5 The electromagnetic torque waveforms for different values of magnet thickness

In the developed parametrical model of the studied machine, the PM dimensions as well as the magnet thickness and the pole embrace can be modified by the introduced number of parameters (Fig. 10.4), particularly,

- the value of PM pole embrace is varied from 0.3 to 0.9.
- the value of magnet thickness is varied 2–4.5 mm.

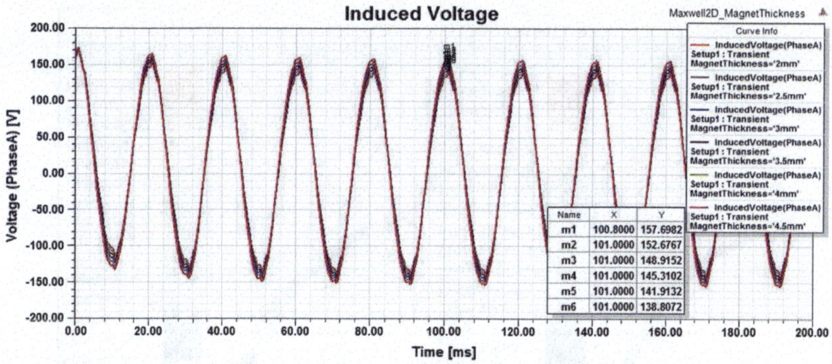

Fig. 10.6 The induced voltage for different values of magnet thickness

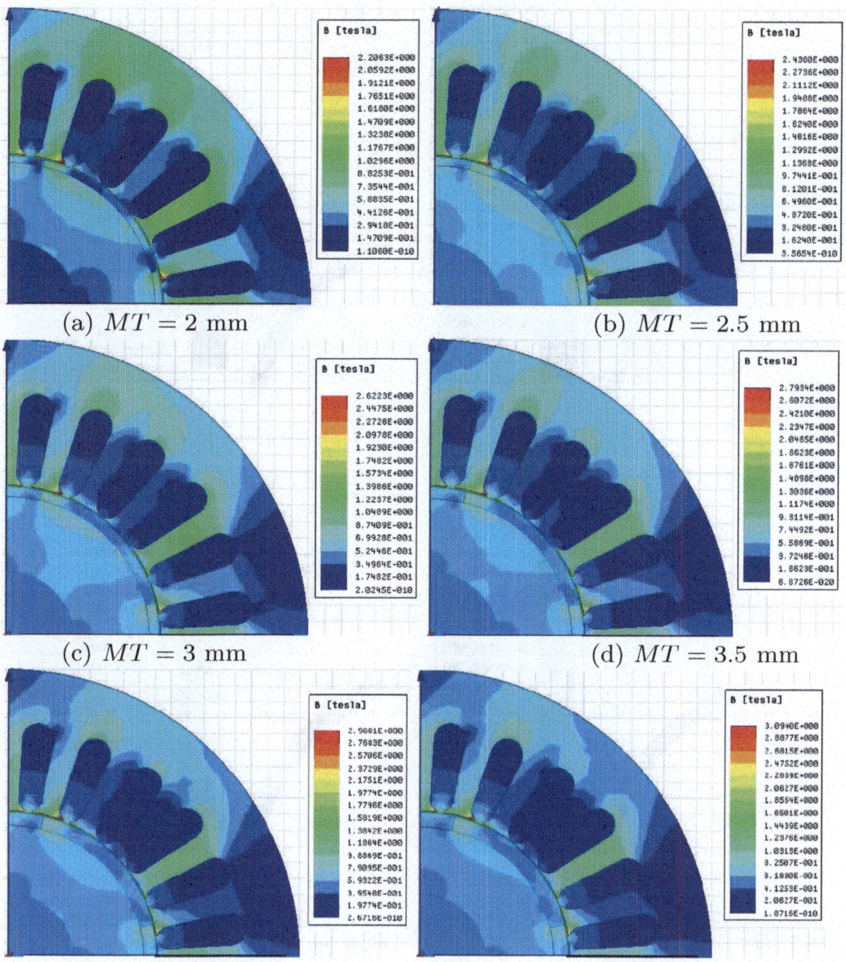

(a) $MT = 2$ mm (b) $MT = 2.5$ mm

(c) $MT = 3$ mm (d) $MT = 3.5$ mm

(e) $MT = 4$ mm (f) $MT = 4.5$ mm

Fig. 10.7 Magnetic flux density distribution of PMSM for different values of magnet thickness

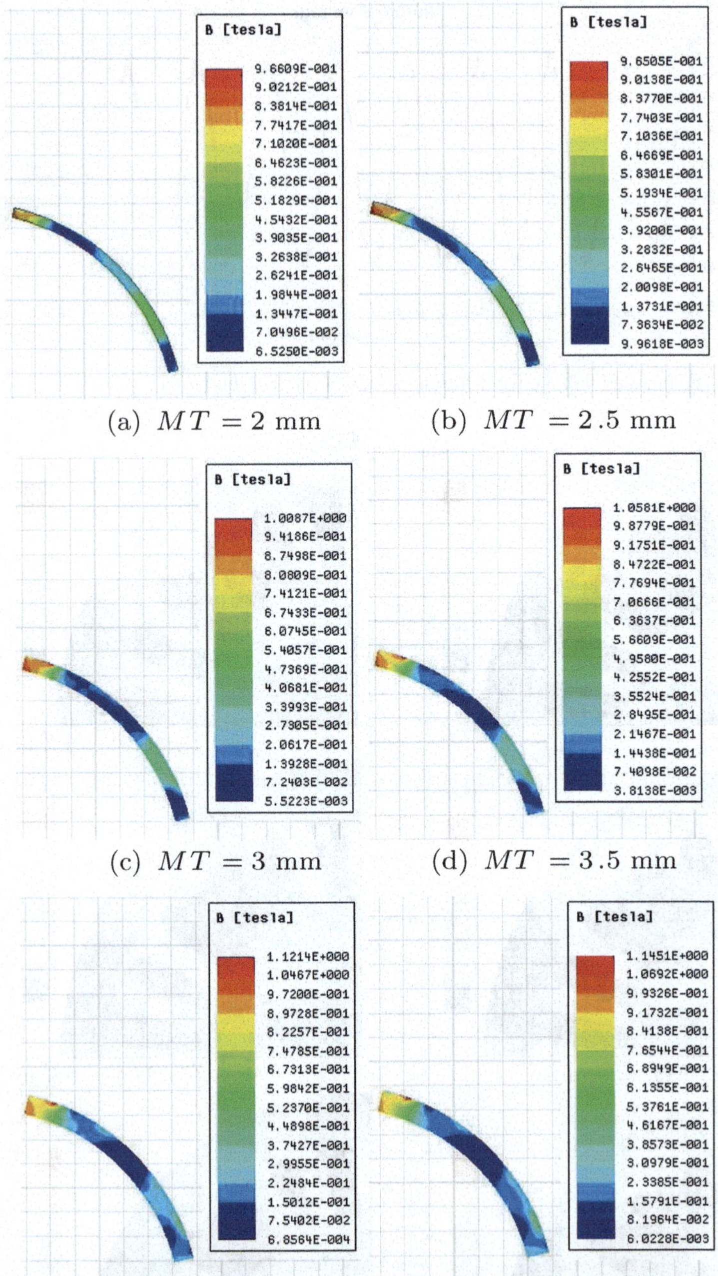

(a) $MT = 2$ mm (b) $MT = 2.5$ mm

(c) $MT = 3$ mm (d) $MT = 3.5$ mm

(e) $MT = 4$ mm (f) $MT = 4.5$ mm

Fig. 10.8 The distribution of flux density in PM for different values of magnet thickness

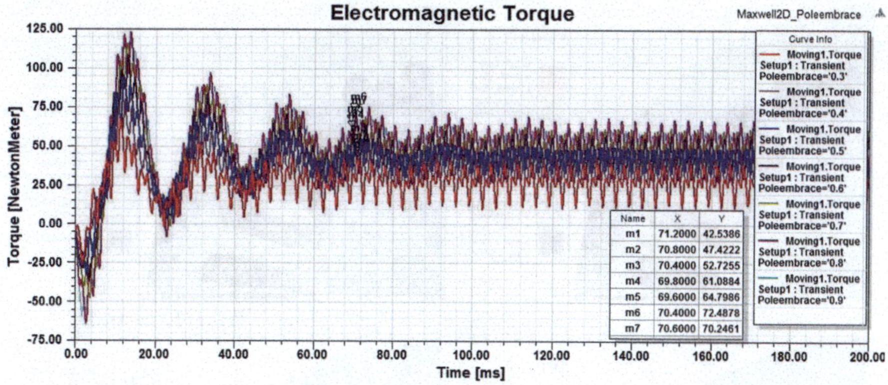

Fig. 10.9 The electromagnetic torque waveforms for different values of pole embrace

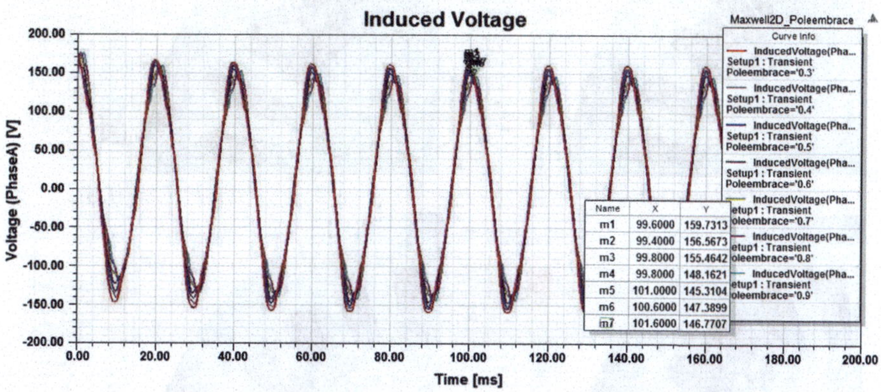

Fig. 10.10 The induced voltage waveforms for different values of pole embrace

By changing the values of the introduced parameters, the geometry of the permanent magnet of the studied machine can be varied. After validation of the model, the analysis of the influence of the imposed parameters on PMSM performance has been conducted.

Using FEA software, rotor magnet thickness was set to constant value from 3.5 mm value while the rotor pole embrace was varied from 0.3 to 0.9 values with step 0.1 value.

The electromagnetic torque and the induced voltage have been analyzed for different values of a PM magnet thickness.

The electromagnetic torque waveforms and the induced voltage were computed by FEM, for several magnet thicknesses (Figs. 10.5 and 10.6). When studying the results presented above, it can be seen that the enlargement of magnet thickness has a significant impact on the performances of the machine. For higher values of magnet thickness, the torque waveforms increases in amplitude; however, the induced voltage decreases.

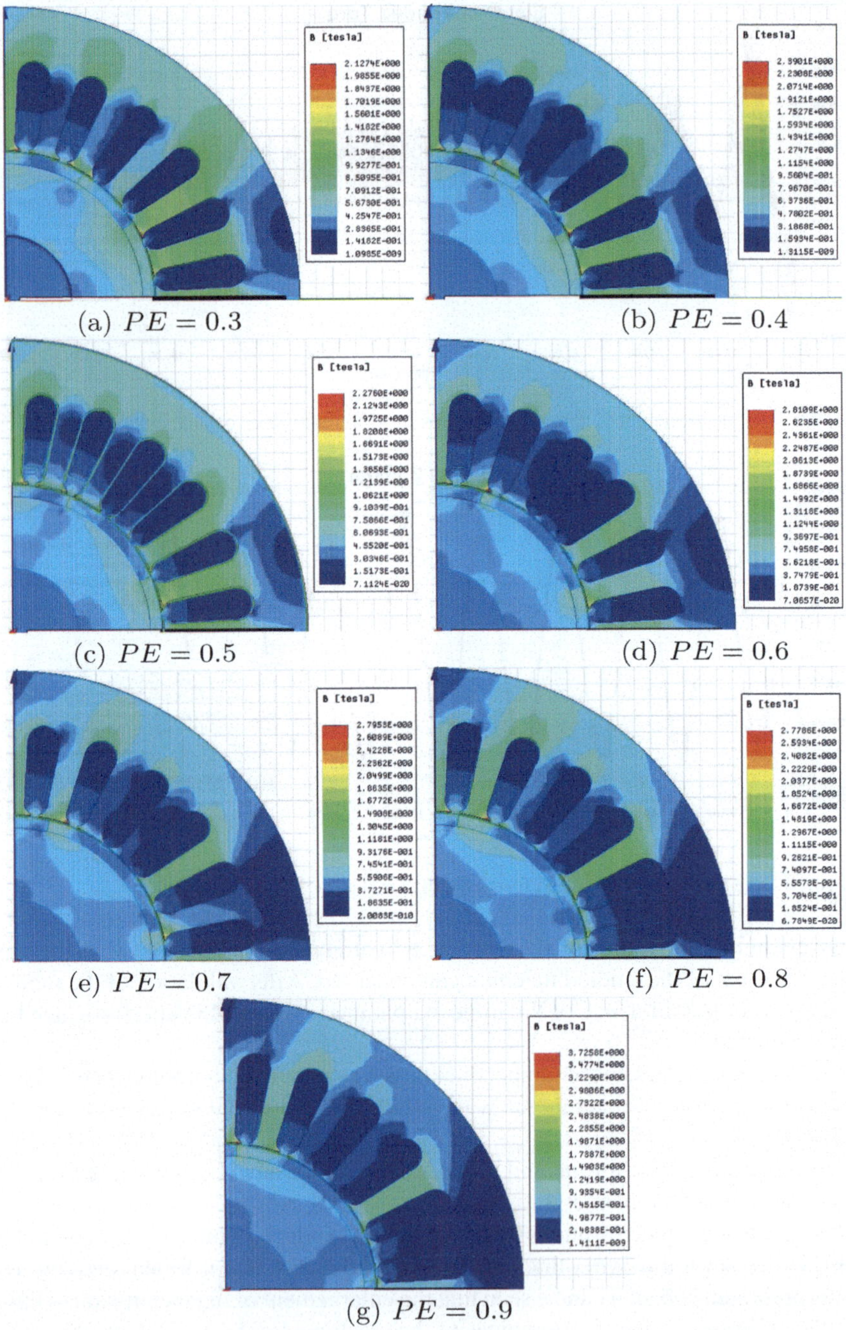

Fig. 10.11 Magnetic flux density distribution of PMSM for different values of pole embrace

Fig. 10.12 The distribution of flux density in PM for different values of pole embrace

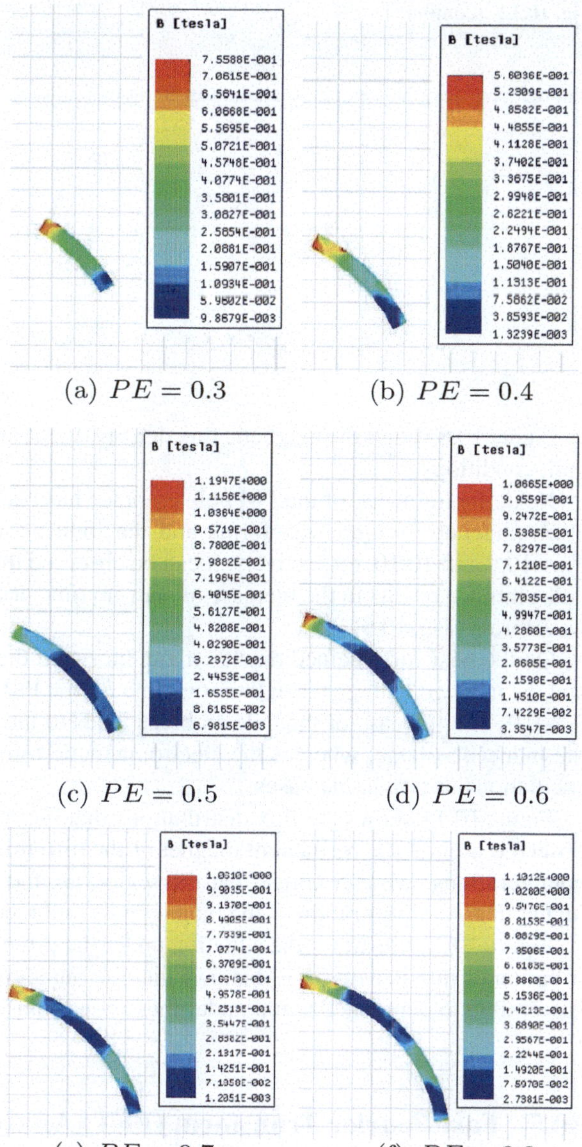

(a) $PE = 0.3$ (b) $PE = 0.4$

(c) $PE = 0.5$ (d) $PE = 0.6$

(e) $PE = 0.7$ (f) $PE = 0.8$

Figure 10.7 shows the magnetic flux density distribution in the PMSM under different fault conditions. It is shown that if the magnet thickness value is taken high, the magnetic flux distribution in permanent magnet shall be low and this causes the core to be saturated, which is undesirable.

Fig. 10.12 (continued)

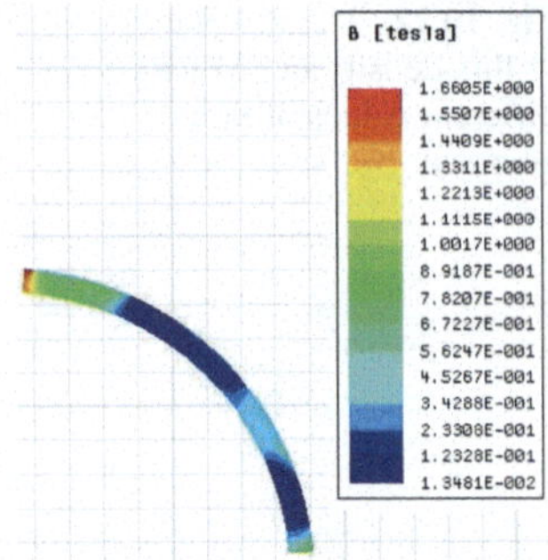

(g) $PE = 0.9$

Figure 10.8 shows the magnetic flux density distribution in the PM under different fault conditions.

Next, the influence of the PM pole embrace has been examined. The determined electromagnetic torque waveforms and the induced voltage have been shown in Figs. 10.9 and 10.10, respectively. It can be observed that the enlargement of the PM pole embrace results in the increase of start-up time and oscillations from the torque waveforms (Fig. 10.9).

On the basis of obtained results, it can be stated that the electromagnetic torque and the induced voltage reach their maxima for the PM pole embrace in the range of 0.3–0.9. The increase of the pole embrace leads to the decrease of the amplitude of the induced voltage, whereas the electromagnetic torque increases and the time of oscillations of torque increases.

Figure 10.11 shows the flux distributions that the magnetic flux concentration is observed around the permanent magnet of the different analyzed PM pole embrace configurations. We can notice that the region around the PM demagnetized has a higher degree of saturation as the magnet pole embrace.

Figure 10.12 illustrates the situation under the different fault conditions. The distribution of flux density in PM shows that the outer corners of the magnets may be susceptible to demagnetization under these conditions.

10.5 Fast Fourier Transform (FFT) Analysis

More frequently, the fault detection in PMSM has been studied by analyzing the stator-induced voltage harmonics by the mean of the well-known fast Fourier transform. This analysis has been extended to the PMSM, in which frequency analysis of a signal highlights many important hidden features and extracts some useful information.

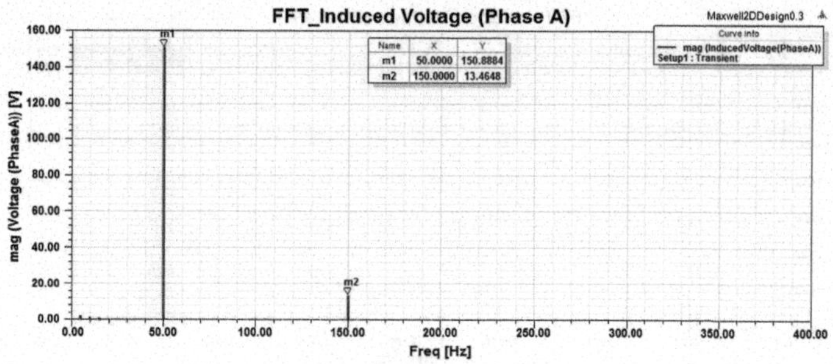

(a) $PE = 0.3$

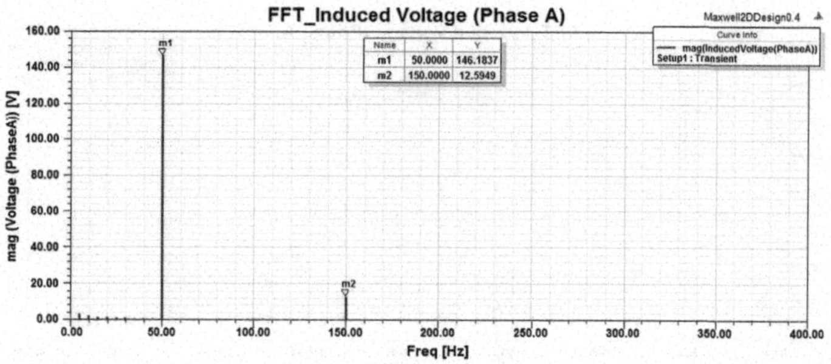

(b) $PE = 0.4$

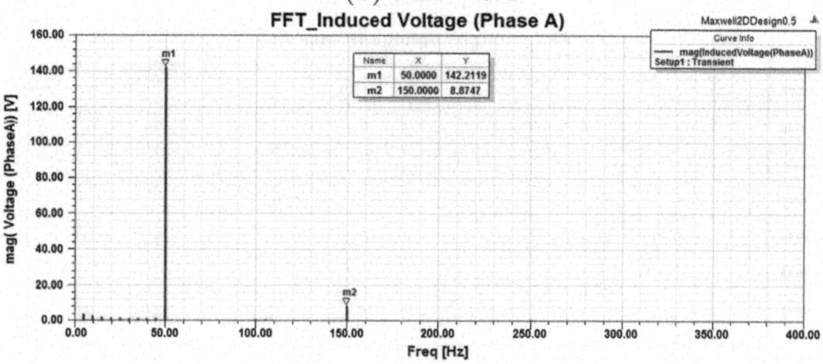

(c) $PE = 0.5$

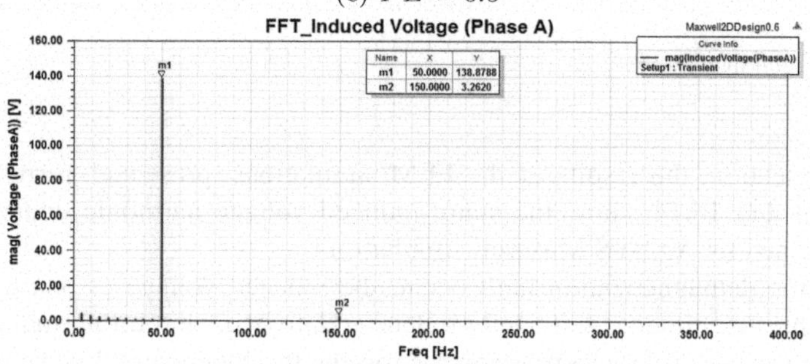

(d) $PE = 0.6$

Fig. 10.13 FFT of induced voltage (Phase A) for different values of pole embrace

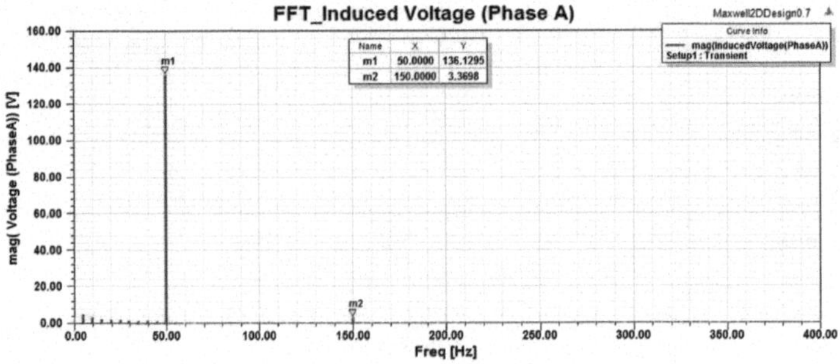

(e) $PE = 0.7$

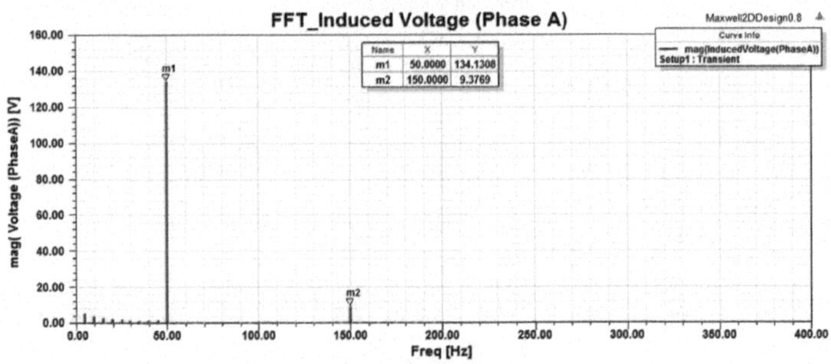

(f) $PE = 0.8$

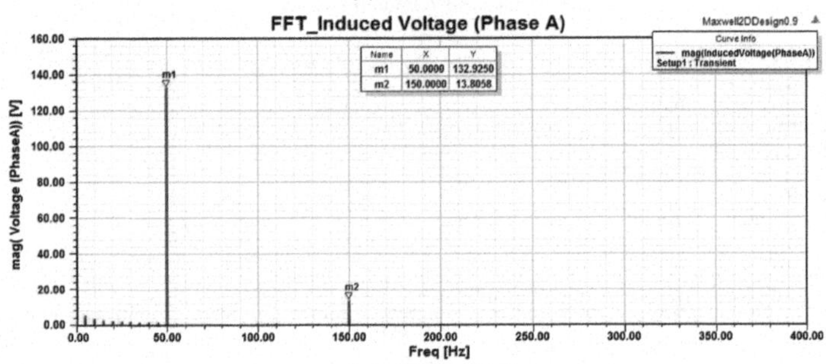

(g) $PE = 0.9$

Fig. 10.13 (continued)

In this section, the results of the FEM simulations, presented, respectively, in Figs. 10.13 and 10.14, show the stator-induced voltage harmonic under different faults obtained by ANSYS Maxwell simulation.

When the demagnetization fault occur, the induced voltage of phase A fundamental wave magnitude significantly reduces relative to the rotor normal condition. With the increase of the electromagnetic torque, the fundamental wave magnitude gets smaller and smaller.

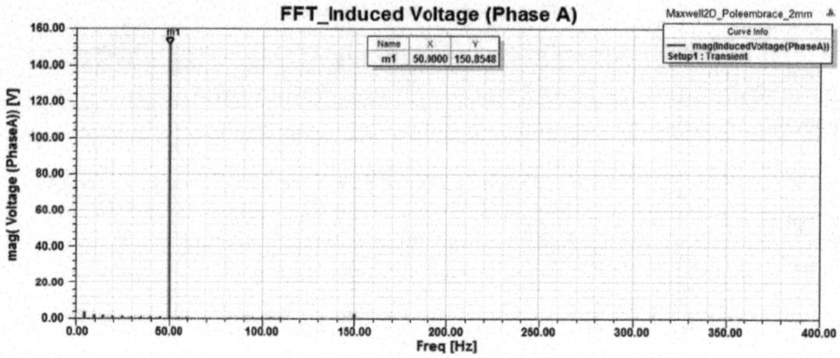

(a) $MT = 2\text{mm}$

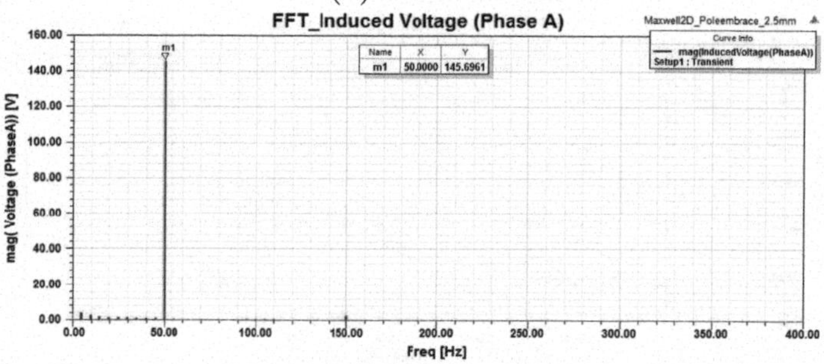

(b) $MT = 2.5\text{mm}$

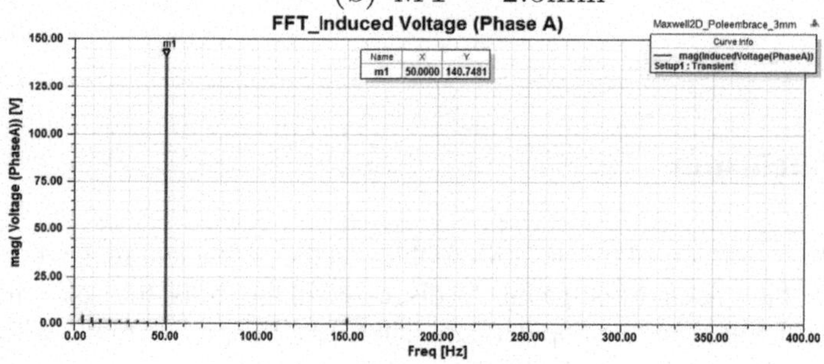

(c) $MT = 3\text{mm}$

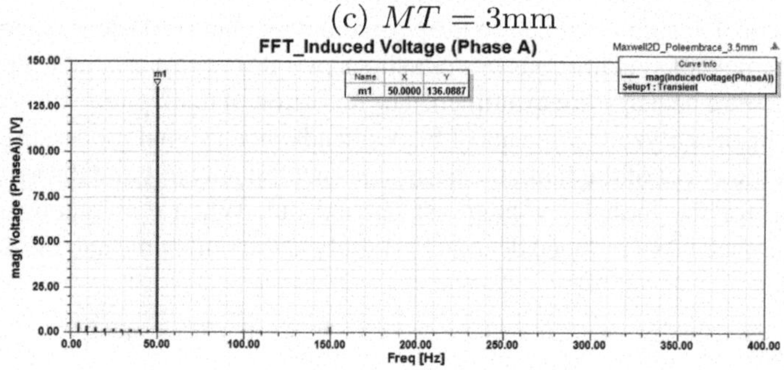

(d) $MT = 3.5\text{mm}$

Fig. 10.14 FFT of induced voltage (Phase A) for different values of magnet thickness

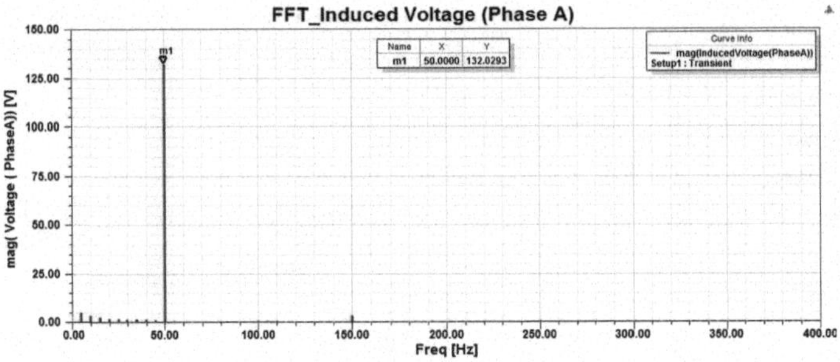

(e) $MT = 4\mathrm{mm}$

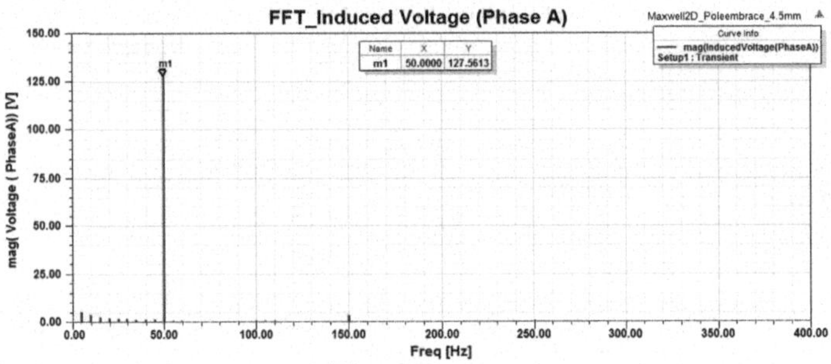

(f) $MT = 4.5\mathrm{mm}$

Fig. 10.14 (continued)

10.6 Conclusion

In this study, the effects of the design variables on the permanent magnet to PMSM's performance have been investigated. As discussed in the study, the pole embrace and the magnet thickness are the popular parameters as a design parameter which effects machine overall performance under demagnetization fault.

The effects of design variables on the permanent magnet (PM) demagnetization of a permanent magnet synchronous motor using a two-dimensional (2D) finite element analysis software package are examined. The detection of demagnetization fault for a PMSM has been analyzed by means of FEA and advanced signal processing. In order to highlight the effect of the fault, the harmonic spectral analysis and the symmetrical components for the induced voltage are carried out. When demagnetization fault happens, the back EMF induced decreases slightly at the same time and the harmonics increase.

Acknowledgements This work was supported by the Ministry of the Higher Education and Scientific Research in Tunisia.

Appendix

See Table 10.1.

Table 10.1 General parameters

Permanent magnet synchronous motor	
Parameters	Values
Number of poles	4
Out put power	15 kW
Rated voltage	127 V
Speed	1500 rpm
Frequency	50 Hz
Rotor inner diameter	40 mm
Length of rotor	101 mm
Stator outer diameter	180 mm
Stator inner diameter	91 mm
Length of stator core	101 mm

References

Ahmed Farooq, J., Djerdir, A., & Miraoui, A. (2009). Identification of demagnetization faults in a permanent magnet synchronous machine by permeance network. *COMPEL-The International Journal for Computation and Mathematics in Electrical and Electronic Engineering*, *28*(6), 1619–1631.

Apostoaia, C. M. (2014). Multi-domain system models integration for faults detection in induction motor drives. In *2014 IEEE International Conference on Electro/Information Technology (EIT)* (pp. 388–393). IEEE.

Bianchi, N., Jahns, T. M., et al. (2004). *Design Analysis and Control of Interior PM Synchronous Machines* (p. 4). Tutorial Course Notes: IEEE-IAS.

Choi, G. & Jahns, T. (2015). Post-demagnetization characteristics of permanent magnet synchronous machines. In *Energy Conversion Congress and Exposition (ECCE), 2015 IEEE* (pp. 1781–1788). IEEE.

Diaz, D., Fernandez, D., Park, Y., Diez, A. B., Lee, S. B., & Briz, F. (2018). Detection of demagnetization in permanent magnet synchronous machines using hall-effect sensors. *IEEE Transactions on Industry Applications*.

Ebrahimi, B. M., Faiz, J., Javan-Roshtkhari, M., & Nejhad, A. Z. (2008). Static eccentricity fault diagnosis in permanent magnet synchronous motor using time stepping finite element method. *IEEE Transactions on Magnetics*, *44*(11), 4297–4300.

Faiz, J., & Mazaheri-Tehrani, E. (2017). Demagnetization modeling and fault diagnosing techniques in permanent magnet machines under stationary and nonstationary conditions: An overview. *IEEE Transactions on Industry Applications*, *53*(3), 2772–2785.

Farooq, J., Djerdir, A., & Miraoui, A. (2008). Analytical modeling approach to detect magnet defects in permanent-magnet brushless motors. *IEEE Transactions on Magnetics*, *44*(12), 4599–4604.

Fitouri, M., BenSalem, Y., & Abdelkrim, M. N. (2016). Analysis and co-simulation of permanent magnet sychronous motor with short-circuit fault by finite element method. In *2016 13th International Multi-Conference on Systems, Signals & Devices (SSD)* (pp. 472–477). IEEE.

Kang, K., Song, J., Kang, C., Sung, S., & Jang, G. (2017). Real-time detection of the dynamic eccentricity in permanent-magnet synchronous motors by monitoring speed and back EMF induced in an additional winding. *IEEE Transactions on Industrial Electronics*, *64*(9), 7191–7200.

Kim, T. H., Choi, S.-K., Ree, C.-J., & Lee, J. (2005). Effect of design variables on irreversible permanent magnet demagnetization in flux-reversal machine. In *2005 International Conference on Electrical Machines and Systems* (Vol. 1, pp. 258–260). IEEE.

Kim, K.-C., Lim, S.-B., Koo, D.-H., & Lee, J. (2006). The shape design of permanent magnet for permanent magnet synchronous motor considering partial demagnetization. *IEEE Transactions on Magnetics*, *42*(10), 3485–3487.

Kim, J.-M., Choi, J.-Y., Lee, S.-H., & Jang, S.-M. (2014). Characteristic analysis and experiment of surface-mounted type variable-flux machines considering magnetization/demagnetization based on electromagnetic transfer relations. *IEEE Transactions on Magnetics*, *50*(11), 1–4.

Kim, K.-S., Kim, K.-S., Lee, B.-H., & Lee, B.-H. (2017). Design of concentrated flux synchronous motor to prevent irreversible demagnetization. In *2017 IEEE International Electric Machines and Drives Conf. (IEMDC)* (pp. 1–6). IEEE.

Kudrjavtsev, O., Kallaste, A., Kilk, A., Vaimann, T., & Orlova, S. (2017). Influence of permanent magnet characteristic variability on the wind generator operation. *Latvian Journal of Physics and Technical Sciences*, *54*(1), 3–11.

Liang, P., Chai, F., Bi, Y., Pei, Y., & Cheng, S. (2016). Analytical model and design of spoke-type permanent-magnet machines accounting for saturation and nonlinearity of magnetic bridges. *Journal of Magnetism and Magnetic Materials*, *417*, 389–396.

Moosavi, S., Djerdir, A., Amirat, Y. A., & Khaburi, D. (2015). Demagnetization fault diagnosis in permanent magnet synchronous motors: A review of the state-of-the-art. *Journal of Magnetism and Magnetic Materials*, *391*, 203–212.

Pacas, M. (2017). Sensorless harmonic speed control and detection of bearing faults in repetitive mechanical systems. In *2017 IEEE 3rd International Future Energy Electronics Conference and ECCE Asia (IFEEC 2017-ECCE Asia)* (pp. 1646–1651). IEEE.

Prasad, B. J. C. & Ram, B. S. (2013). Inter-turn fault analysis of synchronous generator using finite element method (fem). *International Journal of Innovative Technology and Exploring Engineering (IJITEE)*, *3*(7), 170–176.

Ruiz, J.-R. R., Rosero, J. A., Espinosa, A. G., & Romeral, L. (2009). Detection of demagnetization faults in permanent-magnet synchronous motors under nonstationary conditions. *IEEE Transactions on Magnetics*, *45*(7), 2961–2969.

Shin, K.-H., Park, H.-I., Cho, H.-W., & Choi, J.-Y. (2017). Analytical calculation and experimental verification of cogging torque and optimal point in permanent magnet synchronous motors. *IEEE Transactions on Magnetics*, *53*(6), 1–4.

Urresty, J., Riba, J., Saavedra, H., & Romeral, J. (2011). Analysis of demagnetization faults in surface-mounted permanent magnet synchronous motors with symmetric windings. In *2011 IEEE International Symposium on Diagnostics for Electric Machines, Power Electronics & Drives (SDEMPED)* (pp. 240–245). IEEE.

Urresty, J.-C., Riba, J.-R., Delgado, M., & Romeral, L. (2012). Detection of demagnetization faults in surface-mounted permanent magnet synchronous motors by means of the zero-sequence voltage component. *IEEE Transactions on Energy Conversion*, *27*(1), 42–51.

Vaseghi, B. (2009). *Contribution à l'Etude Des Machines Electriques En Présence de Défaut Entre-Spires Modélisation–Réduction du courant de défaut*. PhD thesis, Institut National Polytechnique de Lorraine-INPL.

Chapter 11
Extended Kalman Filtering for Remaining Useful Lifetime Prediction of a Pipeline in a Two-Tank System

M. H. Moulahi and F. Ben Hmida

Abstract In the present chapter, Wiener processes for degradation are used to model faults for two-tank system (clogging or partial blockage in section pipeline). Wiener processes for degradation modeling are appropriate for the case that the degradation processes vary bidirectionally over time with Gaussian noises. In our study, it is assumed that the fault localization is known and remains to analyze the evolution of the blocking particles in the pipeline. Wax can form a solid deposit at the pipe wall, reducing the pipe radius and this phenomenon is also taken into account. The flow through a pipeline is greatly influenced by the formation of a solid deposit, whose evolution depends on the quantity of wax in the fluid. Such phenomenon is important also because, under some circumstances, the deposit can completely obstruct the pipeline stopping the flow. Predicting remaining useful life (RUL) of partial blockages pipeline is necessary to modeling the effect of major preventive maintenance actions within the prediction horizon. In the framework to estimate the non-observable system state, the stochastic filtering approaches are frequently used. The extended Kalman filter (EKF) algorithm is used to estimate the state vector based on all collected measures history; therefore, these methods can give reliable performances for the degradation state estimation. The real data set is not available; we apply Monte Carlo simulation to our predefined Wiener parameters to create the true degradation process.

Keywords Degradation · Clogging · Wiener process · Diagnosis · Prognosis · EKF · RUL estimation

---

M. H. Moulahi (✉) · F. B. Hmida
Laboratory LISIER, ENSIT, Université de Tunis, Tunis, Tunisia
e-mail: mohamedhedia.moulahi@isetn.rnu.tn

F. B. Hmida
e-mail: faycal.benhmida@ensit.rnu.tn

N. Derbel et al. (eds.), *Diagnosis, Fault Detection & Tolerant Control*,
Studies in Systems, Decision and Control 269,
https://doi.org/10.1007/978-981-15-1746-4_11

11.1 Introduction

One of the most serious problems found in subsea pipelines during the flow of crude oil is a major problem in petroleum engineering. The fluid mixture produced from a reservoir is called crud and consists of several hydrocarbon components which can be divided into two main groups, light and heavy hydrocarbon in chemical system. Pipeline is the most efficient way to convey fluids, where it is widely used in transportation of oil, natural gas, industrial plants networks, and water distribution networks. They are required to resist the damaging effects of the surrounding environment, and the potential damage which the material being conveyed might cause. Pipeline managers and operators frequently encounter the need for maintenance actions to clean the pipeline sections. The discussion will concentrate mainly on cross-sectional clogging in the pipeline. Based on the pipeline's circumferential strength, a threshold level is defined a degradation process model will then be developed. The reduction in resistance is assumed to be related directly to the growth of a thickness of wax layer. Therefore, prediction of wax growth and targeted inspection and maintenance actions are required in order to safely manage the operational life of pipeline systems. Predicting remaining useful life is necessary to modeling the effect of major preventive maintenance actions within the prediction horizon. State estimation techniques as stochastic filtering can give reliable performances for the degradation states estimation. The extended Kalman filter requires a model of the pipeline to estimate states or parameters. However, the model must be represented in a state-space form, with the present state being only a function of prior states and inputs. This was required for the implementation of the extended Kalman filter. In literature review, (Ribeiro et al. 1997; Bazán and Beck 2013; Mendes et al. 1999) offer a glimpse of the real deterioration phenomenon in a structural systems. However, the problems were not well understood until recently. In a pioneer study about this problem, (Mendes and Braga 1996; Desmond 1985) worked with the hypothesis that the mechanisms of deposition are molecular diffusion, Brownian diffusion, and shear dispersion. Other researchers followed similar analyses. The presentation is organized as follows. In Sect. 11.2, stochastic degradation modeling in two-tank systems is described with materials and words used in control theory for petrochemical system modeling. In Sect. 11.3, the extended Kalman filter is applied to generate the residual for detection and diagnosis. However, we use Wiener process for degradation modeling and state estimation. As the deterioration states are hidden, we use the maximum likelihood estimation (MLE) to estimate the unknown model parameters. Section 11.4 describes the modeling methodology and develops the first hitting time and RUL distribution model for prognostic. Section 11.5 provides results from simulation-based illustrative example and evaluates the approach. Finally, conclusion drawn from this work is given.

11.2 Stochastic Degradation Modeling in Two-Tank System

In this paper, we address the above issues by utilizing a Wiener-process-based model with extended Kalman filter (EKF) for remaining useful lifetime estimation (RUL). We use a stochastic model for the updating of the RUL estimation. The stochastic filtering approaches are frequently used to recursively update the drift coefficient in the stochastic state-space model representation. In the engineering sciences, it is well recognized that degradation process is uncertain over time. However, the stochastic models are frequently used to describe the evolution of deterioration (Desmond 1985; Si et al. 2013; Lu and Meeker 1993). Once the degradation information of the pipeline system is available by the degradation information, the studies on age- and state-dependent degradation modeling are considered. The RUL can be estimated by using the training data set between the observing time and the failure time. It is an essential part in prognostic; a predictive maintenance is proposed to prevent system's failure. The RUL can be considered as a random variable; it is defined as the time to cross the failure threshold of the monitored degradation data for the first time; we shall denote by the first hitting time (FHT). Pipeline system is considered as a principal component in petrochemical industries. However, a pipeline reliability assessment is described in this work. The life prognostic is vital in their domain since their availability has crucial consequences. The main cause of failure of these systems is the fatigue due to internal pressure-depression variation along the time. Several researchers (Svendsen 1993; Mendes et al. 1999; Mendes and Braga 1996; Benallal 2008; Handal 2008) indicate that molecular diffusion is the best descriptive mechanism to the problem of deposition. Basically, the mathematical model for these degradation processes is based on the assumption of an additive accumulation of degradation with constant wear intensity. Regarding every degradation increment as an additive superposition of a large number of small effects, we can assume that the degradation increases linearly in time with random noise. Therefore, the degradation measure can be described by Wiener process (Kahle and Lehmann 2010). We can assume the degradation process to be a deterministic model (Paris and Erdogan 1963; Mendes et al. 1999). The phenomenon of degradation can be modeled in a probabilistic and a non-probabilistic context (neural network, artificial intelligence, etc…) in the work of Langeron et al. (2015). As a result of the work, we consider the stochastic failure model. Several authors consider that degradation in a stochastic context may be continuous or gradual (Nakagawa 2011; Matta et al. 2012). In the work of Abdel-Hameed (2014), the author proposed the homogeneous gamma process as a model for describing degradation over time.

11.2.1 A Two-Tank System with Damage Modeling

The plant in our example consists of two fluid tanks in cascade as shown schematically in Fig. 11.1a. In this application, the multiplicative fault model corresponds to a variation of the cross section (obstruction phenomenon) over the time of the

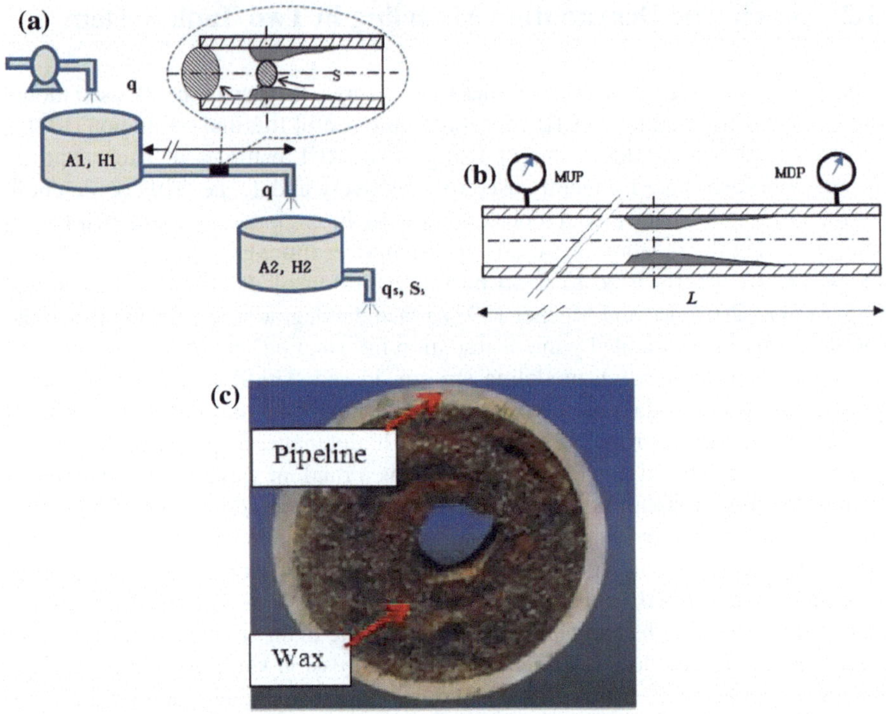

Fig. 11.1 Schematic of pipeline: **a** Showing the deposit profile, **b** control pressure losses, **c** Clogging inside the pipeline

pipe linking the two tanks. Our objective is to control the state of the pipe and to intervene before having the failure. Improving the reliability of the pipeline requires the estimation of RUL.

MUP: Measurement of upstream pressure.
MDP: Measurement of downstream pressure.

Many faults encountered in petrochemical industrial applications; these are

1. Clogging inside the pipeline and increased leakage flow. Two structural internal faults are caused by, respectively, impurities in the fluid and wear (Fig. 11.1c),
2. Performance degradation due to cavitation, and
3. Dry running.

To control pressure losses and obstruction inside the pipeline, we suppose two pressure sensors between the upstream and downstream tank for monitoring the evolution of wax over time, which are represented in Fig. 11.1b. The pipeline has an internal diameter of 0.5640 (m). The model assumes that no leakage occurs at the pipeline. The actual obstruction location is supposed determined and pressure losses are neglected. We suppose that the fluid level in the tank was assumed to remain

Table 11.1 Nomenclature

Parameter	Description
A_i	Tank cross-sectional area (i = 1, 2)
H_i	Fluid level (i = 1, 2)
p_0	Atmospheric pressure (Pa)
p_1	Pressure at the bottom (Pa)
rou	Fluid density (Kg/m^3)
g	Acceleration of gravity (m^2/s)
Z_0	Level fluid in the upper tank (m)
Z_1	Level fluid at the bottom tank (m)
V	Fluid velocity (m/s)
h	State level of fluid at time t in tank (m)
S_s, S_i	Pipeline area (m^2)
S_c	Pipeline area clogging (m^2)
q	Flow rate (m^3/s)

constant because the upstream tank was assumed to have a large cross-sectional area. The fluid is incompressible. The flow is permanent. The Torricelli and Bernoulli theorem are written between two points as follows (Table 11.1):

$$A\frac{dh\,(t)}{dt} = \sum \text{flow} \tag{11.1}$$

$$p_1 - p_0 = (\rho g Z_0 - \rho g Z_1) = \rho g h_1 \tag{11.2}$$

$$V = \sqrt{2gh} \tag{11.3}$$

The fluid flow at each pipeline outlet is given by

$$q = V\ S \tag{11.4}$$

The head and the flow rates of two-tank system outlets are computed with the following equations:

$$\frac{Adh_1}{dt} = S_i V_i - q_c, \quad \frac{Adh_2}{dt} = q_c - q_S \tag{11.5}$$

$$\frac{q_c}{S_c} = \sqrt{2gh_1}, \quad \frac{q_s}{S_s} = \sqrt{2gh_2} \tag{11.6}$$

With Eqs. (11.5) and (11.6), the mathematical model of the two-tank system is given by

$$\frac{dh_1}{dt} = \frac{S_i}{A}Vi - \frac{S_c}{A}\sqrt{2g}\,\sqrt{h_1} \tag{11.7}$$

$$\frac{dh_2}{dt} = \frac{S_c}{A}\sqrt{2g}\sqrt{h_1} - \frac{S_0}{A}\sqrt{2g}\sqrt{h_2} \tag{11.8}$$

Implementing Eqs. (11.7) and (11.8), we can rewrite the discredited model for a sampling period T_e as follows:

$$h_1(k+1) = h_1(k) + T_e\left[\frac{Si}{A}Vi(k) - \frac{Sc(k)}{A}\sqrt{2g}\sqrt{h_1}\right] \tag{11.9}$$

$$h_2(k+1) = h_2(k) + T_e\left[\frac{\sqrt{2g}Sc(k)}{A}\sqrt{h_1(k)} - \frac{\sqrt{2g}S_s}{A}\sqrt{h_2}\right] \tag{11.10}$$

The two-tank system is described by the nonlinear model. If the state $w(k)$ and $v(k)$ measurement noises are taken into account, the model is written in the simplified form state equation $X(k+1)$ and measurement equation $z(k)$.

$$X(k+1) = f_k(X_k, u_k, k) + v(k) \tag{11.11}$$

$$z(k) = HX(k) + w(k) \tag{11.12}$$

where $X_k = \begin{bmatrix} h_1(k) \\ h_2(k) \end{bmatrix}$, $H = [1 \quad 1]$ and $q_i(k) = k_u u_k$.

Where k_u is the amplification gain, and u_k is the control voltage pump. The nonlinear model may be described by the following state equations:

$$\mathrm{X}(\mathrm{k}+1) = \mathrm{X}(\mathrm{k}) + \mathrm{T_e D}(\mathrm{k})\sqrt{\mathrm{X}(\mathrm{k})} + \mathrm{T_e bu}(\mathrm{k}) + \mathrm{v}(\mathrm{k}) \tag{11.13}$$

$$\mathrm{z}(\mathrm{k}) = \mathrm{HX}(\mathrm{k}) + \mathrm{w}(\mathrm{k}) \tag{11.14}$$

where

$$\mathrm{D}(k) = \begin{bmatrix} \frac{-Sc(k)\sqrt{2g}}{A} & 0 \\ \frac{Sc(k)\sqrt{2g}}{A} & \frac{-S_s\sqrt{2g}}{A} \end{bmatrix}, \quad \mathrm{b} = \begin{bmatrix} \frac{1}{\mathrm{A}} \\ 0 \end{bmatrix}$$

11.3 Fault Detection and Diagnosis in Pipelines Using the EKF

The EKF is used to estimate pipeline structural fault, given a pipeline model and set of inputs. Since the equations that represent the conditions in the pipeline are nonlinear, the EKF is required. The extended Kalman filter must linearize the state equations around the most recent state estimate for each time step. The time-update equations are linearized around a posterior state estimates, and the measurement equations are linearized around a priori state estimates. The linearization process is done by using a

Taylor series approximation. The nonlinear model may be described by the following state equation (11.11) and measurement equation (11.12). Since the functions given by Eqs. (11.13) and (11.14) are nonlinear, the Taylor series approximation is used

$$f\left(X_k, u_k\right) \approx f\left(\widehat{X}_k, u_k\right) + dF_k\left(X_k - \widehat{X}_k\right) + dG_k\left(u_k - \overline{u}_k\right) \tag{11.15}$$

such as

$$dF_k = \frac{\partial f\left(X_k, \overline{u}_k\right)}{\partial X}\Big|_{(\widehat{X}_k, \overline{u}_k)} = \begin{bmatrix} 1 - \dfrac{\mathrm{TeSc}}{2\mathrm{A}}\dfrac{\sqrt{2\mathrm{g}}}{\sqrt{\mathrm{h}_1}} & 0 \\ \dfrac{\mathrm{TeSc}}{2\mathrm{A}}\dfrac{\sqrt{2\mathrm{g}}}{\sqrt{\mathrm{h}_1}} & 1 - \dfrac{\mathrm{TeS_s}}{2\mathrm{A}}\dfrac{\sqrt{2\mathrm{g}}}{\sqrt{\mathrm{h}_2}} \end{bmatrix}$$

$$dG_k = \frac{\partial f\left(\widehat{X}_k, u_k\right)}{\partial u}\Big|_{(\widehat{X}_k, \overline{u}_k)} = \begin{bmatrix} Te\dfrac{Si}{A} & 0 \end{bmatrix}$$

$$\mathrm{C_k} = f\left(\widehat{X}_k, \overline{u}\right) - dF_k\widehat{X}_k$$

The EKF equation in the presence of process noise and measurement noise can be obtained (Grewal and Andrews 2001).

The prediction step

$$\widehat{X}_{k|k-1} = f_k\left(\widehat{X}_{k|k-1}, u_k\right) + v_k \tag{11.16}$$

$$P_{k|k-1} = Q_{k|k-1} + dF_k P_{k-1|k-1} dF_k{}^T \tag{11.17}$$

The updating can be implemented iteratively as the following equations:

$$\widehat{X}_{k|k} = \widehat{X}_{k|k-1} + K_k\left[z_k - \left(H_k X_k + w_k\right)\right] \tag{11.18}$$

$$P_{k|k} = P_{k|k-1} - K_k H_k P_{k|k-1} \tag{11.19}$$

$$K_k = P_{k|k-1} H_k{}^T\left[H_k P_{k|k-1} H_k^T + R_k\right]^{-1} \tag{11.20}$$

The expressions of the residues are given by

$$r_1(k) = h_1(k) - h_{1e}(k) \tag{11.21}$$

$$r_2(k) = h_2(k) - h_{2e}(k) \tag{11.22}$$

The simulation data taken from Gomm et al. (1993) and Blanke et al. (2006) are given in Table 11.2.

The measurement and state noise are given by

$w = R\ \mathrm{rand}(2, n)$ and $R = \mathrm{diag}(0.01, 0.01)$.

$v = Q\ \mathrm{rand}(1, n)$ and $Q = \mathrm{diag}(0.002, 0.05, 0.0001)$.

Table 11.2 Physical parameters' system

Physical parameters	Description
$A_1 = A_12 = A = 16\,(\text{m}^2)$	Identical sections for the two tanks
$S_0 = 1/32\,(\text{m}^2)$	Tank outlet section N°2
$S_i = 1/4\,(\text{m}^2)$	Tank outlet section N°2
$g = 9,81\,(\text{m/s}^2)$	Acceleration of gravity
$T = 5\,(\text{s})$	Sampling period
$S_{max} = 1/4\,(\text{m}^2)$	Section of the pipe linking the two tanks
$N = 50$	Time (Year)
$H_1(1) = 0.5\,(\text{m})$	Level of the fluid in the first reservoir
$H_2(1) = 0.01\,(\text{m})$	Level of the fluid in the second reservoir
$q_i = 5\,(\text{m}^3/\text{s})$	Flow rate

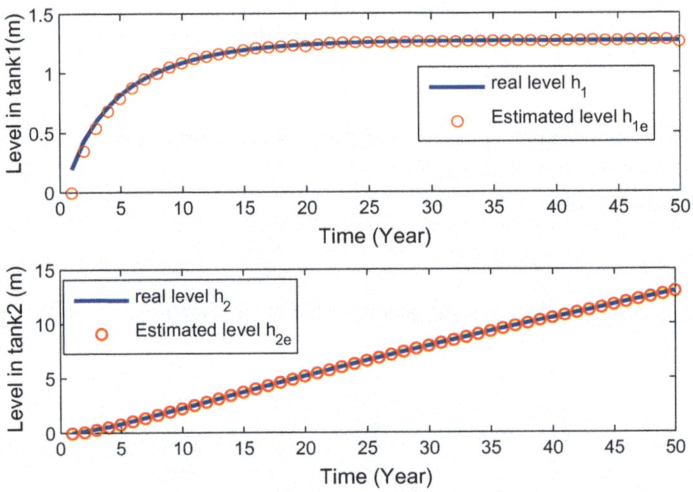

Fig. 11.2 Fluid-level evolution in two tanks

11.3.1 Model Without Faults

The two-tank system is highly nonlinear, and in many applications the operating point is changed frequently.

The algorithm of EKF utilizes level of fluid and flow rate signals to generate the statistical residual life estimates of level tank to quantify the influence of wax deposition action. Of these, the pressure and flow rate signals are commonly measured in petrochemical system applications. Simulations were carried out in the absence of a fault in Figs. 11.2 and 11.3 showing a good estimation of the level heights by the extended Kalman filter method. The EKF is applied to generate the residual for diagnosis. The statistical characteristics of the residual without faults are given in Table 11.3.

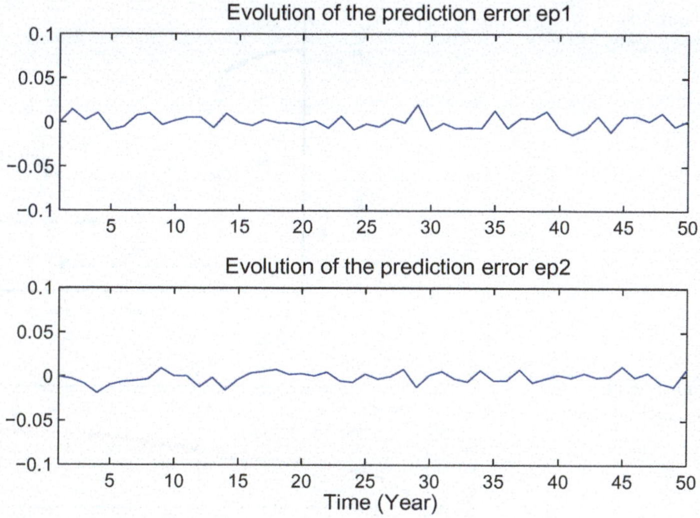

Fig. 11.3 Residuals without faults

Table 11.3 The statistical characteristics of the residual

Residual	Mean	Standard deviation
r_1	$\mu_1 = 0.000537$	$\sigma_1 = 0.0066$
r_2	$\mu_2 = -0.000496$	$\sigma_2 = 0.0074$

11.3.2 Model Including Faults Clogging (20% S_0)

In this industrial application, we will be interested in structural defects. Let us take the case of the obstruction of the pipeline. After a given operating time and a given length L, it is assumed by pressure measurement that there is an increase in the pressure downstream of the obstruction. A degradation of the state of the cross section of the pipeline linking the two reservoirs is considered to occur; at first, clogging is assumed to occur at the given instant $t_1 = 20$ (Years). The problem consists in modeling the evolution of the wax in pipeline and determining the minimum cross section which corresponds to a maximum pressure which can cause the explosion of the pipeline. The illustration in Fig. 11.4 characterizes random evolutions of the internal section Sc of the pipeline for a given length L. The time at which the prognosis is made will be noted in the sequence t_{pro}. The residual life of the system at t_{pro} will be noted $RUL_{t_{pro}}$. It corresponds to the time elapsed between $RUL_{t_{pro}}$ and the moment when the system can no longer perform the required functions satisfactorily. Let T be the date of failure, and then the remaining useful lifetime prediction at the time of prognosis is a random variable defined for $t_{pro} < T$ (see Fig. 11.4).

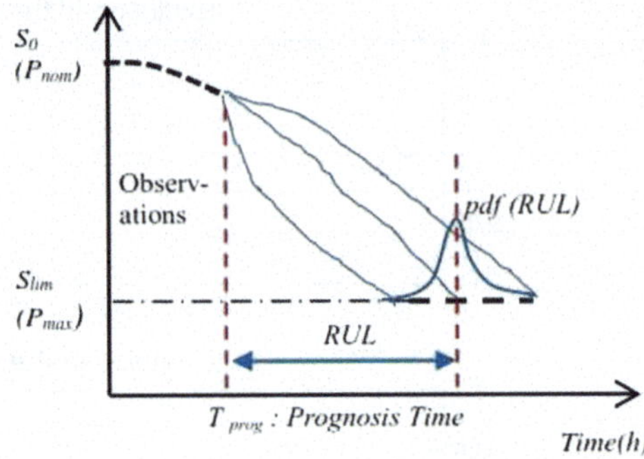

Fig. 11.4 Fluid-level evolution in two tanks

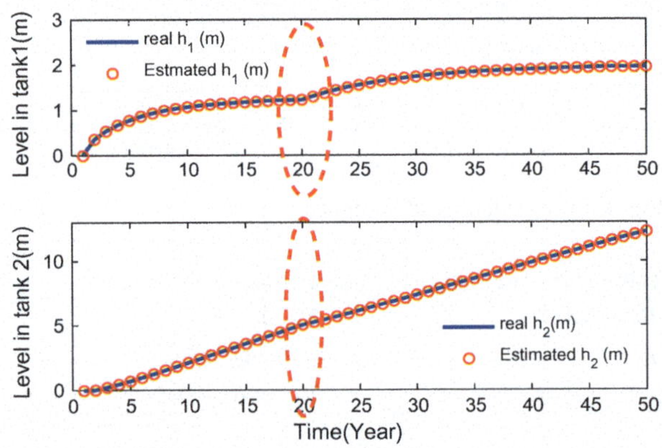

Fig. 11.5 Fluid-level evolution in tank1 and tank2

$$RUL_{t_{pro}} = T - t_{pro} \tag{11.23}$$

S_{lim}: In critical section for a lower value, there will be an explosion in pipeline, according to the literature (Svendsen 1993; Mendes and Braga 1996), who presented models for calculation of the wax appearance at given point. Using the criterion specified in literature (Ahammed and Melchers 1996; Ahammed 1998),

$$P_{\max} = Sy + 68.95\ (\text{MPa}) \tag{11.24}$$

where Sy is the yield strength of the pipe material, and $P_{\max}$ is the flow strength of the pipe material (MPa).

Clogging is assumed to occur at the given instant $t_1 = 20$ (Years). If we consider a reduction of the cross section of $20\% \times Sc\,(t = 0)$ for a given instant, then the dynamic evolution of the heights in tank is affected by this defect as illustrated in Figs. 11.5, 11.6, and 11.7. The residuals are different from the normal mode Figs. 11.2 and 11.3.

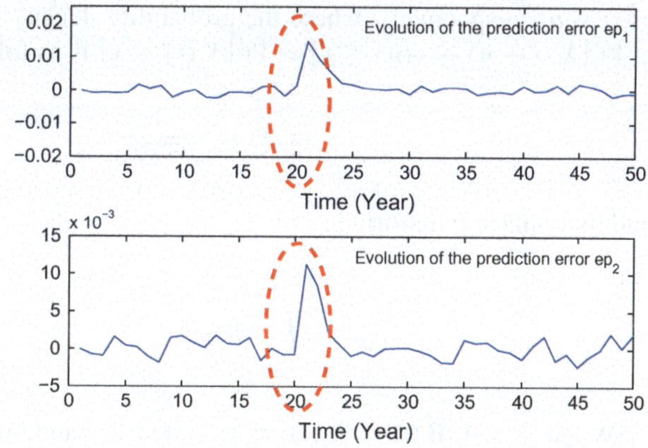

Fig. 11.6 Faulty residuals (Time = 20 Years)

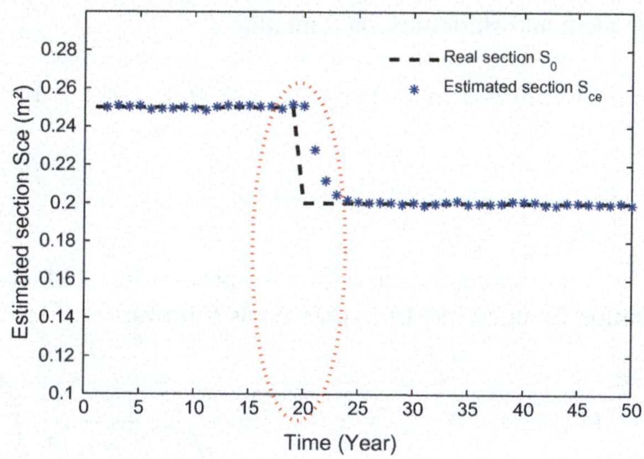

Fig. 11.7 Evolution of section profile with critical case of wax

11.3.3 Wiener Process for Degradation Modeling

In our case, the pressure differential is considered as a degradation indicator chosen (Matta et al. 2012). Degradation modeling attempts to characterize the evolution of damage for example, wear, tear, crack length, and clog. A Wiener process with positive drift is used, due to its mathematical properties and physical interpretations, where a recursive filter was also applied for the updating of the RUL estimation (Ye et al. 2013). We define a Brownian motion process or a Wiener process as follows; a stochastic process is said to be a Brownian motion process (Nakagawa 2011):

1. $X(0) = 0$;
2. The process $\{X(t)\}$ has stationary and independent increment;
3. The process $\{X(t)\}$ is normally distributed with mean 0 and variance $\sigma^2 t$ for any $t > 0$.

For the following time variables $t, u > 0$ the random variables $X(t+u) - X(u)$ and $X(t) - X(u)$ for $t > u$ have a normal density with mean 0, and variance $\sigma^2 t$ and

$\sigma^2(t-u)$, respectively, where the probability density function denoted as $f(x,t)$ of $\Pr\{X(t+u) - x(u) \le x\} = \Pr\{X(t) \le x\}$ is as follows:

$$f(x,t) = \frac{1}{\sqrt{2\pi t}.\sigma} e^{-x^2/(2\sigma^2 t)} \tag{11.25}$$

and its Laplace transform is

$$\int_{-\infty}^{+\infty} f(x,t)\, e^{-sx} = e^{s^2\sigma^2/2} \tag{11.26}$$

When $\sigma = 1$, B (t) $\equiv X(t)/\sigma$ is called a standard Brownian motion because $V\{X(t)/\sigma\} = t$. For any $0 \le t_0 < t_1 < \cdots < t_n < t$, from the properties of independent and stationary increments,

$$\Pr\{X(t) \le x\,|X(t_0) = x_0, \ldots, X(t_n) = x_n\} = \Pr\{X(t) \le x\,|X(t_n) = x_n\} \tag{11.27}$$

$$= \Pr\{X(t) - X(t_n) \le x - x_n\} \tag{11.28}$$

However, we can consider that the process has a Markov property. Hence, its distribution function can be expressed as follows:

$$\Pr\{X(t) - X(t_n) \le x - x_n\} = \frac{1}{\sqrt{2\pi(t-t_n)}\sigma} \int_{-\infty}^{+\infty} e^{-u^2/(2\sigma^2(t-t_n))} du \tag{11.29}$$

where $X(t) - X(t_n)$ has a normal density with mean 0 and variance $\sigma^2(t-t_n)$ for any $t > t_n$, that does not depend on t_n. When

$$Z(t) = \mu t + \sigma B(t) \tag{11.30}$$

where $B(t)$ is a standard Brownian motion (or a Wiener process) representing the stochastic dynamics of the degradation process, then $Z(t)$ is called a Brownian motion with drift parameter μ and variance σ^2. In our case, a Wiener process aims at modeling the heaping and movement of small particles in fluids with tiny fluctuations in pipeline. In the context of reliability, the characteristic of this process is that the pipeline's degradation can increase or decrease gradually and accumulatively over time. The small increase or decrease in degradation pipeline over a small time interval behaves similarly to the random walk of small particles heaping in inside of pipeline section. Therefore, the stochastic processes have been widely used to characterize the path of degradation pipeline where successive fluctuations in degradation can be observed.

11.3.4 Training Date Set and Parameter Estimation

In our example of work, the Wiener process is used for degradation modeling corresponding date set. Assume that $Z_{i,j}$ is a degradation indicator measurements at the ith items at time j, where $i = 1, 2, 3, \ldots N$ and $j = 1, 2, 3, \ldots r$, r is the last observation time. The degradation realization is based on homogeneous Brownian motion with drift of two parameters (μ, σ) that are same for all items. Each increment of degradation $\Delta Z_{i,j} = Z_{i,j+1} - Z_{i,j}$ of each item follows a normal distribution $N\left(\mu \Delta t_{i,j}, \sigma^2 \Delta t_{i,j}\right)$. We consider each increment $\Delta Z_{i,j}$ is independent, identically components, and normally distributed for all. The transition density function of Brownian motion process is given by

$$f_{(\mu \Delta t_{i,j}, \sigma^2 \Delta t_{i,j})}\left(\Delta Z_{i,j}\right) = \frac{1}{\sqrt{2\pi\sigma^2 \Delta t_{i,j}}} \exp\left[-\frac{\left(\Delta Z_{i,j} - \mu \Delta t_{i,j}\right)^2}{2\sigma^2 \Delta t_{i,j}}\right] \tag{11.31}$$

Z is a Markovian process, and then the maximum likelihood estimator (MLE) can be used. The parameter vector $\theta = (\mu, \sigma)$ can be calculated once transition density function of Z is known. Consider the degradation measurements for item i, $\Delta Z_i = \left(\Delta Z_{i,1}, \Delta Z_{i,2}, \Delta Z_{i,3}, \ldots\ldots \Delta Z_{i,r}\right)$. For item i, the likelihood function for item i is

$$L_i(\theta) = \prod_{j=1}^{r} \frac{1}{\sqrt{2\pi\sigma^2 \Delta t_{i,j}}} \exp\left[-\frac{\left(\Delta Z_{i,j} - \mu \Delta t_{i,j}\right)^2}{2\sigma^2 \Delta t_{i,j}}\right] \tag{11.32}$$

For the ith item, the log likelihood is given by

$$l_i(\theta) = \ln L_i(\theta) = \ln\left(\prod_{j=1}^{r} \frac{1}{\sqrt{2\pi\sigma^2 \Delta t_{i,j}}} \exp\left[-\frac{\left(\Delta Z_{i,j} - \mu \Delta t_{i,j}\right)^2}{2\sigma^2 \Delta t_{i,j}}\right]\right) \tag{11.33}$$

The degradation measurement vectors are independent; then, we can write as

$$l(\theta) = \ln\left(\Delta Z_1, \ldots \Delta Z_N\right) = \sum_{\mathrm{i}=1}^{\mathrm{N}} \ln\left(f_i\left(\Delta Z_{i,1}, \ldots \Delta Z_{i,r}\right)\right) \tag{11.34}$$

$$l(\theta) = \sum_{\mathrm{i}=1}^{\mathrm{N}} \ln\left(\prod_{j=1}^{r} \frac{1}{\sqrt{2\pi\sigma^2 \Delta t_{i,j}}} \exp\left[-\frac{\left(\Delta Z_{i,j} - \mu \Delta t_{i,j}\right)^2}{2\sigma^2 \Delta t_{i,j}}\right]\right) \tag{11.35}$$

where $l(\theta) = \sum_{i=1}^{N} l_i(\theta)$, $f_i / l_i(\theta)$ is the probability density function divided by log likelihood of increment s corresponding to each item, and $f/l(\theta)$ is the probability density function divided by log likelihood of increments corresponding to all increments. The maximum likelihood estimator $\widehat{\theta} = (\widehat{\mu}, \widehat{\sigma})$ found by maximizing $l(\theta)$ can be taken as the partial derivative of the log likelihood function of Eq. (11.35); by respecting the derivation variables μ and σ, this gives equations as follows:

$$\frac{\partial l(\theta)}{\partial \mu} = \sum_{i=1}^{N} \sum_{j=1}^{r} \frac{\Delta Z_{i,j} - \mu \Delta t_{i,j}}{\sigma^2} = 0 \tag{11.36}$$

$$\frac{\partial l(\theta)}{\partial \sigma} = -\frac{rN}{\sigma} \sum_{i=1}^{N} \sum_{j=1}^{r} \frac{\left(\Delta Z_{i,j} - \mu \Delta t_{i,j}\right)^2}{\sigma^3 \Delta t_{i,j}} = 0 \tag{11.37}$$

Then, the estimation of the maximum likelihood estimator for $\theta = (\mu, \sigma)$ is given by

$$\widehat{\mu} = \frac{\sum_{i=1}^{N} \sum_{j=1}^{r} \Delta Z_{i,j}}{\sum_{i=1}^{N} \sum_{j=1}^{r} \Delta t_{i,j}} \tag{11.38}$$

$$\widehat{\sigma} = \sqrt{\frac{1}{rN} \sum_{i=1}^{N} \sum_{j=1}^{r} \frac{\left(\Delta Z_{i,j} - \mu \Delta t_{i,j}\right)^2}{\Delta t_{i,j}}} \tag{11.39}$$

11.4 First Hitting Time and RUL Distribution

One of the purposes of studying degradation model is to estimate the RUL of pipeline system and to prevent system's failure. The quantity RUL can be defined as amount of time left before system health falls below a defined failure threshold, refer to Jedrzejewski (2009), Lawler (2018), Klebaner (2012). There are many useful computations about Brownian motions $B(t)$. We need the notion of real-valued stopping time.

11.4.1 First Hitting Time and Strong Markov Property

Let T denote a stopping time for $B(t)$, $t \geq 0$, if for any t it is possible to decide whether T has occurred or not by observing $B(s)$, $0 \leq s \leq t$. More rigorously, for any t the sets $\{T \leq t\} \in F_t$, given a level L, the time of hitting this level could be more than once due to the random nature of Brownian motion process $Z(t)$. The following contents provide the general computation of a first passage time hits level L, hits time T_L, and RUL distribution. Suppose T_L is the first passage time of Brownian motion process $Z(t)$ hits level L, if the maximum of $Z(t)$ at time t is greater than L, then the Brownian motion took value L at some time before t. The mathematical form could be written as follows: let T_L denote the first time, when $Z(t)$ hits level. We can now deduce the distribution of the maximum and the minimum of Brownian motion on $[0, t]$. $M(t) = \max_{0 \leq s \leq t} Z(s)$ and $m(t) = \min_{0 \leq s \leq t} Z(s)$. Similarly, the distribution of the first hitting time of L, $T_L = \inf\{t > 0 : Z(t) = L\}$. By Klebaner theorem (2012), for any $L > 0$,

$$P_0(M(t) \geq L) = 2.P_0(Z(s) \geq L) = 2\left[1 - \Phi\left(\frac{L}{\sqrt{t}}\right)\right] \quad (11.40)$$

Proof Notice that the events $M(t) \geq L$ and $T_L \leq t$ are the same. However, if the maximum of Wiener process at time t is greater than L, then at some time before t Wiener process took value L. On the other hand, if Wiener process took value L at some time before t, therefore the maximum will be at least L. Since $\{Z(t) \geq L\} \subset \{T_L \leq t\}$, therefore, we have also calculated the probability as follows:

$$P(Z(t) \geq L) = P(Z(t) \geq L, T_L \leq t) \quad (11.41)$$

As $Z(T_L) = L$,

$$P(Z(t) \geq L) = P(T_L \leq t, (Z(T_L + (t - T_L)) - Z(T_L)) \geq 0) \quad (11.42)$$

Notice that the T_L is a finite stopping time, and by the strong Markov in Klebaner (2012), the random variable, $\widehat{Z}(s) = Z(T_L + s) - Z(T_L)$, is independent of F_{T_L} (be a collection of subsets of Ω) and has a normal distribution, so we have

$$P(Z(t) \geq L) = P\left(T_L \leq t, \widehat{Z}(t - T_L) \geq 0\right) \quad (11.43)$$

If we had independent of T_L, then

$$\begin{aligned} P\left(T_L \leq t, \widehat{Z}(t) \geq 0\right) &= P(T_L \leq t)\, P\left(\widehat{Z}(s) \geq 0\right) \\ &= P(T_L \leq t)\frac{1}{2} = P(M(t) \geq L)\frac{1}{2} \end{aligned} \quad (11.44)$$

That is to say for any $L > 0$,

$$P(M(t) \geq L) = 2P_0(Z(t) \geq L) = 2\left[1 - \Phi\left(\frac{L - x - \mu t}{\sigma\sqrt{t}}\right)\right] \quad (11.45)$$

In many engineering applications, the failure time T_L for a component is defined as the time at which the degradation path first reaches a predetermined threshold L. The distribution of the first passage time T_L plays an important role in predicting remaining useful life and in determining the optimal maintenance strategies. According to Lovric (2011), the inverse Gaussian distribution describes the distribution of the time a Brownian motion with positive drift takes to reach a given positive level. As what has been denoted before, let T_L be the first passage time for a fixed level $L > 0$ by $Z(t)$. Then T_L can be considered as a random variable governed by inverse Gaussian distribution given by $T_L \sim IG\left(\frac{L}{\mu}, \left(\frac{L}{\sigma}\right)^2\right)$. We should first mention inverse Gaussian distribution (IG) (also known as Wald distribution) as a two-parameter continuous distribution given by its density function

$$f(t, \mu, \lambda) = \sqrt{\frac{\lambda}{2\pi}} t^{-3/2} \exp\left(\frac{-\lambda}{2\mu^2 t}(t-\mu)^2\right), \mathrm{t} > 0 \tag{11.46}$$

The parameter $\mu > 0$ is the mean, and $\lambda > 0$ is the shape parameter. For a random variable X with inverse Gaussian distribution, we write $X \sim IG(\mu, \lambda)$; the inverse Gaussian distribution describes the distribution of the time a Brownian motion with positive drift that takes to reach a given positive level. Then $T_L \sim IG\left(\frac{L}{\mu}, \left(\frac{L}{\sigma}\right)^2\right)$ has inverse Gaussian distribution, so put these parameters into Eq. (11.46). Therefore, we have probability density function given by

$$f(t, \mu, \sigma) = \frac{L}{\sqrt{2\pi\sigma^2 t^3}} \exp\left[-\frac{(L-\mu t)^2}{2\sigma^2 t}\right] \tag{11.47}$$

The first passage time T_L satisfies the following function, which can be written as

$$F(x, \mu, \sigma) = P(T_L \le t) = \int_0^t \frac{L}{\sqrt{2\pi\sigma^2 x^3}} \exp\left[\frac{-(L-\mu x)^2}{2\sigma^2 x}\right] dx \tag{11.48}$$

Given Eq. (11.47), it is easy to obtain residual useful lifetime distribution. Observing the process at time t is at position; the probability that residual useful lifetime is less than a predefined period h is given by

$$P\left(RUL_{z(t)} \le h\right) = \int_0^h \frac{L - z(t)}{\sqrt{2\pi\sigma^2 x^3}} \exp\left[-\frac{[L - z(t) - \mu x]^2}{2\sigma^2 t}\right] dx. \tag{11.49}$$

11.5 Case Study

The main objective of this study is to present an approach to evaluate the reliability and the safety level. However, the remaining useful life is for deteriorating pressurized pipelines at any point (Length L) in time t (Year). This study will focus mainly on longitudinally oriented section in pipeline. Based on the pipeline's circumferential strength, a process models for degradation will then be developed. In the case

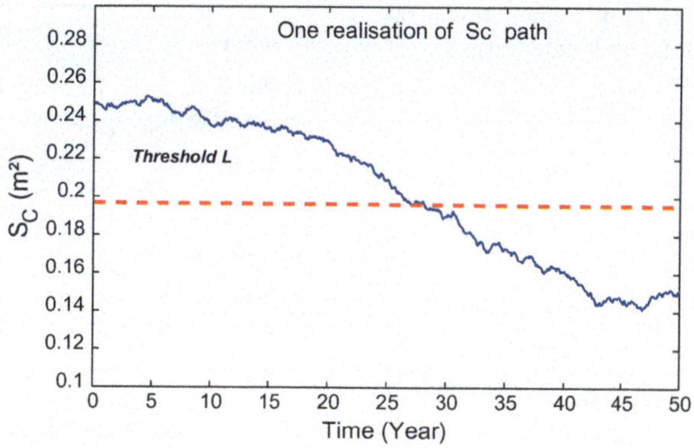

Fig. 11.8 Degradation paths of Sc

of deteriorating structural defects, the damage is defined as Wiener process from Eq. (11.30), and the models for degradation are given by

$$Sc(t) = Sc(t-1) + \mu\, t + \sigma B(t)\,. \tag{11.50}$$

11.5.1 Degradation Path Example, Based on Wiener Process

Now, we provide a numerical simulation to test the performance of the presented approach, including the procedures of initial parameters estimation and RUL estimation. The Wiener process is used for degradation modeling corresponding to training data set. Assume that Sc is a degradation indicator for N = 10 realizations; independent identical tests are based on homogeneous Brownian motion with drift. To illustrate the degradation process, suppose we simulate the testing data set using Matlab. We suppose the first path is simulated with mean and variance parameters, respectively, $\mu = 0.1$, $\sigma^2 = 2.10^{-5}$ and initial condition $Sc\,(0) = 0.25\,\left(\mathrm{m}^2\right)$ (Fig. 11.8).

A different analysis with Matlab demonstrates that when the parameter μ is small in comparison with the parameter σ, drift has a greater impact on the Brownian process; if the parameter σ is small in comparison with the parameter μ, then noise dominates in the behavior of the Brownian process. Ten realizations of paths with initial condition μ and σ are given to visualize the paths of Sc. However, it appears that the paths have regions where motions look like they have trends. Given Eq. (11.38) and (11.39), it is easy to estimate the parameters corresponding to the testing data set. The estimated parameters are obtained as follows: $\widehat{\mu} = 0.1525$ and $\widehat{\sigma} = 0.052$. In the first time, we apply the EKF algorithm in Sect. 11.3. In order to predict the evolution of system's future state, using Eqs. (11.11), (11.12), (11.13), and (11.14) and giving the measurement-update, we can make a posteriori estimation, including measurement value, measurement error, and system noise, to obtain optimal estimation of the system state in the future. Therefore, EKF is also a high-speed recursive algorithm, which only needs to save the system state value and covariance at the last time in

Fig. 11.9 Degradation paths of Sc by Monte Carlo

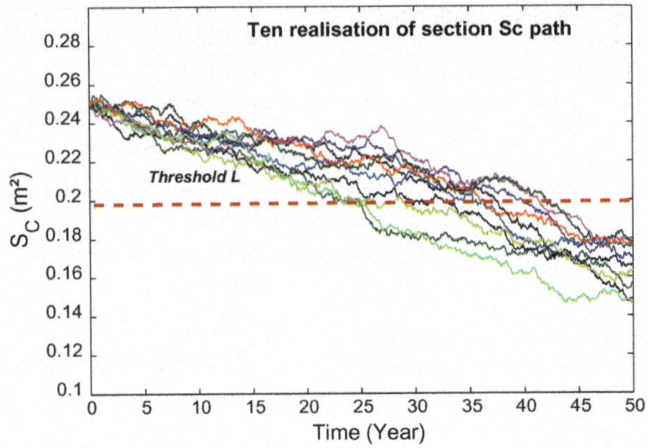

Fig. 11.10 First hitting time estimated by EKF when Sc crossed *L*

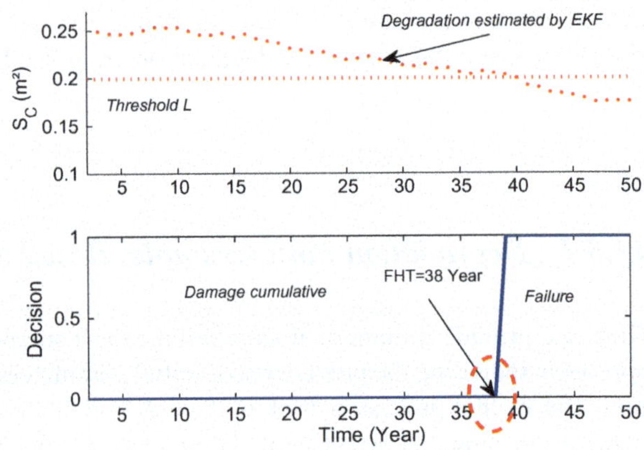

each estimation step. A threshold value of the degradation indicator is fixed for RUL prediction of the degradation process in pipeline system. The threshold value is set to respect the criterion specified in literature (Ahammed and Melchers 1996; Ahammed 1998). Use the criterion in Eq. (11.24) to ensure the reliability and safety level in the pipeline system. Figures 11.12 and 11.13 show the results of prognostication using EKF.

Figure 11.9 shows the EKF prediction of the degradation process Sc state and the detection of the threshold value. Figure 11.10 shows the EKF prediction of the remaining useful life. If the threshold value has been exceeded, the estimation of the remaining useful life can be calculated. If the prognostic begin at the monitoring time of prognostic $t_p = 20$ (Years), then the RUL can be evaluated. In our case, the blue line in Fig. 11.10 is the state estimate of Sc from the EKF. Therefore, the first passage time failure is 37.5 (Years), and then by Eq. (11.23) the $RUL = 17.5$ (Years) (Fig. 11.11).

In the second time, we first simulate the degradation paths of 100 components (measurement value of Sc) based on Wiener process. To illustrate the degradation

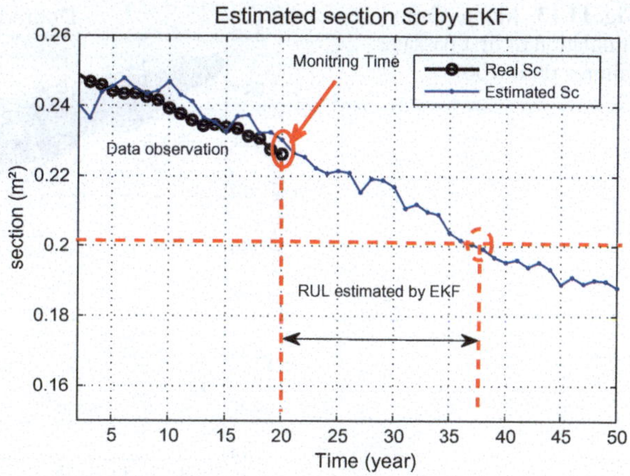

Fig. 11.11 RUL prediction by EKF

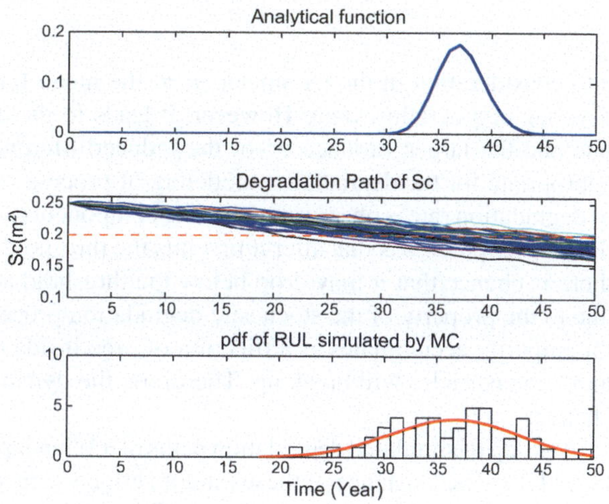

Fig. 11.12 First passage density function ($N = 50$)

process, we apply Monte Carlo simulation to our predefined Wiener parameters to simulate the testing data set, in order to verify that proposed Wiener path model is good enough to capture the true degradation process.

Suppose we want to figure out the time when the process first crosses degradation level $L = Sc_{Lim} = 0.2\left(\mathrm{m}^2\right)$. The random samples are generated, and the distributions of first hitting time(FHT) obtained by Monte Carlo simulations are compared with analytical function of FHT in Eq. (11.48). Therefore, we could say that the analytical function of FHT based on Wiener process roughly captures the true value in degradation process. An advanced evaluation needs to be carried out to judge the accuracy of the proposed Wiener process. In this paper, real data set is not available at the moment and only Monte Carlo simulations are used for N trajectories. So this evaluation process is postponed to future work. As there are random effects taken

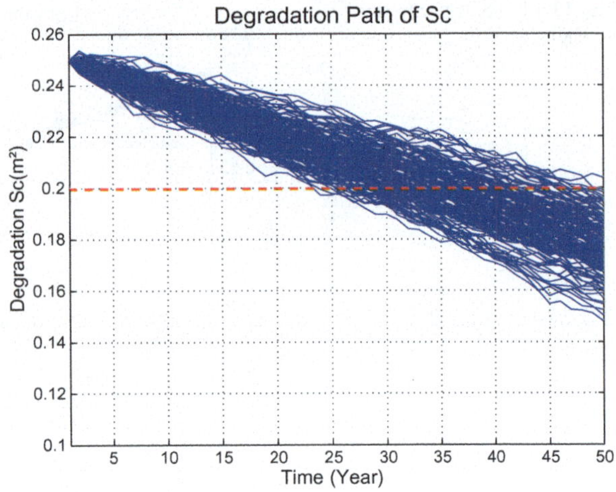

Fig. 11.13 Monte Carlo simulation of first passage failures (N = 100)

into consideration in the parameter μ, σ, the mean function of the Wiener process μ varies in a certain extent. However, it leads to the dispersion of the degradation rate and the larger variance σ^2 of the induced lifetime distribution. This model is appropriate for the degradation modeling of process for which significant variation of degradation rate is observed within the components population. One concern about the Wiener process is that after it first hits the threshold or enters in a threshold zone, there is chance that it may drop below the threshold and return to the normal zone due to the property of the stochastic degradation process. It is logic; in many cases, the impurity accumulates as a function of time inside the pipeline and, on the other hand, the particles will break up. Therefore, the dynamic of the path of degradation changes.

A group of simulated degradation paths of a Wiener process model are presented in Fig. 11.12. The existence of these random effects is involved in the parameter μ. The variance of degradation observations within each unit is significant, and therefore the degradation rates among these units are relevantly coherent. However, the probability density function of failure time is shown in Fig. 11.13.

With Matlab, we can define the statistical properties, the mean, and variance of a random variable RUL which are Tf_{moy}= 36.625 (Year) and standard deviation $\sigma = 13.9$, where $RUL_{estimate}$= $Tf_{moy}-t_0$= 36.625 − 0 = 36.625 (Year).

Therefore, we could say that the analytical function of FHT based on Wiener process roughly captures the true value in degradation process. However, it is not very reasonable to make such deterministic conclusion only according to figures. An advanced evaluation needs to be carried out to judge the accuracy of the proposed Wiener process method. In this paper, real data set is not available at the moment and only Monte Carlo simulations are used.

Suppose we want to find out the mean residual useful lifetime when the unit condition hits critical boundary $Sc_{\lim} = L = 0.2\ (\mathrm{m}^2)$. Given the estimated parameters $\widehat{\mu} = 0.1525$ and $\widehat{\sigma} = 0.052$, where the analytical result of RUL distribution can be

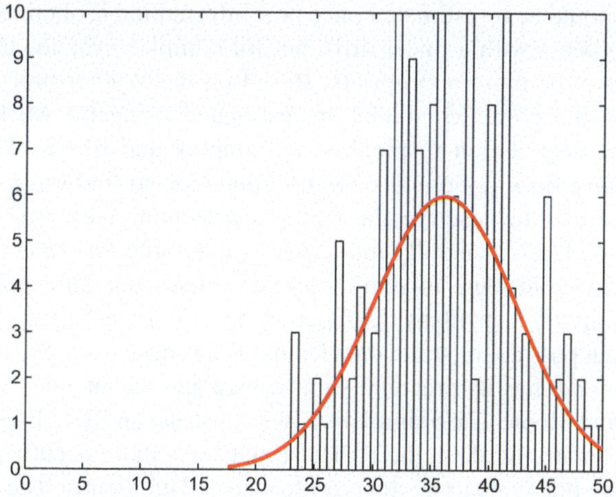

Fig. 11.14 RUL prediction pdf

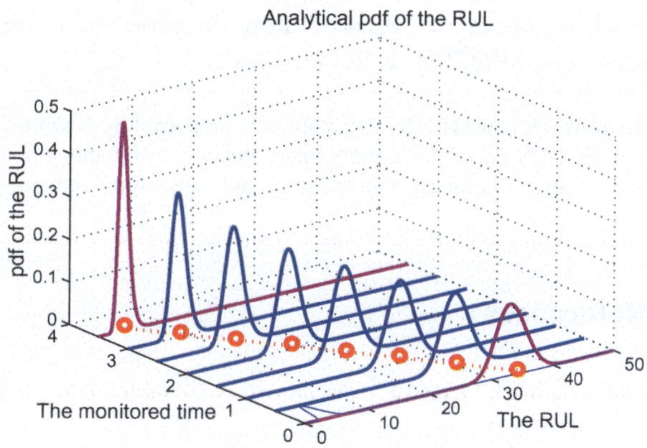

Fig. 11.15 Analytical pdf of the RUL prediction

given based on Eq. (11.49). Correspondingly, we obtain the PDFs of the RUL at different observation times as shown in Fig. 11.14. The first curve in Fig. 11.14 is the RUL distribution when monitoring time is 0.2 time unit, and the last curve shows RUL when monitoring time is 3.4 time unit which is more approaching the mean FHT = L. We draw a new RUL distribution curve for every 0.2 time unit. Clearly, the later the observing time, the sooner that component would fail with an increasing higher possibility (Fig. 11.15).

11.6 Conclusion

We present a Wiener-process-based degradation model with an extended Kalman filter algorithm to estimate the RUL. Moreover, the RUL of a section of pipeline should be re-estimated once new information is obtained. We only consider a Wiener process with a linear drift, but for complex systems in practice, a nonlinear model may be more appropriate. Based on the Wiener process, the unknown parameters in the presented model are estimated using the MLE approach. The mechanism of degradation in pipeline is complex and like a black box. Most efforts which have been done and would be done are to find out a model which could be more precise to describe the true phenomenon. We agree on the fact that degradation process is generally monotone; we assume that in a Wiener process all the factors that contribute to non-monotone effects are pure noise. This assumption is very important, as many degradation data are more or less mixed with noise, and finding out pure degradation data from noise is necessary. So, it is fairly justified that Wiener model has its practical significance and advantages. Based on the concept of FHT and Monte Carlo simulation, we construct an RUL distribution. Therefore, with more monitoring data, the RUL estimation is more accurate. In this practical case study, we find that our model is quite well and efficiently. The application of our approach is higher than the deterministic models and methods in many researches. On the other hand, we verify that incorporating the observation history to date can improve the accuracy of the RUL estimation indeed.

Acknowledgements This research was supported by National high school of engineers of Tunis (LISIER -ENSIT). The authors thank the editor and anonymous reviewers for their valuable and constructive suggestions that led to considerable improvements of this paper.

References

Abdel-Hameed, M. (2014). *Lévy processes and their applications in reliability and storage*. Berlin: Springer.

Ahammed, M. (1998). Probabilistic estimation of remaining life of a pipeline in the presence of active corrosion defects. *International Journal of Pressure Vessels and Piping, 75*(4), 321–329.

Ahammed, M., & Melchers, R. (1996). Reliability estimation of pressurised pipelines subject to localised corrosion defects. *International Journal of Pressure Vessels and Piping, 69*(3), 267–272.

Bazán, F. A. V., & Beck, A. T. (2013). Stochastic process corrosion growth models for pipeline reliability. *Corrosion Science, 74*, 50–58.

Benallal, A. (2008). *Hydrodynamique de l'accumulation des dépôts de paraffines dans les conduites pétrolières*. Ph.D. thesis, École Nationale Supérieure des Mines de Paris.

Blanke, M., Kinnaert, M., Lunze, J., Staroswiecki, M., & Schröder, J. (2006). *Diagnosis and fault-tolerant control* (Vol. 2). Berlin: Springer.

Desmond, A. (1985). Stochastic models of failure in random environments. *Canadian Journal of Statistics, 13*(3), 171–183.

Gomm, J., Williams, D., & Harris, P. (1993). A generic method for fault detection in process control loops by recursive parameter estimation. *European Journal of Diagnosis and Safety in Automation, 3*(1), 47–68.

Grewal, M. S., & Andrews, A. P. (2001). *Kalman filtering: Theory and practice using matlab*. New York: Wiley.

Handal, A. D. (2008). *Analysis of some wax deposition experiments in a crude oil carrying pipe*. Master's thesis.

Jedrzejewski, F. (2009). *Modeles aleatoires et physique probabiliste*. Berlin: Springer Science & Business Media.

Kahle, W., & Lehmann, A. (2010). The wiener process as a degradation model: Modeling and parameter estimation. *Advances in degradation modeling* (pp. 127–146). Berlin: Springer.

Klebaner, F. C. (2012). *Introduction to stochastic calculus with applications*. Singapore: World Scientific Publishing Company.

Langeron, Y., Grall, A., & Barros, A. (2015). A modeling framework for deteriorating control system and predictive maintenance of actuators. *Reliability Engineering & System Safety, 140*, 22–36.

Lawler, G. F. (2018). *Introduction to stochastic processes*. Boca Raton: Chapman and Hall/CRC.

Lovric, M. (2011). *International encyclopedia of statistical science* (pp. 687–688). Berlin: Springer.

Lu, C. J., & Meeker, W. O. (1993). Using degradation measures to estimate a time-to-failure distribution. *Technometrics, 35*(2), 161–174.

Matta, N., Vandenboomgaerde, Y., & Arlat, J. (2012). *Supervision, surveillance et sûreté de fonctionnement des grands systèmes*. Cachan: Lavoisier.

Mendes, P. R. S., & Braga, S. L. (1996). Obstruction of pipelines during the flow of waxy crude oils. *Journal of Fluids Engineering, 118*(4), 722–728.

Mendes, P. S., Braga, A., Azevedo, L., & Correa, K. (1999). Resistive force of wax deposits during pigging operations. *Journal of Energy Resources Technology, 121*(3), 167–171.

Nakagawa, T. (2011). *Stochastic processes: With applications to reliability theory*. Berlin: Springer Science & Business Media.

Paris, P., & Erdogan, F. (1963). A critical analysis of crack propagation laws. *Journal of Basic Engineering, 85*(4), 528–533.

Ribeiro, F. S., Mendes, P. R. S., & Braga, S. L. (1997). Obstruction of pipelines due to paraffin deposition during the flow of crude oils. *International Journal of Heat and Mass Transfer, 40*(18), 4319–4328.

Si, X.-S., Wang, W., Hu, C.-H., Chen, M.-Y., & Zhou, D.-H. (2013). A wiener-process-based degradation model with a recursive filter algorithm for remaining useful life estimation. *Mechanical Systems and Signal Processing, 35*(1–2), 219–237.

Svendsen, J. A. (1993). Mathematical modeling of wax deposition in oil pipeline systems. *AIChE Journal, 39*(8), 1377–1388.

Ye, Z.-S., Wang, Y., Tsui, K.-L., & Pecht, M. (2013). Degradation data analysis using wiener processes with measurement errors. *IEEE Transactions on Reliability, 62*(4), 772–780.

Chapter 12
Fault-Tolerant Control of Two-Time-Scale Systems

A. Tellili, N. Abdelkrim, A. Challouf, A. Elghoul and M. N. Abdelkrim

Abstract This work presents fault-tolerant control of two-time-scale systems in both linear and nonlinear cases. An adaptive approach for fault-tolerant control of singularly perturbed systems is used in linear case, where both actuator and sensor faults are examined in the presence of external disturbances. For sensor faults, an adaptive controller is designed based on an output-feedback control scheme. The feedback controller gain is determined in order to stabilize the closed-loop system in the fault-free case and vanishing disturbance, while the additive gain is updated using an adaptive law to compensate for the sensor faults and the external disturbances. To correct the actuator faults, a state-feedback control method based on adaptive mechanism is considered. The both proposed controllers depend on the singular perturbation parameter ε leading to ill-conditioned problems. A well-posed problem is obtained by simplifying the Lyapunov equations and subsequently the controllers using the singular perturbation method and the reduced subsystems yielding to an ε-independent controller. In the nonlinear case, an additive fault-tolerant control for nonlinear time-invariant singularly perturbed system against actuator faults based on Lyapunov redesign principle is presented. The full-order two-time-scale system is decomposed into reduced slow and fast subsystems by time-scale decomposition using singular perturbation method. The time-scale reduction is carried out by setting

A. Tellili (✉) · A. Elghoul
Higher Institute of Technological Studies at Djerba (ISETJB), Djerba, Tunisia
e-mail: Adel.Tellili@isetjb.rnu.tn

A. Elghoul
e-mail: aymengoul@hotmail.fr

N. Abdelkrim
Higher Institute of Industrial Systems at Gabès, University of Gabes, Gabes, Tunisia
e-mail: nouceyba.nacer@laposte.net

A. Challouf
Research Unit MACS, Higher National Engineering School at Gabes, Gabes, Tunisia
e-mail: amina_chalouf@yahoo.fr

M. N. Abdelkrim
National Engineering School at Gabes, University of Gabes, Gabes, Tunisia
e-mail: naceur.abdelkrim@enig.rnu.tn

N. Derbel et al. (eds.), *Diagnosis, Fault Detection & Tolerant Control*,
Studies in Systems, Decision and Control 269,
https://doi.org/10.1007/978-981-15-1746-4_12

the singular perturbation parameter to zero, which permits to avoid the numerical stiffness due to the interaction of two different dynamics. The fault-tolerant controller is computed in two steps. First, a nominal composite controller is designed using the reduced subsystems. Second, an additive part is appended to the nominal controller to compensate for the effect of an actuator fault. In both cases, the Lyapunov stability theory is used to prove the stability provided the singular perturbation parameter is sufficiently small. The designed control schemes guarantee asymptotic stability in the presence of additive faults. Finally, the effectiveness of the theoretical results is illustrated using numerical examples.

Keywords Fault-tolerant control · Nonlinear singularly perturbed systems · Time-scale decomposition · Adaptive control · Additive fault
Lyapunov equations · Singular perturbation method

12.1 Introduction

Systems, where slow and fast dynamic phenomena arise, are called singularly perturbed systems. They model many control systems like robotic systems, motor control systems, chemical processes, convection–diffusion systems, and electric circuits. Those systems are distinguished by the existence of a small positive parameter called singular perturbation parameter. It indicates the degree of separation between "slow" and "fast" modes of the system. Such small parameters can be used for modeling machine reactance in power systems, capacitance in electronic and wire inductance control systems. Singular perturbation parameter leads frequently to ill-conditioned results in the system analysis and synthesis methods. In order to handle such problems, a reduction technique, called singular perturbation method, is proposed in the literature. Thus, the full-order system is decoupled into slow and fast subsystems, which makes possible to deal with lower order systems and consequently to simplify the analysis and synthesis problems (El Hachemi et al. 2012; Kokotovic et al. 1999; Khalil 1987). The complexity of the singularly perturbed systems makes them vulnerable to faults being able to corrupt the controller, the sensors, the process itself, or the actuators. In this case, an adequate control scheme, known as reconfigurable control or fault-tolerant control, is needed to provide system stability even in the presence of defects. Otherwise, the occurrence of such faults may cause production to stop and threats human and material safety. The main purpose of the fault-tolerant control is to maintain stability and performance in the event of malfunctions in sensors, actuators, or other system components and to inhibit the restrictions of conventional feedback control (Bustan et al. 2014; Wang et al. 2014; Noura et al. 2000; Patton 1997). Fault-tolerant design methods can be classified into active and passive approaches. The active control technique requires a fault diagnosis block to detect and identify the faults in real time, and then a mechanism to adapt the controller to the new faulty situation according to the information recovered about malfunctions (Richter et al. 2011; Corradini and Orlando 2007; Ye and Yang 2006; Wang et al. 2007). In contrast, the passive approach, also called reliable control, offers a fixed-parameter controller in order to maintain (at least) stability in occurrence of presumed faults without the need for controller reconfiguration and fault diagnosis scheme (Noura et al. 2000; Corradini and Orlando 2007; Wang et al. 2007; Stoustrup and Blondel 2004).

12.2 Linear Case

Diverse methods are proposed to design reconfigurable controllers in order to ensure closed-loop stability, even in faulty case. A reliable fault-tolerant control method designed for linear systems using simultaneous stabilization is carried out by Stoustrup and Blondel (2004). Tellili et al. (2007) developed a reliable H_∞ controller for linear time-invariant multiparameter singularly perturbed system to tolerate sensor faults and to ensure H_∞ performance. The full-order system controller is then simplified to three reduced reliable H_∞ sub-controllers based on the fast and slow problems through the manipulation of the algebraic Riccati equations. Liu et al. (2012) proposed a H_∞ fault-tolerant controllers to overcome actuator faults. The authors handle three cases of such faults: normal, loss of efficiency, and outage. The designed H_∞ controllers can reduce the degree of conservatism compared with existing methods. Richter and Lunze (2010) developed the fault-hiding approach for linear and Hammerstein systems and used virtual actuators and sensors, respectively, for actuator and sensor faults. The results are extended to piecewise affine systems in Richter et al. (2011). Another widely studied method consists in the design of fault-tolerant control scheme based on adaptive control principles. Such approaches can be applied without using control restructuring and fault diagnosis procedures (Wang and Wen 2011; Chen and Saif 2007). Many authors were interested in this subject. In particular, Chen and Saif (2007) designed an adaptive scheme to diagnose and to accommodate actuator faults in linear multi-input single-output (MISO) systems with unknown system parameters. The fault-tolerant control problem is resolved using the remaining operative actuators. In Tao et al. (2002), a class of adaptive control methods based on output feedback approach in order to stabilize linear systems with complete actuator failures is developed. Jin and Yang (2009) designed a direct adaptive state feedback control approach based on Lyapunov stability theory. The resulting closed-loop system is then asymptotically stable in the presence of actuator faults and external disturbances. The method is extended and improved in Fan et al. (2010). Wang et al. (2013) developed an adaptive output feedback control to accommodate actuator faults including outage, loss of efficiency, and stuck. An adaptive fault-tolerant controller including a fault estimation error minimization problem is designed by Casavola and Garone (2010). The considered fault in this case is assumed to be piecewise constant with a slowly varying behavior. For the singularly perturbed systems, their control in the presence of actuator failures is investigated by some authors. Liu (2012) developed a controller scheme for singularly perturbed systems in the presence of actuator saturation under the assumption that the fast subsystem is stable. Xin et al. (2008, 2010) proposed the reduced-order adjoint systems, by which some methods to estimate the basin of attraction of singularly perturbed systems were developed. In Zhou et al. (2013), an optimal controller based on state-feedback approach is designed to control nonlinear singularly perturbed systems subject to actuator saturation. However, the abovementioned approaches were limited to actuator saturation and did not consider the loss of efficiency by sensors and actuators.

The main goal of this part is to design an adaptive fault-tolerant control scheme for singularly perturbed systems in the presence of external disturbances and additive faults characterized by a loss of effectiveness in sensors and actuators. The remaining part of this section is organized as follows. The system description and preliminaries are presented in Sect. 12.2.1. The fault model and the control problem are formulated in case of sensor faults in Sect. 12.2.2. In Sect. 12.2.3, an adaptive fault-tolerant control approach against actuator faults and external disturbances is established. An example of application is developed in Sect. 12.2.4.

12.2.1 System Description and Preliminaries

Consider the following time-invariant two-time-scale singularly perturbed system under external disturbances described by

$$\begin{cases} \begin{bmatrix} \dot{x}(t) \\ \varepsilon \dot{z}(t) \end{bmatrix} = \begin{bmatrix} A_{11} & A_{12} \\ A_{21} & A_{22} \end{bmatrix} \begin{bmatrix} x(t) \\ z(t) \end{bmatrix} + \begin{bmatrix} B_{1x} \\ B_{1z} \end{bmatrix} w(t) + \begin{bmatrix} B_{2x} \\ B_{2z} \end{bmatrix} u(t) \\ y(t) = C \begin{bmatrix} x(t) \\ z(t) \end{bmatrix} = \begin{bmatrix} C_1 & C_2 \end{bmatrix} \begin{bmatrix} x(t) \\ z(t) \end{bmatrix} \end{cases} \tag{12.1}$$

where $x \in \mathbb{R}^{n_1}$ and $z \in \mathbb{R}^{n_2}$ are state vectors, $u \in \mathbb{R}^m$ is the control vector, $y \in \mathbb{R}^l$ is the output, $w \in \mathbb{R}^q$ models piecewise continuous bounded external disturbances acting on the system and verifying $\|w\| \leq \overline{w}$ with $\overline{w}$ being an unknown positive constant. $A_{11} \in \mathbb{R}^{n_1 \times n_1}$, $A_{12} \in \mathbb{R}^{n_1 \times n_2}$, $A_{21} \in \mathbb{R}^{n_2 \times n_1}$, $A_{22} \in \mathbb{R}^{n_2 \times n_2}$, $B_{1x} \in \mathbb{R}^{n_1 \times q}$, $B_{2x} \in \mathbb{R}^{n_1 \times m}$, $B_{1z} \in \mathbb{R}^{n_2 \times q}$, $C_1 = \begin{bmatrix} C_{11} \\ C_{12} \end{bmatrix} \in \mathbb{R}^{l \times n_1}$, $C_2 = \begin{bmatrix} C_{21} \\ C_{22} \end{bmatrix} \in \mathbb{R}^{l \times n_2}$, and $B_{2z} \in \mathbb{R}^{n_2 \times m}$ are constant matrices. The matrix A_{22} is assumed to be nonsingular (standard singularly perturbed systems). The parameter ε, called singular perturbation parameter, is a positive scalar taking values between 0 and 1. Denote throughout the paper: $B_i = \begin{bmatrix} B_{ix}^T & B_{iz}^T \end{bmatrix}^T$ for $i = 1, 2$ and $\|(.)\|$ the Euclidean norm of (.).

According to the time-scale property of the singularly perturbed system, the slow and the fast subsystems of full-order system (12.1) can be derived by formally setting the singular perturbation parameter ε to zero (Kokotovic et al. 1999; Datta and Rai Chaudhuri 2010). The slow subsystem is obtained as

$$\begin{cases} \dot{x}_s = A_s x_s + B_{1s} w_s + B_{2s} u_s \\ y_s = C_s x_s + D_s u_s \end{cases} \tag{12.2}$$

where
$A_s = A_{11} - A_{12}A_{22}^{-1}A_{21}$, $C_s = C_1 - C_2A_{22}^{-1}A_{21}$, $D_s = -C_2A_{22}^{-1}B_2$, $B_{is} = B_{ix} - A_{12}A_{22}^{-1}B_{iz}$ for $i = 1, 2$; x_s, u_s, and w_s are respectively, the slow part of the states, the control input u, and the disturbance input w.

If (A_s, B_{2s}) is stabilizable and (C_s, A_s) is detectable, then there exists a symmetric and positive definite matrix P_s satisfying the following slow Lyapunov equation:

$$(A_s + B_{2s}K_sC_s)^T P_s + P_s (A_s + B_{2s}K_sC_s) = -Q_s \tag{12.3}$$

where Q_s is any given positive definite symmetric matrix and K_s is a static output feedback gain stabilizing the slow subsystem such that $u_s = K_s y_s$. The closed-loop slow subsystem is then defined by

$$\dot{x}_s = (A_s + B_{2s}K_sC_s)x_s + B_{1s}w_s \tag{12.4}$$

Most often in literature (Tellili et al. 2007; Khalil and Chen 1992; Tan et al. 1998), the following approximation is used: $A_s = A_{11}$, $B_{1s} = B_{1x}$, $B_{2s} = B_{2x}$, and $C_s = C_1$. Consequently, the slow subsystem (12.2), the slow Lyapunov equation (12.3), and the closed-loop slow subsystem (12.4) can be approached, respectively, by Eqs. (12.5), (12.6), and (12.7):

$$\dot{x}_s = A_{11}x_s + B_{1x}w_s + B_{2x}u_s \tag{12.5}$$

$$(A_{11} + B_{2x}K_sC_1)^T P_s + P_s (A_{11} + B_{2x}K_sC_1) = -Q_s \tag{12.6}$$

$$\dot{x}_s = (A_{11} + B_{2x}K_sC_1)x_s + B_{1x}w_s \tag{12.7}$$

Since C is of full rank, it is assumed, without loss of generality, that the output matrix can be transformed to block-diagonal form Fridman (Fridman and Shaked 2000), $C = \begin{bmatrix} C_1 & 0 \\ 0 & C_2 \end{bmatrix}$, where C_1x and C_2z describe, respectively, the slow and the fast part of the output.

The fast subsystem is given by

$$\begin{cases} \varepsilon\dot{z}_f = A_{22}z_f + B_{1z}w_f + B_{2z}u_f \\ y_f = C_2x_f \end{cases} \tag{12.8}$$

where z_f, u_f, and w_f are the fast parts, respectively, of the states, the control input u and the disturbance input w. The fast subsystem can be rewritten in the stretching (fast) time scale t/ε as follows:

$$\begin{cases} \dot{z}_f = A_{22}z_f + B_{1z}w_f + B_{2z}u_f \\ y_f = C_2x_f \end{cases} \tag{12.9}$$

Assuming that (A_{22}, B_{2z}) is stabilizable and (C_2, A_{22}) is detectable, there exists a symmetric and positive definite matrix P_f satisfying the following fast Lyapunov equation:

$$(A_{22} + B_{2z} K_f C_2)^T P_f + P_f (A_{22} + B_{2z} K_f C_2) = -Q_f \tag{12.10}$$

where Q_f is any given positive definite symmetric matrix and K_f is a static output feedback gain stabilizing the fast subsystem such that $u_f = K_f y_f$. The closed-loop fast subsystem is then defined by

$$\dot{z}_f = (A_{22} + B_{2z} K_f C_2) z_f + B_{1z} w_f \tag{12.11}$$

12.2.2 Controller Design with Respect to Sensor Fault

This section will concentrate on the design of an adaptive fault-tolerant controller to handle sensor faults in the presence of external disturbances.

12.2.2.1 Failure Model and Problem Formulation

The sensor faults which have been taken into account in this section involve loss of efficiency. For the output y_i, $i = 1, \ldots, l$, let y_i^f be the signal issued from the ith faulty sensor; accordingly, the sensor-fault model is expressed as follows:

$$y_i^f(t) = \rho_i \, y_i(t) \tag{12.12}$$

ρ_i represents the sensor efficiency factor and verifies $0 \le \underline{\rho}_i \le \rho_i \le \overline{\rho}_i \le 1$. $\underline{\rho}_i$ and $\overline{\rho}_i$ indicate the known lower and upper bounds of ρ_i, respectively. The case $\underline{\rho}_i = \overline{\rho}_i = 0$ describes the completely interruption of the sensor i. $\underline{\rho}_i > 0$ denotes the case of partial failure of y_i. If $\underline{\rho}_i = \overline{\rho}_i = 1$, then $y_i^f(t) = y_i(t)$, which depicts the case of no failure. Denoting $\rho = diag(\rho_i)$, $i = 1, \ldots, l$, the uniform sensor-fault model becomes

$$y^f(t) = \rho \, y(t) \tag{12.13}$$

The assumption that the control signals and disturbances use identical channels is used by many authors (Jin and Yang 2009; Wang et al. 2013) to solve robust control problems. Consequently, the following supposition will be held: $B_{1x} = B_{2x} F$ and $B_{1z} = B_{2z} F$, where F is a matrix with appropriate dimension.

The problem under consideration is to design a control law such that the closed-loop singularly perturbed system is asymptotically stable for any $\varepsilon \in]0, \varepsilon^*]$ despite the sensor fault occurrence and disturbance.

12.2.2.2 Controller Proposal for the Full-Order Singularly Perturbed System

Let us introduce the following notations:

$$X = \begin{bmatrix} x(t) \\ z(t) \end{bmatrix}, A(\varepsilon) = \begin{bmatrix} A_{11} & A_{12} \\ \frac{1}{\varepsilon}A_{21} & \frac{1}{\varepsilon}A_{22} \end{bmatrix},$$

$B_i(\varepsilon) = \begin{bmatrix} B_{ix} \\ \frac{1}{\varepsilon}B_{iz} \end{bmatrix}$ for $i = 1,\ 2$.

System (12.1) can be expressed as

$$\begin{cases} \dot{X}(t) = A(\varepsilon)\, X(t) + B_1(\varepsilon)\, w(t) + B_2(\varepsilon)\, u(t) \\ y(t) = C\, X(t) \end{cases} \tag{12.14}$$

By using the sensor-fault model (12.13) and the hypothesis assumed for the disturbances, system (12.14) is transformed to

$$\begin{cases} \dot{X}(t) = A(\varepsilon)\, X(t) + B_2(\varepsilon) F\, w(t) + B_2(\varepsilon)\, u(t) \\ y(t) = \rho\, C\, X(t) \end{cases} \tag{12.15}$$

The proposed fault-tolerant controller to stabilize the system (12.14) is described by

$$u\,(t) = K_1\; y\,(t) + K_2(t) \tag{12.16}$$

where K_1 is chosen such that $(A(\varepsilon) + B_2(\varepsilon)\, K_1 C)$ is Hurwitz and $K_2(t)$ is governed by the following update law:

$$K_2(t) = -\frac{B_2^T(\varepsilon) P(\varepsilon) X}{\|B_2^T(\varepsilon) P(\varepsilon) X\|}\,(\hat{k}_3(t) + \|K_1 C X\|) \tag{12.17}$$

where $\hat{k}_3$ is adjusted using the following adaptive law:

$$\frac{d\hat{k}_3(t)}{dt} = \varepsilon\,\gamma\,\left\|X^T P(\varepsilon) B_2(\varepsilon)\right\| \tag{12.18}$$

where γ is a suitable positive constant and ε is the singular perturbation parameter. Let $\tilde{k}_3\,(t) = \hat{k}_3\,(t) - k_3$. Then the expression (12.18) can be transformed to $\dot{\tilde{k}}_3(t) = \varepsilon\,\gamma\,\left\|X^T P(\varepsilon) B_2(\varepsilon)\right\|$.

As mentioned in Fan et al. (2010), to avoid discontinuity which can be caused by the term $\left\|X^T P(\varepsilon) B_2(\varepsilon)\right\|$ in the case where $X = 0$, it is sufficient to add a small constant in the denominator of the control law (12.17).

The following theorem is proposed to solve the fault-tolerant control problem (12.15).

Theorem 1 *Consider the singularly perturbed system described by Eq. (12.14). Suppose the following assumptions are satisfied:*

1. *There exists a singular perturbation parameter $\varepsilon^* > 0$ such that $(A(\varepsilon), B_2(\varepsilon))$ is stabilizable and $(C, A(\varepsilon))$ is detectable for all $\varepsilon \in]0, \varepsilon^*]$;*
2. *There exists a symmetric and positive definite matrix $P(\varepsilon)$ satisfying the following Lyapunov equation:*

$$(A(\varepsilon) + B_2(\varepsilon)\, K_1 C)^T\, P(\varepsilon) + P(\varepsilon)\,(A(\varepsilon) + B_2(\varepsilon)\, K_1 C) = -Q \tag{12.19}$$

where Q is any given positive definite symmetric matrix;

3. *K_1 is designed such that $(A(\varepsilon) + B_2(\varepsilon)\, K_1 C)$ is Hurwitz.*

Then the fault-tolerant controller (12.16) verifying the adaptive laws (12.17) and (12.18) stabilizes asymptotically the system (12.14) subject to sensor fault (12.13) and external disturbances for any $\varepsilon \in]0, \varepsilon^*]$.

Proof From Eqs. (12.15) and (12.16), it follows for the closed-loop fault-tolerant control system:

$$\begin{cases} \dot{X}(t) = A(\varepsilon)\, X(t) + B_2(\varepsilon) F\, w(t) + B_2(\varepsilon)\, K_1\, \rho\, C\, X(t) + B_2(\varepsilon)\, K_2(t) \\ y(t) = \rho\, C\, X(t) \end{cases} \tag{12.20}$$

which can be rewritten as

$$\begin{aligned} \dot{X}(t) = {} & (A(\varepsilon) + B_2(\varepsilon)\, K_1 C)\, X(t) + B_2(\varepsilon)\, K_1 (\rho - I)\, C X(t) \\ & + B_2(\varepsilon)\, K_2(t) + B_2(\varepsilon)\, F\, w(t) \end{aligned} \tag{12.21}$$

Define an ε-dependent Lyapunov function candidate,

$$V(\varepsilon) = X^T\, P(\varepsilon)\, X + \varepsilon^{-1} \gamma^{-1} \tilde{k}_3^2 > 0 \tag{12.22}$$

Computing the derivative of $V(\varepsilon)$ along the trajectories of system (12.21) and taking into account the assumptions about the disturbances leads to

$$\begin{aligned} \dot{V}(\varepsilon) = {} & X^T [(A(\varepsilon) + B_2(\varepsilon)\, K_1 C)^T\, P(\varepsilon) + P(\varepsilon)\,(A(\varepsilon) + B_2(\varepsilon)\, K_1 C)]\, X \\ & + 2 \varepsilon^{-1} \gamma^{-1} \tilde{k}_3\, \dot{\tilde{k}}_3 + 2 X^T P(\varepsilon) B_2(\varepsilon)\, K_2 + 2 X^T P(\varepsilon) B_2(\varepsilon)\, F\, w \\ & + 2 X^T P(\varepsilon) B_2(\varepsilon) K_1 (\rho - I) C X \end{aligned} \tag{12.23}$$

Let k_3 be a constant used to limit the unknown bounded constant $\overline{w}$ such that (Wang et al. 2013)

$$\left\| X^T P(\varepsilon) B_2(\varepsilon) \right\|\, \|F\|\, \bar{w} \le \left\| X^T P(\varepsilon) B_2(\varepsilon) \right\|\, k_3 \tag{12.24}$$

Since $\overline{\rho}_i \le 1$, it can be shown that

$$X^T P(\varepsilon) B_2(\varepsilon) K_1 (\rho - I) C X \le \|K_1 C X\|\, \left\| X^T P(\varepsilon) B_2(\varepsilon) \right\| \tag{12.25}$$

Substituting the adaptive law (12.17) and expression (12.19) into Eq. (12.23) yields the following form:

$$\dot{V}(\varepsilon) \leq -2\,X^T Q\,X + 2\,\left\|X^T P(\varepsilon) B_2(\varepsilon)\right\|\,k_3 + 2\,\|K_1 C X\|\,\left\|X^T P(\varepsilon) B_2(\varepsilon)\right\| \\ -2\,\frac{X^T P(\varepsilon) B_2(\varepsilon)\, B_2^T(\varepsilon) P(\varepsilon) X}{\left\|B_2^T(\varepsilon) P(\varepsilon) X\right\|}\,(\hat{k}_3(t) + \|K_1 C X\|) + 2\,\varepsilon^{-1}\,\gamma^{-1}\tilde{k}_3\,\dot{\tilde{k}}_3 \tag{12.26}$$

This can be transformed into the following form,

$$\dot{V}(\varepsilon) \leq -2\,X^T Q X + 2\,\left\|X^T P(\varepsilon) B_2(\varepsilon)\right\| \\ (k_3 - \hat{k}_3(t) + \|K_1 C X\| - \|K_1 C X\|) + 2\,\varepsilon^{-1}\,\gamma^{-1}\tilde{k}_3\,\dot{\tilde{k}}_3 \tag{12.27}$$

In the light of the adaptation law (12.18), it is easy to see that $\dot{V}(\varepsilon) \leq 0$. Therefore, the faulty closed-loop system is asymptotically stable for any singular perturbation parameter $\varepsilon \in]0, \varepsilon^*]$.

12.2.2.3 Controller Simplification

The fault-tolerant controller (12.16) involves an output feedback gain K_1 designed such that $(A(\varepsilon) + B_2(\varepsilon)\,K_1 C)$ is Hurwitz. Under mild technical conditions, asymptotic stability of the full-order singularly perturbed system is guaranteed through the asymptotic stability of both the slow subsystem and fast subsystem for sufficiently small values of the singular perturbation parameter ε (Kokotovic et al. 1999; Glielmo and Corless 2010).

The conception of the static output feedback gain K_1 can be achieved through the simultaneous design of static output feedback controllers for the reduced slow and fast subsystems for small ε and some constraints on the system (Khalil 1987; Glielmo and Corless 2010; Wang et al. 1993). Thus, a simplified ε-independent feedback controller is designed to stabilize the full-order system.

The approximations taken in Sect. 12.2 (see Eqs. (12.5) and (12.9)) permit to simplify $B_2(\varepsilon)$ in the control law (12.17) through $B_2 = \begin{bmatrix} B_{2x} \\ B_{2z} \end{bmatrix}$. The ε-dependent $P(\varepsilon)$ will be simplified in the following section.

12.2.2.4 Solving Lyapunov Equation Using Reduced-Order Models

In order to alleviate the numerical stiffness caused by the simultaneous occurrence of slow and fast phenomena and characterized by the presence of the small singular perturbation parameter ε, the full-order Lyapunov equation (12.19) will be formulated using slow and fast subsystem components.

The structure of $P(\varepsilon)$ is assumed to be of the form

$$P(\varepsilon) = \begin{bmatrix} P_1 & P_2 \\ P_2^T & P_3 \end{bmatrix} \tag{12.28}$$

The solution $P(\varepsilon)$ is ε-dependent, because Eq. (12.19) contains ε^{-1}-order matrices. Let Q be of the form $Q = \begin{bmatrix} I_{n_1} & 0 \\ 0 & \varepsilon^{-1} I_{n_2} \end{bmatrix}$ where I is identity matrix and $K_1 = \begin{bmatrix} K_{11} & K_{12} \end{bmatrix}$. Expanding the Lyapunov equation (12.19) after the substitution of $P(\varepsilon)$ leads to the following partitioned three equations:

$$a_{11}^T P_1 + \varepsilon^{-1} a_{21}^T P_2^T + P_1 a_{11} + \varepsilon^{-1} P_2 a_{21} = -I_{n_1} \tag{12.29}$$

$$\varepsilon^{-1} a_{21}^T P_3 + a_{11}^T P_2 + P_1 a_{12} + \varepsilon^{-1} P_2 a_{22} = 0 \tag{12.30}$$

$$a_{12}^T P_2 + \varepsilon^{-1} a_{22}^T P_3 + P_2^T a_{12} + \varepsilon^{-1} P_3 a_{22} = -\varepsilon^{-1} I_{n_2} \tag{12.31}$$

where $a_{11} = A_{11} + B_{2x} K_{11} C_1$, $a_{12} = A_{12} + B_{2x} K_{12} C_2$, $a_{21} = A_{21} + B_{2z} K_{11} C_1$, and $a_{22} = A_{22} + B_{2z} K_{12} C_2$.

Pre- and post-multiplying equations and (12.31) by the singular perturbation parameter ε and letting $\varepsilon \to 0$ lead to the following zero-order equations:

$$a_{11}^T P_1 + P_1 a_{11} = -I_{n_1} \tag{12.32}$$

$$a_{22}^T P_3 + P_3 a_{22} = -I_{n_2} \tag{12.33}$$

$$a_{21}^T P_2^T + P_2 a_{21} = 0 \tag{12.34}$$

Since K_{11} and K_{12} can be approximated, respectively, by K_s and K_f (Noura et al. 2000; Richter and Lunze 2010), it is clear that Eqs. (12.32) and (12.33) correspond, respectively, to the Lyapunov equations of the slow subsystem approximation described by Eq. (12.5) and the fast subsystem given by Eq. (12.9) of the singularly perturbed system (12.1). Hence, P_1 and P_3 correspond, respectively, to the solution P_s of the slow Lyapunov equation (12.3) (approximated by (12.6)) and the solution P_f of the fast Lyapunov equation (12.10). It follows that the solution $P(\varepsilon)$ of Eq. (12.19) can be approximated by $P(\varepsilon) \approx \begin{bmatrix} P_s & 0 \\ 0 & P_f \end{bmatrix}$. Thus, the ill-defined controller (12.16) is simplified and made ε-independent using the corresponding reduced order well-defined problem.

12.2.3 Controller Design with Respect to Actuator Fault

In this section, a fault-tolerant controller will be designed to compensate actuator fault and external disturbances and then the singular perturbation method will be used to simplify the ε-dependent controller to avoid the numerical stiffness caused by the presence of time scales.

12.2.3.1 Preliminaries and Failure Model

Consider the LTI singularly perturbed system (12.1) with its slow subsystem (12.2) and fast subsystem (12.9).

If (A_s, B_{2s}) is stabilizable, then the state feedback controller $u_s = G_s\, x_s$ is considered to stabilize the slow subsystem (12.2). G_s is the slow controller gain satisfying the following slow Lyapunov equation:

$$(A_s + B_{2s}G_s)^T L_s + L_s (A_s + B_{2s}G_s) = -M_s \tag{12.35}$$

with M_s any given positive definite symmetric matrix. The closed-loop slow subsystem is then obtained

$$\dot{x}_s = (A_s + B_{2s}G_s)\, x_s + B_{1s}w_s \tag{12.36}$$

Employing the same approximation used in Sect. 12.2.1 for the system matrices of the slow subsystem, the slow Lyapunov equation (12.35) and the closed-loop slow subsystem (12.36) can be approximated, respectively, by Eqs. (12.37) and (12.38).

$$(A_{11} + B_{2x}G_s)^T L_s + L_s (A_{11} + B_{2x}G_s) = -M_s \tag{12.37}$$

$$\dot{x}_s = (A_{11} + B_{2x}G_s)\, x_s + B_{1x}w_s \tag{12.38}$$

Assuming that (A_{22}, B_{2z}) is stabilizable, there exists a feedback controller $u_f = G_f\, z_f$ stabilizing the fast subsystem (12.9). G_f is the fast controller gain satisfying the fast Lyapunov equation (12.39).

$$(A_{22} + B_{2z}G_f)^T L_f + L_f (A_{22} + B_{2z}G_f) = -M_f \tag{12.39}$$

with M_f any given positive definite symmetric matrix. The closed-loop fast subsystem is then described by

$$\dot{z}_f = (A_{22} + B_{2z}G_f)\, z_f + B_{1z}w_f \tag{12.40}$$

The faults dealt with in this section are given by a decrease in effectiveness. For the control input u_i, $i = 1, \ldots, m$, consider u_i^F the signal from the faulty actuator; consequently, the failure model is adopted as follows:

$$u_i^F = \alpha_{ai}\, u_i \tag{12.41}$$

where α_{ai} represents the actuator efficiency factor and verifies $0 \leq \underline{\alpha}_{ai} \leq \alpha_{ai} \leq \overline{\alpha}_{ai} \leq 1$. $\underline{\alpha}_{ai}$ and $\overline{\alpha}_{ai}$ indicate the known lower and upper bounds on α_{ai}, respectively. Thus, if $\underline{\alpha}_{ai} = \overline{\alpha}_{ai} = 1$, the ith actuator is working perfectly, whereas if $\underline{\alpha}_{ai} > 0$, partial loss of effectiveness is present. The case $\underline{\alpha}_{ai} = \overline{\alpha}_{ai} = 0$ describes the completely failing of the actuator i. Denoting $\alpha_a = diag(\alpha_{ai}), i = 1, \ldots, m$, the uniform actuator-fault model becomes

$$u^F(t) = \alpha_a\, u(t) \tag{12.42}$$

The problem under consideration is to develop a fault-tolerant controller making the closed-loop singularly perturbed system asymptotically stable in the presence of actuator faults and external disturbances.

12.2.3.2 Controller Synthesis for the Global Singularly Perturbed System

Considering the system (12.14), the actuator fault (12.42) and the assumption held for the disturbances, system (12.14) can be rewritten as

$$\dot{X}(t) = A(\varepsilon)\, X(t) + B_2(\varepsilon)\, F\, w(t) + B_2(\varepsilon)\, \alpha_a\, u(t) \tag{12.43}$$

The proposed controller model to stabilize the system (12.14) under actuator faults and external disturbances is given by

$$u(t) = G_1(t)\; X(t) + G_2(t) \tag{12.44}$$

where G_1 is chosen such that $(A(\varepsilon) + B_2(\varepsilon)\, G_1)$ is Hurwitz and $G_2(t)$ is designed using the following update law:

$$G_2(t) = -\frac{\beta\, B_2^T(\varepsilon)L(\varepsilon)X}{\alpha \left\| B_2^T(\varepsilon)L(\varepsilon)X \right\|}\,(\hat{k}(t) + \|G_1 X\|) \tag{12.45}$$

where α and β are appropriate positive constants satisfying (Jin and Yang 2009; Fan and Song 2010; Wang et al. 2013).

$$\alpha \left\| B_2^T(\varepsilon)L(\varepsilon)X \right\|^2 \leq \beta \left\| B_2^T(\varepsilon)L(\varepsilon)X\sqrt{\alpha_a} \right\|^2 \tag{12.46}$$

and $\hat{k}$ obeys the following adaptive law:

$$\frac{d\hat{k}(t)}{dt} = \varepsilon\gamma \left\| X^T L(\varepsilon) B_2(\varepsilon) \right\| \tag{12.47}$$

where γ is an appropriate positive constant and ε is the singular perturbation parameter. Let $\tilde{k}(t) = \hat{k}(t) - k$, where k is a constant verifying

$$\left\| X^T L(\varepsilon) B_2(\varepsilon) \right\| \; \|F\| \; \bar{w} \leq \left\| X^T L(\varepsilon) B_2(\varepsilon) \right\| \; k \tag{12.48}$$

then the update law (12.47) can be expressed as

$$\dot{\hat{k}}(t) = \varepsilon \gamma \left\| X^T L(\varepsilon) B_2(\varepsilon) \right\| \tag{12.49}$$

As main result, the following theorem will be proposed to solve the fault-tolerant control problem (12.43).

Theorem 2 *Consider the system described by Eq. (12.14). Suppose that*

1. *There exists a positive singular perturbation parameter* $\varepsilon^* > 0$ *such that* $(A(\varepsilon), B_2(\varepsilon))$ *is stabilizable for all* $\varepsilon \in]0, \varepsilon^*]$;
2. *For any given positive definite symmetric matrix M, there exists a symmetric and positive definite matrix* $L(\varepsilon)$ *satisfying the following Lyapunov equation:*

$$(A(\varepsilon) + B_2(\varepsilon) G_1)^T L(\varepsilon) + L(\varepsilon) (A(\varepsilon) + B_2(\varepsilon) G_1) = -M \tag{12.50}$$

3. G_1 *is chosen such that* $(A(\varepsilon) + B_2(\varepsilon) G_1)$ *is Hurwitz.*

Then there exists $\varepsilon^* > 0$ such that for every $\varepsilon \in]0, \varepsilon^*]$, the controller described by Eq. (12.44), with the update laws (12.45) and (12.47) and the controller gain G_1, stabilize asymptotically the system (12.43) subject to actuator fault (12.42) and external disturbances.

Proof The Lyapunov-based proof of stability can be done using the following Lyapunov function candidate:

$$V(\varepsilon) = X^T \; L(\varepsilon) \; X + \varepsilon^{-1} \gamma^{-1} \tilde{k}^2 > 0 \tag{12.51}$$

The other steps are similar to Theorem 1 and are detailed in Tellili et al. (2016).

12.2.3.3 Design of the ε-Independent Controller

G_1 represents the state-feedback controller gain stabilizing the full-order singularly perturbed system (12.1) in the absence of faults and perturbations. It is well known (Dragan 2011; Kokotovic et al. 1999) that G_1 can be designed using the locale feedback gains G_s and G_f provided that (A_s, B_s) and (A_f, B_f) are controllable and A_{22}^{-1} exists. Accordingly, the composite control is of the form $u(t) = G_1(t) X(t) = \begin{bmatrix} G_{11} & G_{12} \end{bmatrix} \begin{bmatrix} x \\ z \end{bmatrix}$.

Where $G_{12} = G_f$ and $G_{11} = (I_f + G_f A_{22}^{-1} B_2) G_s + G_f A_{22}^{-1} A_{21}$. Thus, the simplification of the feedback controller gain G_1 conserves the stability of the full-order singularly perturbed system. To simplify the adaptive laws (12.45) and (12.47), it is adequate to use the slow parts of the command matrix $B_2(\varepsilon)$ and the simplified form of the Lyapunov matrix $L(\varepsilon)$ which will be derived in the following.

12.2.3.4 Simplifying of the Lyapunov Equation

To remove the numerical stiffness in the Lyapunov equation given by expression (12.50), the latter will be decomposed in slow and fast parts (Koskie et al. 2010). The structure of $L(\varepsilon)$ is assumed to be of the form $L(\varepsilon) = \begin{bmatrix} L_1 & L_2 \\ L_2^T & L_3 \end{bmatrix}$. The solution $L(\varepsilon)$ is dependent on the singular perturbation parameter ε because Eq. (12.50) contains ε^{-1}-order matrices. A positive definite symmetric matrix M will be chosen of the form $M = \begin{bmatrix} I_{n_1} & 0 \\ 0 & \varepsilon^{-1} I_{n_2} \end{bmatrix}$, where I is identity matrix. Substituting $L(\varepsilon)$ into Lyapunov equation (12.50) yields the following partitioned three equations:

$$a_{11}^T L_1 + \varepsilon^{-1} a_{21}^T L_2^T + L_1 a_{11} + \varepsilon^{-1} L_2 a_{21} = -I_{n_1} \tag{12.52}$$

$$\varepsilon^{-1} a_{21}^T L_3 + a_{11}^T L_2 + L_1 a_{12} + \varepsilon^{-1} L_2 a_{22} = 0 \tag{12.53}$$

$$a_{12}^T L_2 + \varepsilon^{-1} a_{22}^T L_3 + L_2^T a_{12} + \varepsilon^{-1} L_3 a_{22} = -\varepsilon^{-1} I_{n_2} \tag{12.54}$$

where $a_{11} = A_{11} + B_{2x} G_{11}, a_{12} = A_{12} + B_{2x} G_{12}, a_{21} = A_{21} + B_{2z} G_{11}$, and $a_{22} = A_{22} + B_{2z} G_{12}$.

Using the same method as in Sect. 12.2.1 yields the following equations:

$$a_{11}^T L_1 + L_1 a_{11} = -I_{n_1} \tag{12.55}$$

$$a_{22}^T L_3 + L_3 a_{22} = -I_{n_2} \tag{12.56}$$

$$a_{21}^T L_2^T + L_2 a_{21} = 0 \tag{12.57}$$

It is well known that the controller gains G_{11} and G_{12} can be approximated, respectively, by the slow controller gain G_s and the fast controller gain G_f (Shao and Sawan 2006). Hence, expressions (12.55) and (12.56) match, respectively, the slow and the fast Lyapunov equations until $M_s = I_{n_1}$ and $M_f = I_{n_2}$. Furthermore, considering Eq. (12.57), the approximate solution of the Lyapunov matrix becomes $L(\varepsilon) \approx \begin{bmatrix} L_s & 0 \\ 0 & L_f \end{bmatrix}$, which removes the numerical stiffness in the solution of Eq. (12.50).

12.2.4 *Example of Application*

To illustrate the effectiveness of the proposed method, the following numerical example is considered.

12.2.4.1 Fault-Tolerant Control in the Presence of Sensor Faults and Disturbance

Consider the singularly perturbed system (12.1) with parameters given by

$$A_{11} = \begin{bmatrix} -5 & 0.2 \\ -0.5 & 6 \end{bmatrix}, A_{12} = \begin{bmatrix} 0 & 0.1 \\ -1 & 1 \end{bmatrix}, A_{21} = \begin{bmatrix} -9 & -8 \\ 0.3 & 0.1 \end{bmatrix}, A_{22} = \begin{bmatrix} -7 & 1 \\ -0.5 & -6 \end{bmatrix},$$

$$B_2 = B_1 = \begin{bmatrix} 1 & 1.55 \\ 1 & -0.5 \\ 0.9 & 0.8 \\ 0.2 & 0.11 \end{bmatrix}, C = \begin{bmatrix} -2 & 0.5 & 0 & 0 \\ 1 & -2 & 0 & 0 \end{bmatrix} \text{ and } w = \begin{bmatrix} 0.01 & \sin(5t) \\ & 0.5 \end{bmatrix}$$

The full-order system is open-loop unstable (the system has one positive pole). The considered faulty model is a 50% loss of effectiveness in the first sensor and 70% in the second sensor, that is, $\alpha_a = diag(0.5, 0.7)$. The singular perturbation parameter is set to $\varepsilon = 0.05$. In the fault-free case, the output feedback controller is computed using the method described in Glielmo et al. (2010) based on the reduced order models. The state trajectories starting from $X_0 = [0.2,\ 0.03,\ 0.02,\ 0.01]^T$, the output, and the simplified controller gain are shown in Fig. 12.1. It is clear from the abovementioned figure that the full-order system is stabilized using the output feedback controller. However, the occurrence of sensor faults at time instance 50 s causes instability of the full-order system (see Fig. 12.2). Next, an adaptive fault-tolerant controller will be designed in the presence of sensor faults and external disturbances. The simulation results are shown in Fig. 12.3. They indicate that the designed adaptive fault-tolerant controller stabilizes asymptotically the singularly perturbed system subject to sensor faults and external disturbances.

Next, the case of actuator faults will be examined.

12.2.4.2 Fault-Tolerant Control in the Presence of Actuator Faults and Disturbance

The same singularly perturbed system treated below is considered with 10% loss of efficiency in the first actuator and 50% in the second actuator, that is, $\rho = \text{diag}(0.1, 0.5)$. In the fault-free case, the full-order system is stabilized through a composite controller based on the slow and fast subsystems (see Fig. 12.4). The appearance of defects at time 50 s yields a loss of the actuators performances which is indicated in Fig. 12.5. To compensate the fault effect, an adaptive fault-tolerant controller is designed. Figure 12.6 shows the trajectories of the states, the output, the controller $u(t)$, and the gain $k_3(t)$ after fault occurrence at time 50 s by initial values $X_0 = [0.2,\ 0.03,\ 0.02,\ 0.01]^T$ and $\varepsilon = 0.05$. It can be deduced that the computed adaptive controller compensates the actuator faults in the presence of external disturbances.

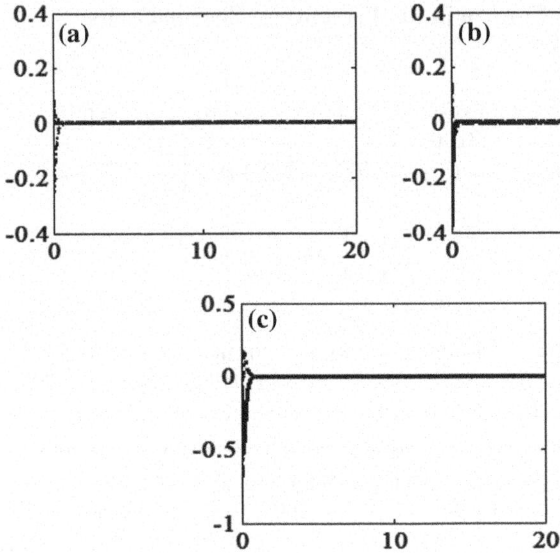

Fig. 12.1 **a** States, **b** Output, and **c** Controller $u(t)$ in the fault-free case by output feedback control and by $\varepsilon = 0.05$

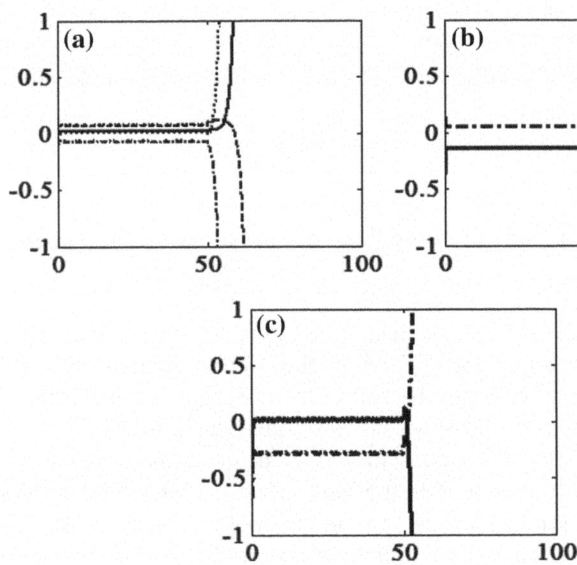

Fig. 12.2 **a** States, **b** Output, and **c** Controller $u(t)$ in the faulty case by output feedback control and by $\varepsilon = 0.05$

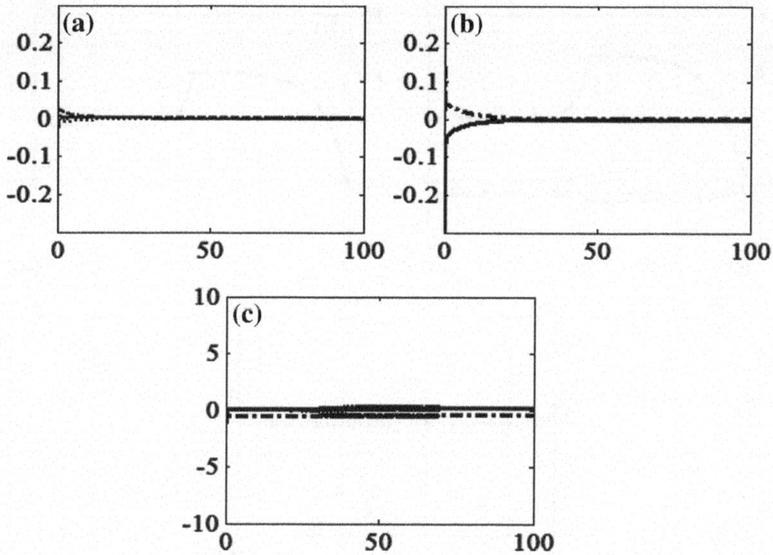

Fig. 12.3 **a** States, **b** Output, and **c** Controller $u(t)$ in the faulty case by adaptive fault-tolerant control and by $\varepsilon = 0.05$

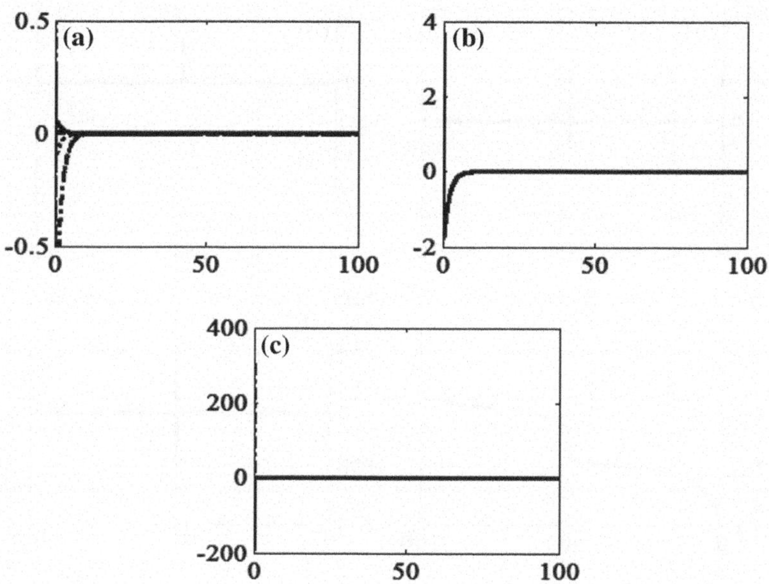

Fig. 12.4 **a** States, **b** Output, and **c** Controller $u(t)$ in the fault-free case by composite control

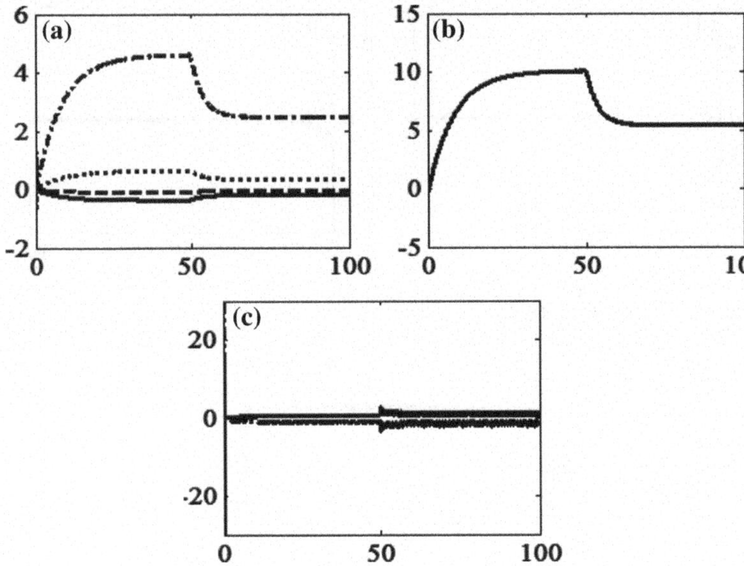

Fig. 12.5 **a** States, **b** Output, and **c** Controller $u(t)$ in the faulty case by composite control

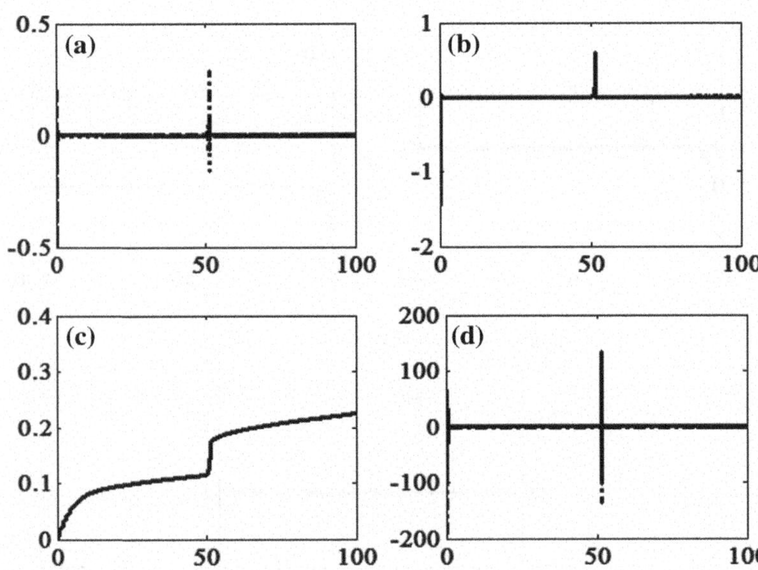

Fig. 12.6 **a** States, **b** Output, **c** Gain $k_3(t)$, and **d** Controller $u(t)$ in the faulty case by adaptive FTC control

12.3 Nonlinear Case

For nonlinear systems, meaningful results are presented in the literature. Two different methods are presented. Passive fault-tolerant controller design methods, where the controller parameters remain unchangeable throughout the faulty and the fault-free case, use techniques like Hamilton–Jacobi inequality approach and robust pole region assignment method (Yang et al. 1998; Zhao and Jiang 1998). In contrast to the passive methods, active fault-tolerant control systems, like sliding mode-based control methods and adaptive back-stepping compensation control, need a fault diagnosis block to the fault diagnosis in real time, and then a mechanism to ensure the compensation of detected faults (Corradini and Orlando 2007; Wang and Wen 2011). Diverse techniques are proposed in the literature to design fault-tolerant controllers for nonlinear systems. Liang and Xu (2006) proposed a variable structure stabilizing control law to tolerate the presence of actuator fault in a nonlinear system. The derived control scheme does not require the solution of Hamilton–Jacobi inequality. Benosman and Lum (2010) developed a Lyapunov-based passive feedback controller to guarantee stability of nonlinear affine system with actuator faults. Ma and Yang (2012) designed a fault diagnosis procedure and an active fault-tolerant control scheme for nonlinear uncertain dynamic system against time-varying actuator fault. An extended high-gain observer is applied to provide more online information to generate the reconfigurable controller.

On the other hand, several approaches to control singularly perturbed systems have been proposed. In particular, in linear case, Li (2002) derived a reliable linear-quadratic state feedback control for singularly perturbed systems in the presence of actuator failures. The designed controller is based on controllers of slow and fast subsystems so that it becomes independent of the singular perturbation parameter. Tellili et al. (2015) proposed a robust adaptive fault-tolerant control method to compensate for actuator and sensor faults in the presence of external perturbation. In both cases, a controller for the original system was designed and then simplified using singular perturbation method. In nonlinear case, reconfigurable control was investigated by some authors in case of systems subject to actuator saturation (Liu 2001; Yang et al. 2013; Zhou et al. 2013). However, the abovementioned methods for nonlinear singularly perturbed systems are limited to actuator saturation and did not consider actuator additive faults.

The main objective of this part is to develop a fault-tolerant control scheme for nonlinear singularly perturbed systems in the presence of additive actuator fault. The remaining part of this section is organized as follows. The system description and problem formulation are presented in Sect. 12.3.1. The main results are formulated in Sect. 12.3.2. An example of application is developed in Sect. 12.3.3.

12.3.1 System Description and Problem Formulation

Consider the following time-invariant nonlinear two-time-scale singularly perturbed system of the form: (Fridman 2000; Aliyu and Boukas 2011).

$$\begin{cases} \dot{x}(t) = f_1(x, y) + g_1(x, y)\, u(t) \\ \varepsilon\, \dot{y}(t) = f_2(x, y) + g_2(x, y)\, u(t) \end{cases} \tag{12.58}$$

where $x \in \mathrm{B}_x \subset \mathbb{R}^{n_1}$ and $y \in \mathrm{B}_y \subset \mathbb{R}^{n_2}$ are state vectors, $u \in \mathbb{R}^m$ is the control vector. For $i = 1, 2, f_i$ and g_i are locally Lipschitz in a domain that contains the origin and f_i satisfy $f_i(0, 0) = 0$. The scalar ε is a singular perturbation parameter taking values between 0 and 1; it gives the speed ratio between the slow and the fast dynamics. It is supposed that $(x, y) = (0, 0)$ is an isolated equilibrium state.

The singular perturbation theory will be used to approximate the slow and fast dynamics by setting $\varepsilon = 0$ in the $\dot{y}$-equation and solving for y in terms of x (Kokotovic et al. 1999; Saberi and Khalil 1985; Siddarth and Valasek 2011). The second equation in system (12.58) can be expressed as

$$0 = f_2(x, y) + g_2(x, y)\, u(t) \tag{12.59}$$

The singularly perturbed system is supposed to be standard, which means Eq. (12.59) has a unique solution $y = h(x, u_s)$, where u_s is the slow part of the control u. Substituting this solution into the first equation of system (12.58), the following reduced slow subsystem is obtained:

$$\dot{x}(t) = f_1(x, h(x, u_s)) + g_1(x, h(x, u_s))\, u_s(t) \tag{12.60}$$

It is assumed that the origin of the closed-loop slow subsystem (12.60) is the unique asymptotically stable equilibrium, so there exists a feedback slow control law $u_s(t) = p_s(x)$ that renders the slow dynamics asymptotically stable and a positive definite Lyapunov function $V_s(x)$ guaranteeing for all $x \in \mathrm{B}_x$,

$$\frac{\partial V_s(x)}{\partial x} [f_1(x, h(x, u_s)) + g_1(x, h(x, u_s))\, u_s] \leq -a\, L_s^2(x) \tag{12.61}$$

where $a \succ 0$, $L_s(x)$ is a positive definite function and p_s is locally Lipschitz vector function. The reduced fast subsystem, called boundary layer system, is given by

$$\left\{ \frac{dy}{d\tau} = f_2(x, y) + g_2(x, y)\, (u_s + u_f) \right. \tag{12.62}$$

in which the fast time scale is defined as $\tau = t/\varepsilon$ and x is assumed to be a constant parameter equal to its initial value. There exists a feedback fast control law $u_f(t) = p_f(x, y)$ which asymptotically stabilize the fast dynamics, such that $y = h(x, u_s)$ is supposed asymptotically stable equilibrium of the closed-loop fast subsystem uniformly in $x \in \mathbb{R}^{n_1}$. The fast controller satisfies $p_f(x, h(x, p_s(x)) = 0$, where p_f is locally Lipschitz vector function. The composite control of the overall system is expressed as a sum of the slow and fast sub-controllers (Kokotovic et al. 1999; Siddarth and Valasek 2011; Chow and Kokotovic 1978).

$$u = u_s(x) + u_f(x, y) \tag{12.63}$$

The composite feedback control is designed so that the origin is an asymptotically stable equilibrium of the singularly perturbed closed-loop system (12.58). Let $V_f(x, y)$ be a positive definite Lyapunov function for the fast subsystem (12.62) such that for all $(x, y) \in B_x \times B_y$:

$$\frac{\partial V_f}{\partial y}\left[f_2(x, y) + g_2(x, y)\,(p_s + p_f)\right] \leq -b\,L_f^2(y - h(x, p_s)) \tag{12.64}$$

where $b \succ 0$ and L_f is a positive definite function.

A composite Lyapunov function candidate for the singularly perturbed system (12.58) is defined by a weighted sum of the Lyapunov functions for the reduced slow and fast subsystems, so that

$$W(x, y) = (1-d)\,V_s + d\,V_f \tag{12.65}$$

where $0 \prec d \prec 1$ is a free parameter, which is to be to be chosen.

The derivative of this composite Lyapunov function along the trajectories of (12.58) gives

$$\frac{\partial W(x, y)}{\partial t} = (1-d)\,\frac{\partial V_s(x)}{\partial x}\,\frac{\partial x}{\partial t} + d\,\left(\frac{\partial V_f(x, y)}{\partial x}\,\frac{\partial x}{\partial t} + \frac{\partial V_f(x, y)}{\partial y}\,\frac{\partial y}{\partial t}\right) \tag{12.66}$$

Using Eq. (12.58), we get

$$\begin{aligned}\frac{\partial W(x, y)}{\partial t} &= (1-d)\,\frac{\partial V_s(x)}{\partial x}\,[f_1(x, y) + g_1(x, y)\,u] \\ &+ \frac{d}{\varepsilon}\,\frac{\partial V_f(x, y)}{\partial y}\,[f_2(x, y) + g_2(x, y)\,u] \\ &+ d\,\frac{\partial V_f(x, y)}{\partial x}\,[f_1(x, y) + g_1(x, y)\,u]\end{aligned} \tag{12.67}$$

After some algebraic manipulations, Eq. (12.67) can be expressed as

$$\begin{aligned}\frac{\partial W(x, y)}{\partial t} &= \frac{d}{\varepsilon}\,\frac{\partial V_f(x, y)}{\partial y}\,[f_2(x, y) + g_2(x, y)\,u] \\ &+ (1-d)\frac{\partial V_s(x)}{\partial x}\,[f_1(x, h(x, u_s)) \\ &+ g_1(x, h(x, u_s))\,u_s] + T(x, y, u, u_s)\end{aligned} \tag{12.68}$$

The first two terms are the derivatives of V_f and V_s along the trajectories of the fast and slow subsystems. These two terms are negative definite in y and x, respectively, by inequalities (12.64) and (12.61).

The last term represents the effect of the interconnection between the slow and fast dynamics. Its upper bound is assumed to depend on comparison functions L_s and L_f.

Using inequalities (12.61) and (12.64), and some algebraic manipulations, Eq. (12.68) can be rearranged further to get (Khalil 2015; Chen 1998).

$$\frac{\partial W(x,y)}{\partial t} \leq -N(x,y) \tag{12.69}$$

where N(x, y) is positive definite. Consequently, the composite control (12.63) ensures that the origin is an asymptotically stable equilibrium of the closed-loop system (12.58) for a given interval of the singular perturbation parameter ε and $W(x,y)$ is a Lyapunov function of this system (Kokotovic et al. 1999; Saberi and Khalil 1985; Narang-Siddarth and Valasek 2014).

12.3.2 Principal Results

In the following, the essential results will be presented. Suppose that both slow and fast subsystems are stabilizable in $B_x \times B_y$ and the state (x, y) is available for feedback, we seek for a control law which asymptotically stabilizes the equilibrium state $(x = 0, y = 0)$ of the closed-loop singularly perturbed system despite the actuator fault occurrence.

The system (12.58), in the presence of additive actuator fault, can be expressed as

$$\begin{cases} \dot{x}(t) = f_1(x,y) + g_1(x,y)\,(u + D(t,x,y)) \\ \varepsilon\,\dot{y}(t) = f_2(x,y) + g_2(x,y)\,(u + D(t,x,y)) \end{cases} \tag{12.70}$$

where $D(t, x, y)$ represents an actuator fault which verifies $\|D(t,x,y)\| \leq B(t,x,y)$ and $B(t, x, y)$ is a nonnegative continuous function. The controller takes the following form:

$$u = u_{nom} + u_{add} \tag{12.71}$$

where u_{nom} represents the nominal controller that stabilizes the overall system in the fault-free case, it corresponds to the composite controller (12.63). u_{add} denotes the additive part to be designed to compensate for the effect of the actuator fault. The proposed fault-tolerant controller to stabilize the faulty system (12.70) is expressed as

$$u = u_{nom} - B\,\frac{(1-d)\frac{\partial V_s(x)}{\partial x}\,g_1 + d\,\frac{\partial V_f(x,y)}{\partial y}\,g_2}{\left\|(1-d)\frac{\partial V_s(x)}{\partial x}\,g_1 + d\,\frac{\partial V_f(x,y)}{\partial y}\,g_2\right\|} \tag{12.72}$$

where $(.)^T$ denotes the transpose of (.) and $\|(.)\|$ the Euclidian norm of (.).

The following theorem is proposed to accomplish the fault-tolerant control of the nonlinear singularly perturbed system (12.70) even when some actuators operate abnormally:

Theorem 1 *Consider the faulty nonlinear singularly perturbed system given by Eq. (12.70). Suppose that the slow and fast subsystems are stabilizable, and the actuator fault is bounded. There exists a singular perturbation parameter* $\varepsilon^* > 0$ *such that for all* $\varepsilon \in]0, \varepsilon^*]$*, the origin* $(x, y) = (0, 0)$ *of the faulty overall system is locally asymptotically stable under the fault-tolerant control law (12.72) even when the actuators undergo abnormal operation.*

Proof From Eqs. (12.70) and (12.71), it follows for the closed-loop faulty nonlinear singularly perturbed system:

$$\begin{cases} \dot{x}(t) = f_1(x, y) + g_1(x, y)\,(u_{nom} + u_{add} + D(t, x, y)) \\ \varepsilon\,\dot{y}(t) = f_2(x, y) + g_2(x, y)\,(u_{nom} + u_{add} + D(t, x, y)) \end{cases} \tag{12.73}$$

Define an ε-dependent Lyapunov function candidate,

$$W(\varepsilon, x, y) = (1 - d)\,V_s + \varepsilon\, d\, V_f \tag{12.74}$$

Computing the derivative of $W\,(\varepsilon, x, y)$ along the trajectories of system (12.73) leads to

$$\begin{aligned} \frac{\partial W(\varepsilon, x, y)}{\partial t} = {} & d\,\frac{\partial V_f(x, y)}{\partial y}\,[f_2 + g_2\,(u_{nom} + u_{add} + D)] \\ & + (1 - d)\frac{\partial V_s(x)}{\partial x}[f_1 + g_1(u_{nom} + u_{add} + D)] \\ & + d\,\varepsilon\,\frac{\partial V_f(x, y)}{\partial x}\,[f_1 + g_1\,(u_{nom} + u_{add} + D)] \end{aligned} \tag{12.75}$$

The isolation of the terms depending of the nominal control yields the following system:

$$\begin{aligned} \frac{\partial W(\varepsilon, x, y)}{\partial t} = {} & d\,\frac{\partial V_f(x, y)}{\partial y}(f_2 + g_2 u_{nom}) + (1 - d)\,\frac{\partial V_s(x)}{\partial x}\,(f_1 + g_1\,u_{nom}) \\ & + d\,\varepsilon\,\frac{\partial V_f(x, y)}{\partial x}\,(f_1 + g_1\,u_{nom}) \\ & + (1 - d)\,\frac{\partial V_s(x)}{\partial x}\,g_1(u_{add} + D) + d\,\frac{\partial V_f(x, y)}{\partial y}\,g_2(u_{add} + D) \\ & + d\,\varepsilon\,\frac{\partial V_f(x, y)}{\partial x}\,g_1(u_{add} + D) \end{aligned} \tag{12.76}$$

set $R_1^T = \frac{\partial V_s(x)}{\partial x} g_1$, $R_2^T = \frac{\partial V_f(x,y)}{\partial y} g_2$, $R_3^T = \frac{\partial V_f(x,y)}{\partial x} g_1$ and tacking in account inequality (12.69), Eq. (12.77) takes the following form:

$$\frac{\partial W(\varepsilon,x,y)}{\partial t} \le -N(x,y) + (1-d)\,R_1^T\,(u_{add}+D) + d\,R_2^T\,(u_{add}+D) + d\,\varepsilon\,R_3^T\,(u_{add}+D) \tag{12.77}$$

which can be rearranged in the following form:

$$\frac{\partial W(\varepsilon,x,y)}{\partial t} \le -N(x,y) + ((1-d)\,R_1^T + d\,R_2^T + d\,\varepsilon\,R_3^T)\,u_{add} + ((1-d)\,R_1^T + d\,R_2^T + d\,\varepsilon\,R_3^T)D \tag{12.78}$$

Using the singular perturbation method by setting the singular perturbation parameter ε to zero and considering the upper bound of the fault leads to the following expression:

$$\frac{\partial W(0,x,y)}{\partial t} \le -N(x,y) + ((1-d)\,R_1^T + d\,R_2^T)\,u_{add} + \left\|((1-d)\,R_1^T + d\,R_2^T)\right\|\,B \tag{12.79}$$

The first term on the right-hand side is due to the nominal composite control of the fault-free overall system; this term is negative definite in x and y by Eq. (12.69), whereas the second and third terms designate, respectively, the effect of the additive control u_{add} and the fault $D(t, x, y)$ on $\dot{W}(x, y)$. Consequently, u_{add} will be chosen to compensate for the effect of the actuator fault $D(t, x, y)$ on $\dot{W}(x, y)$.

In view of the control law (12.72) and taking into account the assumptions about the fault, it is obvious that $\dot{W}(x, y) \le 0$. Thus, it can be concluded that the origin $(x, y) = (0, 0)$ of the faulty overall system is a locally asymptotically stable equilibrium point for the system (12.73) under the fault-tolerant control law (12.72) for any singular perturbation parameter $\varepsilon \in\,]0, \varepsilon^*]$.

Remark The discontinuity of the control law (12.72) may engender chattering effect. This problem is usually solved by approximating the discontinuous function by a saturation function.

12.3.3 Example of Application

To illustrate the effectiveness of the proposed method, the following nonlinear singularly perturbed system is considered:

$$\begin{cases} \dot{x} = -x + y \\ \varepsilon\,\dot{y} = -x - e^y + 1 + u \end{cases} \tag{12.80}$$

The reduced slow subsystem is established by taking $\varepsilon = 0$ in (12.80). It takes the following form:

$$\dot{x} = \log_e(-x + 1 + u_s) - x \tag{12.81}$$

and the slow part of the state y is expressed as

$$y_s = \log_e(-x + 1 + u_s) \tag{12.82}$$

It is clear that the slow control $u_s = x - 1 + e^{-x}$ ensures that the closed-loop reduced slow subsystem (12.81) becomes asymptotically stabilizing about the origin. A corresponding slow Lyapunov function candidate is $V_s = 0.5\,x^2$. The reduced fast subsystem is given by

$$\frac{dy}{d\tau} = -x - e^y + 1 + u_s + u_f \tag{12.83}$$

Taking into account the abovementioned slow control, the fast control $u_f = e^y - e^{-x} - y_f$ stabilizes the state y about the origin. y_f is the fast part of the state y, such that $y_f = y - y_s$. A corresponding fast Lyapunov function candidate is $V_f = 0.5\,y_f^2 = 0.5\,(x + y)^2$. The composite control, considered as the nominal control of the fault-free overall system, will be designed as the sum of the slow and fast reduced-order controls, taking the following form:

$$u_{comp} = u_s + u_f = e^y - y - 1 \tag{12.84}$$

The resulting closed-loop overall system, after substituting of (12.84) in (12.80) is

$$\begin{cases} \dot{x} = -x + y \\ \varepsilon\,\dot{y} = -x - y \end{cases} \tag{12.85}$$

where the real parts of the eigenvalues remain negative. This concludes that the origin of system (12.85) becomes an asymptotically stable equilibrium of the closed-loop system. The simulation results in Fig. 12.7 show the composite control and the states trajectories in the fault-free case, starting from $(x_0, y_0) = (-2, 2)$. It is clear that the composite control (12.84) ensures the asymptotical stability of the origin.

However, the occurrence of constant additive actuator fault of amplitude 1, at time instance 10 s, yields a loss of the actuator performance. Consequently, the states drive to another stationary point (see Fig. 12.8a). This means that the composite controller failed to stabilize the origin equilibrium point in the faulty case.

Next, a fault-tolerant controller will be designed, according to Eq. (12.72), to compensate for the actuator fault, and it takes the form:

$$u = u_{comp} + u_{add} = e^y - y - 1 - 0.7\,\frac{x + y}{\|x + y\|} \tag{12.86}$$

As shown in Fig. 12.8b, the states' deviation is corrected using the fault-tolerant control (12.86) and the singularly perturbed system stabilizes at the origin equilibrium point, despite the presence of actuator faults. The zoom in Fig. 12.8b illustrates the

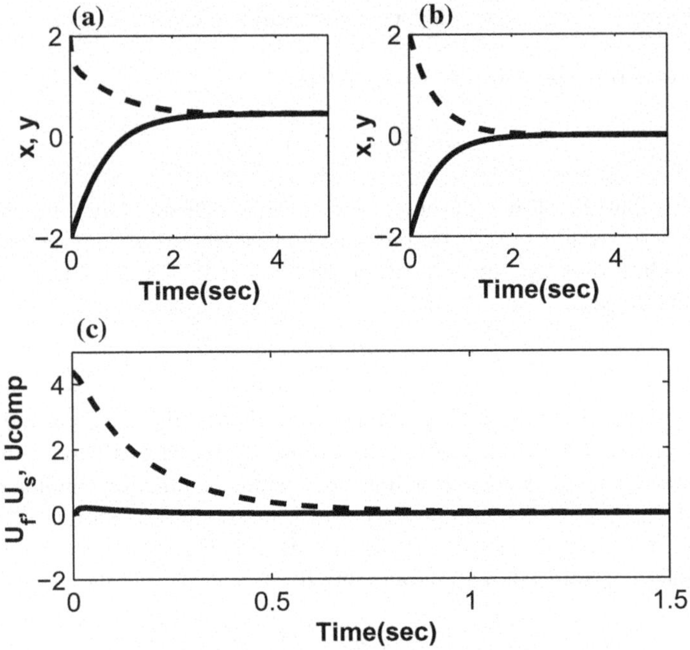

Fig. 12.7 States trajectories in fault-free case. **a** x (dashed line) and y (solid line) in open loop. **b** States in closed loop with composite control. **c** Fast (-.), slow (–) and composite (-) controllers by $\varepsilon = 0.01$

effect of fault on the states when the fault-tolerant control is used. Figure 12.8c depicts the corresponding fault-tolerant control which presents high chattering effects.

To solve this problem, the discontinuous function will be changed with a saturation function. Simulation results in Fig. 12.9a and b represent, respectively, the states and the controller after the substitution of the discontinuity in the control law. It is clear that the chattering effect is reduced, and the states remain at the same equilibrium point.

12.4 Conclusion

In this work, the stabilization problem of singularly perturbed systems subject to additive faults and external disturbances is investigated. ε-dependent controllers are first proposed to handle sensor and actuator faults covering the normal operation and faulty cases. The control scheme includes a feedback controller to handle the fault-free case and an adaptive part to compensate additive faults and external disturbances. The resulting ill-conditioned Lyapunov equation in both cases is solved using the singular perturbation method and the simultaneous design of Lyapunov equations of the

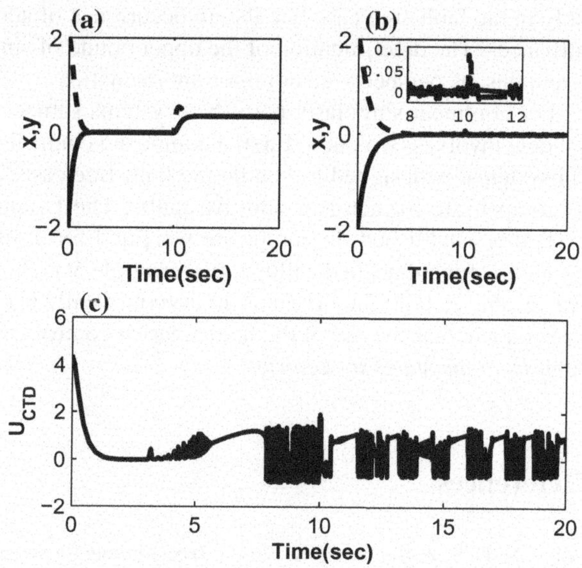

Fig. 12.8 States trajectories after the appearance of the actuator fault. **a** x (dashed line) and y (solid line) in case of nominal control. **b** States in case of fault-tolerant control. **c** Fault-tolerant control controller by $\varepsilon = 0.2$

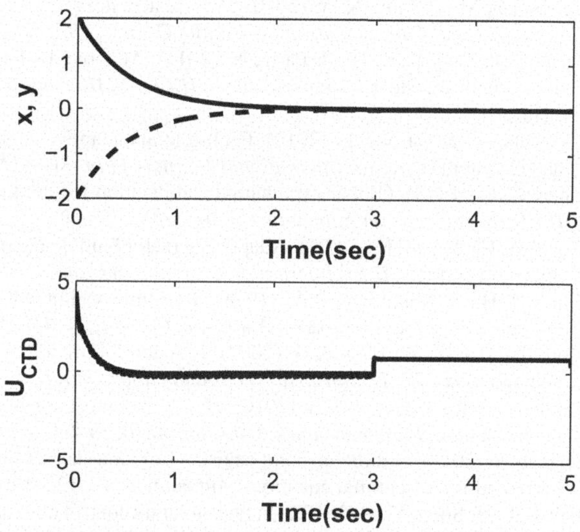

Fig. 12.9 States trajectories (**a**) fault-tolerant control (**b**) after the attenuation of chattering effect

slow and fast subsystems. In the case of sensor faults, the first part of the controller, designed by an output feedback controller, is approximated by gains stabilizing the reduced-order systems. While, in occurrence of actuator faults, the state-feedback controller is approximated by a composite controller which depends on the gains stabilizing the slow and fast subsystems. The adaptive laws are simplified using ε-independent matrices. The synthesis of the reconfigurable control system guarantees the desired asymptotic stability of the full-order singularly perturbed system not only in the fault-free case but also in occurrence of additive faults and external disturbances. The determination of the upper bound of singular perturbation parameter ε remains, as perspective, an important thematic.

For nonlinear singularly perturbed systems subject to actuator fault, the control scheme involves two parts. First, a composite controller, based on the slow and fast subsystems, is designed to handle the fault-free case. The second part is generated to compensate for actuator additive faults. The Lyapunov function for the overall system is established, in composite weighted form, using the slow and fast locale Lyapunov functions. In the illustrative example, it is shown that the composite control was unable to hold the origin as an asymptotically stable equilibrium of the overall system, whereas the use of the fault-tolerant control eliminates the effect of actuator fault from the states trajectories.

References

Aliyu, M. D. S., & Boukas, E. K. (2011). $H\infty$-filtering for singularly perturbed nonlinear systems. *International Journal of Robust and Nonlinear Control, 21*, 218–236.

Benosman, M., & Lum, K. Y. (2010). Passive actuators' fault-tolerant control for affine nonlinear systems. *IEEE Transactions on Automatic Control, 18*, 152–163.

Bustan, D., Sani, S. K. H., & Pariz, N. (2014). Adaptive fault-tolerant spacecraft attitude control design with transient response control. *IEEE/ASME Transactions on Mechatronics, 19*, 1404–1411.

Casavola, A., & Garone, E. (2010). Fault-tolerant adaptive control allocation schemes for overactuated systems systems. *International Journal of Robust and Nonlinear Control, 20*, 1958–1980.

Chen, C. C. (1998). Global exponential stabilization for nonlinear singularly perturbed systems. *IEEE Transactions on Automatic Control, 145*, 377–382.

Chen, W. H., & Saif, M. (2007). Adaptive actuator fault detection, isolation and accommodation in uncertain systems. *IEEE International Journal of Control, 80*, 45–63.

Chow, J. H., & Kokotovic, P. V. (1978). Two-time-scale feedback design of a class of nonlinear systems. *IEEE Transactions on Automatic Control, 23*, 438–443.

Corradini, M. L., & Orlando, G. (2007). Actuator failure identification and compensation through sliding modes. *IEEE Transactions on Automatic Control, 15*, 184–190.

Datta, K., & Rai Chaudhuri, A. (2010). $H_2/H\infty$ control of singularly perturbed systems: The state feedback case. *European Journal of Control, 10*, 54–69.

Dragan, V. (2011). Stabilizing composite control for a class of linear systems modeled by singularly perturbed Ito differential equations. *Automatica, 47*, 122–126.

Fan, L. L., & Song, Y. D. (2010). On fault-tolerant control of dynamic systems with actuator failures and external disturbances. *Acta Automatica Sinica, 36*, 1620–1625.

Fridman, E., & Shaked, U. (2000). Robust $H\infty$ minimum entropy static output-feedback control of singularly perturbed systems. *Automatica, 36*, 1181–1188.

Fridman, E. (2000). Exact slow-fast decomposition of the nonlinear singularly perturbed optimal control problem. *Systems & Control Letters, 40*, 121–131.

Glielmo, L., & Corless, M. (2010). On output feedback control of singularly perturbed systems. *Applied Mathematics and Computation, 217*, 1053–1070.

El Hachemi, F., Sigalotti, M., & Daafouz, J. (2012). Stability analysis of singularly perturbed switched linear systems. *IEEE Transactions on Automatic Control, 57*, 2116–2121.

Jin, X.-Z., & Yang, G.-H. (2009). Robust adaptive fault-tolerant compensation control with actuator failures and bounded disturbances. *Acta Automatica Sinica, 35*, 305–309.

Khalil, H. K. (1987). Output feedback control of linear two-time-scale systems. *IEEE Transactions on Automatic Control*, *32*, 784–792.

Khalil, H. K., & Chen, F. C. (1992). $H\infty$ control of two-time-scale systems. *Systems and Control Letters*, *19*, 35–42.

Khalil, H. K. (2015). *Nonlinear Control* (Global ed.). New York: Pearson Education.

Kokotovic, P., Khalil, H. K., & O'Reilly, J. (1999). *Singular perturbation methods in control-Analysis and design*. New York: Academic, Harcourt Brace Jovanovich publishers.

Koskie, S., Coumarbatch, C., & Gajic, Z. (2010). Adaptive output feedback control using fault compensation and fault estimation for linear system with actuator failure. *Applied Mathematics and Computation*, *216*, 1404–1411.

Liang, Y. W., & Xu, S. D. (2006). Reliable control of nonlinear systems via variable structure scheme. *IEEE Transactions on Automatic Control*, *51*, 1721–1726.

Li, Y., Wang, J. L., & Yang, G. H. (2002). Reliable linear quadratic control for singularly perturbed systems. *International Journal of Systems Science*, *33*, 949–958.

Liu, P. L. (2001). Stabilization of singularly perturbed multiple-time-delay systems with a saturating actuator. *International Journal of Systems Science*, *32*, 1041–1045.

Liu, P. L., Shen, Y., Dowell, E. H., & Zhu, C. (2012). A general $H\infty$ fault tolerant control and management for a linear system with actuator faults. *Automatica*, *48*, 1676–1682.

Ma, H. J., & Yang, G. H. (2012). Detection and adaptive accommodation for actuator faults of a class of non-linear systems. *IET Control Theory and Applications*, *6*, 2292–2307.

Narang-Siddarth, A., & Valasek, J. (2014). *Nonlinear time scale systems in standard and nonstandard forms-analysis and control*. Philadelphia: Siam.

Noura, H., Sauter, D., Hamelin, F., & Theilliol, D. (2000). Fault tolerant control in dynamic systems: Application to a winding machine. *IEEE Control Systems*, *20*, 33–49.

Patton, R.J. (1997). Fault-tolerant control systems: The 1997 situation. In *Proceedings of the Safe'Process*, UK

Richter, J. H., & Lunze, J. (2010). Reconfigurable control of Hammerstein systems after actuator failures: Stability, tracking, and performance. *International Journal of Control*, *47*, 1612–1630.

Richter, J. H., Heemels, W. P. M. H., van de Wouw, N., & Lunze, J. (2011). Reconfigurable control of piecewise affine systems with actuator and sensor faults: Stability and tracking. *Automatica*, *83*, 678–691.

Saberi, A., & Khalil, H. K. (1985). Stabilization and regulation of nonlinear singularly perturbed systems-composite control. *IEEE Transactions on Automatic Control*, *30*, 739–747.

Shao, Z. H., & Sawan, M. E. (2006). Stabilization of uncertain singularly perturbed systems. *IEE Proceedings: Control Theory and Applications*, *153*, 99–103.

Siddarth, A., & Valasek, J. (2011). Kinetic state tracking for a class of singularly perturbed systems. *Journal of Guidance, Control, and Dynamics*, *34*, 734–749.

Stoustrup, J., & Blondel, V. D. (2004). Fault tolerant control: A simultaneous stabilization result. *IEEE Transactions on Automatic Control*, *49*, 305–310.

Tan, W., Leung, T., & Tu, Q. (1998). $H\infty$ control for singularly perturbed systems. *Automatica*, *34*, 255–260.

Tao, G., Chen, S. H., & Joshi, S. M. (2002). An adaptive actuator failure compensation controller using output feedback. *IEEE Transactions on Automatic Control*, *47*, 506–511.

Tellili, A., Abdelkrim, N., Challouf, A., & Abdelkrim, M. N. (2015). Reconfigurable control of two-time scale systems in presence of additive faults. *Information Technology and Control*, *44*, 357–366.

Tellili, A., Abdelkrim, N., Challouf, A., & Abdelkrim, M. N. (2016). Adaptive fault tolerant control of singularly perturbed systems with actuator faults. *International Journal Automation and Control*, *10*, 104–119.

Tellili, A., Abdelkrim, M. N., & Benrejeb, M. (2007). Reliable $H\infty$ control of multiple time scales singularly perturbed systems with sensor failure. *International Journal of Control*, *80*, 659–665.

Wang, M. S., Li, T. H. S., & Sun, Y. Y. (1993). Robustness of dynamic output feedback control for singular perturbation systems. *Systems and Control Letters*, *10*, 353–361.

Wang, R., Jin, G., & Zhao, J. (2007). Robust fault-tolerant control for a class of switched nonlinear systems in lower triangular form. *Asian Journal of Control*, *9*, 68–72.

Wang, W., & Wen, C. Y. (2011). Adaptive compensation for infinite number of actuator failures or faults. *Automatica*, *47*, 2197–2210.

Wang, J., Pei, H. L., & Wang, N. Z. (2013). Adaptive output feedback control using fault compensation and fault estimation for linear system with actuator failure. *International Journal of Automation and Computing*, *10*, 463–471.

Wang, Y., Zhou, P., Wang, Q., & Duan, D. (2014). Reliable robust sampled-data h-infinity output tracking control with application to flight control. *Information Technology and Control*, *43*, 175–182.

Xin, H., Wu, D., Gan, D. Q., & Qin, J. J. (2008). A method for estimating the stability region of singular perturbation systems with saturation nonlinearities. *Acta Automatica Sinica*, *34*, 1549–1555.

Xin, H., Gan, D., Huang, M., & Wang, K. (2010). Estimating the stability region of singular perturbation power systems with saturation nonlinearities: An linear matrix inequality based method. *IET Control Theory and Applications*, *4*, 351–361.

Yang, G. H., Zhang, S. Y., Lam, J., & Wang, J. L. (1998). Reliable control using redundant controllers. *IEEE Transactions on Automatic Control*, *43*, 1588–1593.

Yang, C., Sun, J., & Ma, X. (2013). Stabilization bound of singularly perturbed systems subject to actuator saturation. *Automatica*, *43*, 1588–1593.

Ye, D., & Yang, G. H. (2006). Adaptive fault-tolerant tracking control against actuator faults with application to flight control. *IEEE Transactions on Control Systems Technology*, *14*, 1088–1096.

Zhao, Q., & Jiang, J. (1998). Reliable state feedback control system design against actuator failures. *Automatica*, *34*, 1267–1272.

Zhou, L., Yang, C., & Zhang, W. (2013). Control for a class of non-linear singularly perturbed systems subject to actuator saturation. *IET Control Theory and Applications*, *7*, 1415–1421.

Chapter 13
Reliable Control of Power Systems

H. M. Soliman and J. Ghommam

Abstract Power systems are subject to spontaneous disturbances (due to, e.g., faults, topology changes, etc.), which result in system oscillations. If such oscillations are not well damped, it may increase to cause system separation and consequently great loss in the national economy. This chapter describes the authors' research work on different techniques of power system stabilization: excitation control, power system stabilizers (PSS), governors control, and flexible AC transmission (FACTS) control. Reliable control means a system stabilized by two controllers, rather than one, to achieve better control grip if either controller fails or both are unfaulty. Different designs are presented in the frequency domain and time domain. The designs consider two cases: system without uncertainty and with uncertainty due to load changes. The solution employs Kharitonov's theorem, linear matrix inequalities (LMI). It also tackles a more difficult case when one controller is fast, whereas the other is slow due to its inherent time delay. Lyapunov–Krasovski method is used to handle this case.

Keywords Power system · Reliable control · Fault-tolerant control · LMI · Stabilization

List of Main Symbols

All quantities are in per unit (p.u.) unless otherwise stated:

P_m Mechanical input power of the generator,
P_e Electrical output power of the generator,
X_e Transmission line+ transformer reactance,

H. M. Soliman (✉) · J. Ghommam
Department of Electrical and Computer Engineering, Sultan Qaboos University, Muscat, Oman
e-mail: hsoliman1@yahoo.com

J. Ghommam
e-mail: jawhar.ghommam@gmail.com

N. Derbel et al. (eds.), *Diagnosis, Fault Detection & Tolerant Control*, Studies in Systems, Decision and Control 269,
https://doi.org/10.1007/978-981-15-1746-4_13

X_d, X_q	Direct and quadrature-axis synchronous reactances, respectively,
T'_{d0}	d-axis open circuit field time constant (sec),
M	Inertia constant (sec),
δ	Torque angle (rad),
ω	Angular velocity,
E_{fd}	Field voltage,
E'_q	q-axis voltage behind transient reactance,
u_1	Stabilizing signal of controller 1,
u_2	Stabilizing signal of controller 2,
K_A, T_A	Exciter–AVR gain and time constant (sec),
T_g	Governor time constant (sec),
R	Speed regulation,
K_C, T_C	TCSC gain and time constant,
V	Infinite bus voltage,
X_C^{ref}	TCSC reference reactance,
P, Q	Machine loading at the infinite bus, and
s	The Laplace operator.

13.1 Introduction

Stability of power systems is of paramount importance as it guarantees the continuity of service to consumers. It is observed that there are blackouts because of small signal oscillations in the power systems. There are three types of power system stability: rotor-angle stability, frequency stability, and voltage stability. The utility companies encounter the problem of voltage stability and small signal stability, particularly, during heavy-loading periods. In this chapter, the following are the central topics that will be discussed: angle reliable control and frequency reliable control.

To improve oscillation damping, the synchronous generators are equipped with power system stabilizers (PSSs) which provide feedback stabilizing signals in the excitation systems. Design techniques for the lead-type conventional PSS are found in Mondal et al. (2014). The changes in the power system caused by loading and varying operating conditions can be treated as model uncertainty. The controller which keeps stability in face of various operating conditions is termed robust control (Soliman and El-Metwally 2017) and references therein. Other PSS designs to tackle load changes have been proposed, e.g., adaptive control and intelligent control (Hussein et al. 2010). For large-scale power systems, using a hub computer to achieve centralized control is impractical, because any failure in the hub computer will cause a system failure. In addition, centralized control needs transmitting all system's states using a costly communication network which is associated with time delay. Time delays result in system's performance degradation, and even instability. The decentralized control is an effective method for stabilizing large power systems. In this approach, the system is decomposed to subsystems with smaller sizes. For each subsystem, a controller is installed which uses only the local states and is termed decentralized.

Decentralized robust control, against load changes, is employed for large power systems which avoids costly communications, and its associated time delays are given in Soliman et al. (2018). The ellipsoidal approximation to the reachable sets is used in the latter approach. Practically imposed control constraints is also considered in that reference.

Moreover, it should be mentioned that abrupt changes often emerge in the structure and parameters of many real systems due to the phenomena such as component failures as well as repairs, changing subsystem interconnections and abrupt environmental disturbances. Considering these facts, power systems encountering abrupt changes may be treated as systems which contain finite modes and the modes may jump from one to another at different times so it is not wondering that these kinds of systems can be represented as Markov jump linear systems (MJLSs). Generally speaking, MJLS is a class of stochastic linear systems subject to abrupt variations. Therefore, researchers have used this class of systems to model the physical systems with abrupt structural changes such as failure in power systems and communication systems. Stabilizing such systems is called fault-tolerant control. Stabilizing power systems subject to a series of lightning strokes and the consequent transmission lines switching caused by the auto-enclosure of circuit breakers protection is studied in Soliman et al. (2015). The case of unknown failure rate of controllers, actuators, considering the system's time delays is given in Kaviarasan et al. (2016).

Besides PSSs, flexible AC transmission (FACTS) controllers are also applied to enhance system stability (Mondal et al. 2014). Particularly, in multi-machine systems, using only conventional PSS may not provide sufficient damping for inter-area oscillations. In these cases, FACTS power oscillation damping controllers are effective solutions. Furthermore, in recent years, with the deregulation of the electricity of market, the traditional concepts and practices of power systems have changed. Better utilization of the existing power system to capacities by installing FACTS devices becomes imperative. FACTS devices are playing an increasing and major role in the operation and control of competitive power systems.

FACTS devices can be (1) series connected, for example, thyristor-controlled series capacitors (TCSC) or thyristor-controlled phase angle regulators (TCPAR); or (2) shunt-connected device such as static VAR compensators (SVC). TCSC devices are the key devices of the FACTS family and they are recognized as effective and economical means to damp power system oscillation. Therefore, in this chapter, a series FACTS device, the thyristor-controlled series capacitor (TCSC), is employed for damping system oscillations.

13.2 Motivation, Related Work, and Objectives

Unlike the fault-tolerant control which tackles the case of partial failure of sensors, the present chapter deals with the problem of complete failure of sensors or actuators. Note that stabilizing a single plant by two controllers is better than one since it provides tighter control grip. Stability has to be kept if either controller is faulty or both are sound. Such an approach is termed reliable control. There are different tools for oscillation damping: excitation control, power system stabilizers (PSSs), governors control, and flexible AC transmission (FACTS) control. Different combinations of the previously mentioned controls will be presented in this chapter to achieve reliable stabilization.

It should be pointed out that redundancy is the key attribute in reliable control systems. The fundamental difference between the robustness and reliability of control systems could be explained as follows. A robust controller can function acceptably with small- and medium-size parameter variations and/or model uncertainties, plausibly due to different loading conditions. Meanwhile, a reliable controller accommodates more drastic changes in system configurations, probably caused by component failures and/or outages. Reliable control systems are extremely vital in practice, e.g., in avionics (Jin et al. 2014).

Although there are at least one hundred research papers dealing with PSS for excitation channel, it seems that very little effort is done to use additional redundant controller, e.g., governor channel, for reliable stabilization of power systems. Loss of one control signal is equivalent to actuator failure. Such faults may be attributed to a loss of signal, a communication channel, a controller malfunction, or a combination of these.

The remaining parts of this chapter are organized as follows. Section 13.3 presents the problem statement, and Sect. 13.4 presents the problem solution techniques. Section 13.5 presents the numerical simulation results by illustration the developed concepts a benchmark example for power system reliable control. Section 13.6 presents conclusions and future research directions of the work presented in this chapter.

Notation and Facts

$\mathbb{R}^n$, $\mathbb{R}^{n\times m}$ denote the n-dimensional Euclidean space and the set of $n \times m$ real matrices, respectively. In the sequel, W' and W^{-1} denote the transpose, and the inverse of any square matrix W, respectively. The notation $W > 0$, ($W < 0$) is used to denote a symmetric positive (negative) definite matrix; I denotes the identity matrix of appropriate dimension. The symbol $\star$ is as an ellipsis for terms in matrix expressions that are induced by symmetry, e.g.,

$$\begin{bmatrix} L + (W + N + W' + N') & N \\ N' & R \end{bmatrix} = \begin{bmatrix} L + (W + N + \star) & N \\ \star & R \end{bmatrix}$$

- Fact 1: For any real matrices W_1, W_2 and $\Delta(t)$ with appropriate dimensions and $\Delta\Delta' < I$, it follows that (Liu and Yao 2016), $\exists\varepsilon > 0$:

$$W_1\Delta(t)W_2 + \star < \varepsilon W_1 W_1' + \varepsilon^{-1} W_2' W_2$$

- Fact 2: The congruence transformation zWz does not change the definiteness of W (Liu and Yao 2016).
- Fact 3: (Schur complement) This fact is used to transform a nonlinear matrix inequality to a linear one. Given constant matrices W_1, W_2, and W_3 where $W_1 = W_1'$ and $W_2 = W_2' > 0$. Then (Liu and Yao 2016),

$$W_1 + W_3' W_2^{-1} W_3 < 0 \Longleftrightarrow \begin{bmatrix} W_1 & \star \\ W_3 & -W_2 \end{bmatrix} < 0$$

- Fact 4: (Kharitonov's theorem for robust stability) Checking the stability of a polynomial with fixed coefficients can be easily done using Routh's criterion. However, if the coefficients are randomly changing within intervals having upper and lower bounds, infinite number of Routh's criterion has to be applied for checking the stability of such interval or uncertain polynomial. Kharitonov proved that only four polynomials are needed to check the stability of an uncertain polynomial (Datta et al. 2009).
  Consider the uncertain polynomial with independently changing coefficients

$$p = \sum_{i=0}^{n} a_i s^i$$

with $a_i \in [a_i^-, a_i^+]$ denotes the a priori bounding interval of the ith component of uncertainty (the ith uncertain coefficient) a_i. The above uncertain polynomial is stable if and only if the following four Kharitonov's polynomials are stable:

$$\begin{aligned}
p_1 &= a_0^- + a_1^- s + a_2^+ s^2 + a_3^+ s^3 + a_4^- s^4 + a_5^- s^5 + a_6^+ s^6 + \cdots \\
p_2 &= a_0^+ + a_1^+ s + a_2^- s^2 + a_3^- s^3 + a_4^+ s^4 + a_5^+ s^5 + a_6^- s^6 + \cdots \\
p_3 &= a_0^+ + a_1^- s + a_2^- s^2 + a_3^+ s^3 + a_4^+ s^4 + a_5^- s^5 + a_6^- s^6 + \cdots \\
p_4 &= a_0^- + a_1^+ s + a_2^+ s^2 + a_3^- s^3 + a_4^- s^4 + a_5^+ s^5 + a_6^+ s^6 + \cdots
\end{aligned}$$

Kharitonov's test for robust stability can significantly be simplified if the interval polynomial studied is of degree 5, 4, or 3. In these cases, the Kharitonov polynomials necessary for performing the test are 3, 2, or 1 in number, respectively, as against the four polynomials of the general case. More precisely, we have the following.

- An interval polynomial p with degree 5 is robustly stable if and only if the Kharitonov polynomials p_1, p_2, and p_3 are stable.
- An interval polynomial p with degree 4 is robustly stable if and only if the Kharitonov polynomials p_2, and p_3 are stable.
- An interval polynomial p with degree 3 is robustly stable if and only if the Kharitonov polynomial p_3 is stable.

13.3 Problems' Statement

The reliable control problem is stated as follows. Design controllers such that the system stability is retained when either controller, actuator, fails, or both are sound, Fig. 13.1. The case of simultaneous failure of both controllers is excluded since it is very unlikely in practice.

The case study system dealt with in this chapter is single machine infinite bus system (SMIB). The infinite bus represents the Thévenin equivalent for a large network, Fig. 13.2.

Different combinations of stabilization tools (excitation control, PSS, governor control, and fact controllers) will be presented in progressive complexity to achieve reliable stabilization of SMIB as follows.

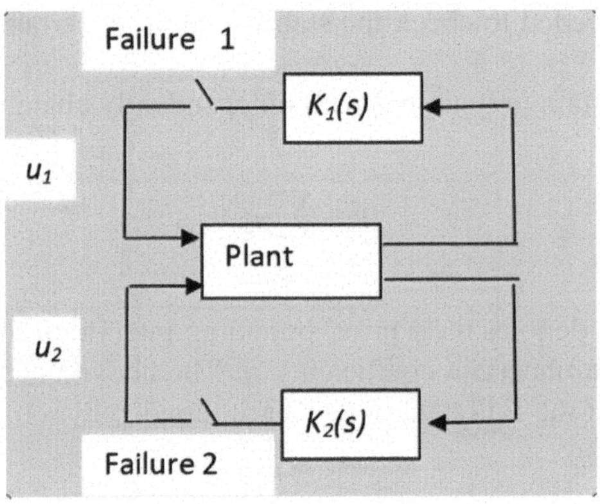

Fig. 13.1 Reliable control

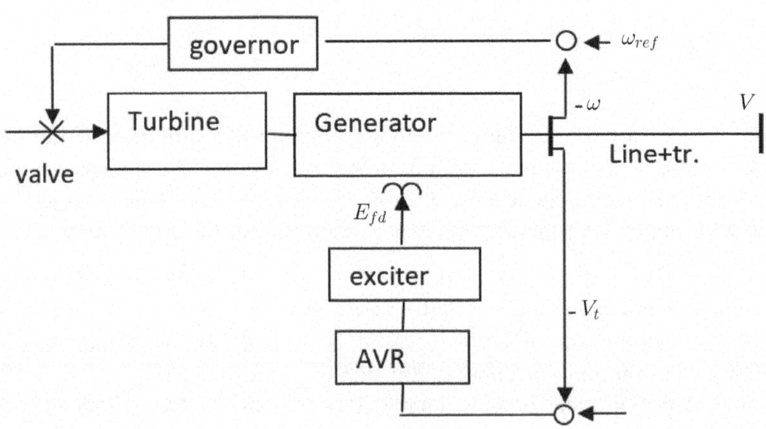

Fig. 13.2 SMIB with excitation and governor channels

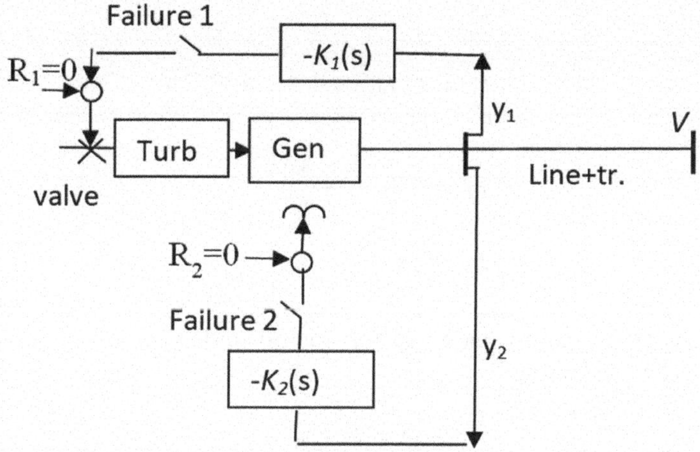

Fig. 13.3 Excitation/governor reliable decentralized control

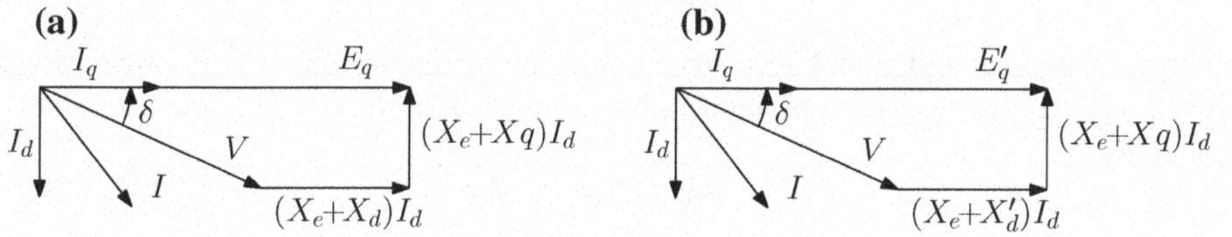

Fig. 13.4 Phasor diagram during steady state and dynamic periods. **a** steady state, **b** dynamic state

13.3.1 Problem1: Excitation/Governor Reliable Control (Soliman et al. 2008a)

The additional signals generated by the excitation and governor controllers are shown in Fig. 13.3.

The phasor diagram for an SMIB during steady state and dynamic periods is shown in Fig. 13.4.

Using Fig. 13.4, the system dynamics is described by the following nonlinear differential equations (Mondal et al. 2014):

$$\begin{cases} \dot{\delta} = \omega_0 \Delta\omega \\ \dot{\omega} = \dfrac{1}{M}(P_m - P_e) \\ \dot{E}'_q = \dfrac{1}{T'_{d0}}(E_f - E_q) \end{cases} \tag{13.1}$$

where

$$P_e = E_q I_q$$
$$E_q = [E'_q + (X_q - X'_d)I_d$$
$$I_d = \frac{E'_q - V\cos\delta}{X'_d + X_e}$$
$$I_q = \frac{V\sin\delta}{X'_d + X_e}$$

Equations (13.1) can be linearized for small disturbances around an operating point ($\delta_0, \omega_0, E'_{q0}$). Letting

$$\Delta\delta = \delta - \delta_0$$
$$\Delta\omega = \omega - \omega_0$$
$$\Delta E_q = E'_q - E'_{q0}$$

one gets the linearized model. Linearization can be carried out using hand calculations or numerical differentiation by the Matlab command "linmod". The linear model is given by

$$\dot{x} = Ax + Bu, \quad y = Cx \tag{13.2}$$

where

$$x = [\Delta\delta, \ \ \Delta\omega, \ \ \Delta E'_q]', \ \ u = [\Delta P_m, \ \ \Delta E_f]^T$$

and

$$A = \begin{bmatrix} 0 & \omega_0 & 0 \\ -\frac{k_1}{M} & 0 & -\frac{k_2}{M} \\ -\frac{k_3 k_4}{T} & 0 & -\frac{1}{T} \end{bmatrix}, \quad B = \begin{bmatrix} 0 & 0 \\ \frac{1}{M} & 0 \\ 0 & \frac{1}{T'_{d0}} \end{bmatrix} = [B_1, B_2], \ \ C = \begin{bmatrix} 1\,0\,0 \\ 0\,1\,0 \end{bmatrix} = \begin{bmatrix} C_1 \\ C_2 \end{bmatrix}$$

with $T = T'_{d0}k_3$, $\omega_0 = 2\pi f_0$, $f_0 = 50$ Hz.

It is to be noticed that the machine loading conditions affect the constants k_l, k_2, and k_4, considerably, whereas k_3 is not affected (Mondal et al. 2014). Machine loading P and Q, as well as system data, are given in the Appendix.

It is desired to stabilize the system in a decentralized manner, that is, each of the two controllers (excitation and governor) uses its local output only. This assumption of using an independent output for each controller is crucial to increase the chance that at least one of the controllers survives in case of a sensor failure.

Accordingly, system (13.2) is decomposed as follows:

$$u = \begin{bmatrix} u_1 \\ u_2 \end{bmatrix}, \quad B = [B_1, B_2], \ \ C = \begin{bmatrix} C_1 \\ C_2 \end{bmatrix} \tag{13.3}$$

where

$$B_1 = \begin{bmatrix} 0 \\ \frac{1}{M} \\ 0 \end{bmatrix}, \quad B_2 = \begin{bmatrix} 0 \\ 0 \\ \frac{1}{T'_{d0}} \end{bmatrix} \quad C_1 = \begin{bmatrix} 1\,0\,0 \end{bmatrix}, \quad C_2 = \begin{bmatrix} 0\,1\,0 \end{bmatrix}$$

Thus, the system (13.2) will be treated as a two-channel linear system

$$\dot{x} = Ax + B_1 u_1 + B_2 u_2, \quad y_1 = C_1 x, \quad y_2 = C_2 x \tag{13.4}$$

It is assumed that the triple (A, B_1, C_1) and (A, B_2, C_2) are both stabilizable and detectable. Therefore, the system can be stabilized either from u_1 using y_1 only or from u_2 using y_2 only (decentralized control). From (13.4), we obtain the transfer matrix of the system as

$$Y(s) = G(s)U(s) \Longleftrightarrow \begin{bmatrix} Y_1 \\ Y_2 \end{bmatrix} = \begin{bmatrix} G_{11} & G_{12} \\ G_{21} & G_{22} \end{bmatrix} \begin{bmatrix} U_1 \\ U_2 \end{bmatrix} \tag{13.5}$$

where the Laplace variable "s" is dropped for simplicity, and where

$$\begin{aligned} G_{11} &= C_1(sI - A)^{-1} B_1 \\ G_{12} &= C_1(sI - A)^{-1} B_2 \\ G_{21} &= C_2(sI - A)^{-1} B_1 \\ G_{22} &= C_2(sI - A)^{-1} B_2 \end{aligned}$$

Now we connect two dynamic output feedback controllers:

$$U_1 = R_1 - K_1(s)Y_1, \quad U_2 = R_2 - K_2(s)Y_2 \tag{13.6}$$

Our objective is to design reliable excitation and governor controllers that maintain the desired degree of stability of power system G over a wide range of operation ($0.4 \leq P \leq 1.1$ and $0.2 \leq Q \leq 0.4$). Variations in the operating point cause G to be uncertain. The reliable design is to stabilize G with possible failure of either one of the two controllers or with the two controllers acting together.

The problem is summarized as follows.

Given a power System (13.5) operating over a wide range, design K_1 and K_2 such that

- Condition (1): when K_1 is the only active controller (K_2 is off or fails), it stabilizes G.
- Condition (2): K_2 stabilizes G when K_1 is either on (active) or off (fails). Of course, we could have started with K_2 instead of K_1. Out of conditions (1) and (2), we may consider K_1 as the main controller and K_2 as a redundant (backup) controller.

It should be pointed out that redundancy is the key ingredient in any reliable control system.

13.3.2 Problem 2: PSS/Governor Reliable Control (Soliman et al. 2008b)

In problem 1, the exciter/governor dynamics are neglected and the analysis is carried out in the frequency domain to design output feedback reliable control. The high gain used in the AVR loop to adjust the terminal voltage might cause system instability. For this reason, a PSS is installed to provide an additional stabilizing signal in the excitation channel. Here, the problem of PSS/governor state feedback reliable control is presented and tackled in the time domain taking into consideration the governor dynamics, Fig. 13.5.

The synchronous machine under study is provided with a thyristor-based excitation system of transfer function $T_1(s)$, while the governor is represented by the transfer function $T_2(s)$, Appendix. Out of $T_1(s)$, $T_2(s)$, the exciter and governor dynamics are added to Eq. (13.1) to get

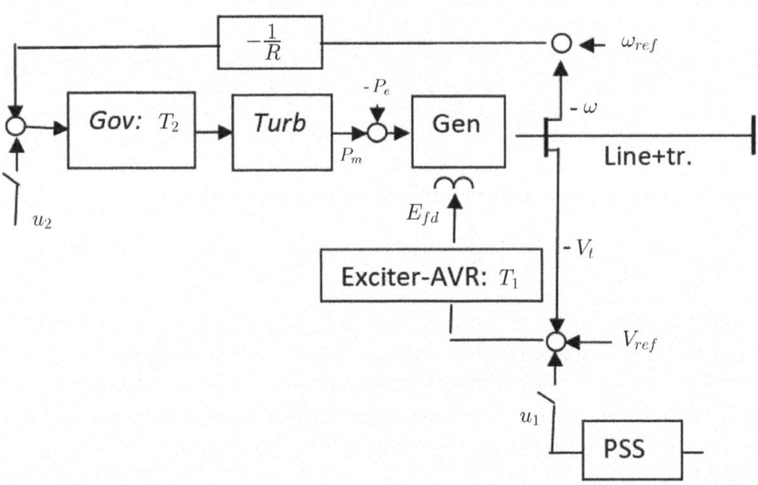

Fig. 13.5 PSS/governor reliable control

$$\begin{cases} \dot{\delta} = \omega_0 \Delta\omega \\ \dot{\omega} = \dfrac{1}{M}(P_m - P_e) \\ \dot{E}'_q = \dfrac{1}{T'_{d0}}[E_f - E'_q - (X_d - X'_d)I_d] \\ \dot{E}_{fd} = \dfrac{1}{T_E}[-E_{fd} + K_E(V_{ref} - V_t + u_1)] \\ \dot{P}_m = \dfrac{1}{T_g}\Big[- P_m - \frac{1}{R}(\omega_{ref} - \omega) + u_2 \Big] \end{cases} \tag{13.7}$$

where:

$$P_e = [E'_q + (X_q - X'_d)I_d]I_q$$
$$V_t = \sqrt{(X_q I_q)^2 + (E'_q - X'_d I_d)^2}$$

Linearizing (13.7) around an operating point, the resulting 5th-order model is

$$\dot{x} = Ax + B_1 u_1 + B_2 u_2$$

$$A = \begin{bmatrix} 0 & \omega_0 & 0 & 0 & 0 \\ -\dfrac{k_1}{M} & 0 & -\dfrac{k_2}{M} & 0 & \dfrac{1}{M} \\ -\dfrac{k_4}{T'_{d0}} & 0 & -\dfrac{1}{T} & -\dfrac{1}{T'_{d0}} & 0 \\ -\dfrac{k_5 K_E}{T_E} & 0 & -\dfrac{k_6 K_E}{T_E} & -\dfrac{1}{T_E} & 0 \\ 0 & -\dfrac{1}{T_g R} & 0 & 0 & -\dfrac{1}{T_g} \end{bmatrix}, \quad B = \begin{bmatrix} 0 & 0 \\ 0 & 0 \\ 0 & 0 \\ \dfrac{K_E}{T_E} & 0 \\ 0 & \dfrac{1}{T_g} \end{bmatrix} = [B_1, B_2] \tag{13.8}$$

with

$$x = [\Delta\delta,\ \Delta\omega,\ \Delta E'_q,\ \Delta E_{fd},\ \Delta P_m]^T, \quad u = [u_1,\ u_2]^T$$

For a severe load condition, $P = 1.2$, $Q = -0.5$ p.u., the resulting system matrices are

$$A = \begin{bmatrix} 0 & 314 & 0 & 0 & 0 \\ -0.0889 & 0 & -0.1389 & 0 & 0.1 \\ -0.2963 & 0 & -0.4633 & -0.1667 & 0 \\ 131.473 & 0 & -169.6247 & -20 & 0 \\ 0 & -20 & 0 & 0 & -20 \end{bmatrix}, \quad B = \begin{bmatrix} 0 & 0 \\ 0 & 0 \\ 0 & 0 \\ 500 & 0 \\ 0 & 1 \end{bmatrix}$$

The problem is to determine state feedback:

$$u = -Kx, \quad K = \begin{bmatrix} K_1 \\ K_2 \end{bmatrix}$$

For excitation and governor controllers $u = [u_1 \ u_2]'$, which ensure stability when either one fails or when both are sound. It is assumed that (A, B_1) and (A, B_2) are both controllable. Therefore, the system can be stabilized from u_1 only, from u_2 only, or from both acting together. Given the system described via Eq. (13.7), design K_1 and K_2 such that

(i) when only the excitation controller (K_1) is active, it stabilizes the system, and
(ii) the governor controller (K_2) stabilizes the system when K_1 is either on or off.

There is no mathematical preference on the order by which the two controllers have to be designed. However, based on the above conditions, K_1 may be considered the main controller and K_2 a redundant (backup) controller. In practice, the exciter channel provides rapid control, so it is selected as the main controller due to the considerable difference in the quickness of response. The governor channel provides slow control due to the time delay of the mechanical system, so it is selected as the backup controller. Nevertheless, the evolution of electronic governors that contribute to the actuation of the intercept valve has reduced the time delay in governors to 3–5 s. After such a period, both the excitation and governor controllers will damp the small frequency oscillations (0–2 Hz) in a reliable manner.

13.3.3 Problem 3: PSS/Governor Reliable Control with Governor Time Delay (Soliman et al. 2016)

Steam valve control is usually discarded in power system stability due to belief in its slow response. The present section makes use of it as a backup control in the case of failure of the main fast excitation control. The model describing system dynamics as a function of the two controllers, with wide-range loading conditions, is derived in a norm-bounded format. Linear matrix inequalities are derived as a sufficient condition to obtain reliable controllers that provide good oscillation damping when both controllers are sound or even in the case of failure of either one. The design scheme is robust in the sense that it keeps reliable stability against wide load changes as well.

The time delay d in measuring and transmitting ω to the governor controller is included in the last equation of (13.7) as follows:

$$\begin{cases} \dot{\delta} = \omega_0 \Delta\omega \\ \dot{\omega} = \dfrac{1}{M}\big[P_m - [E'_q + (X_q - X'_d)I_d]I_q\big] \\ \dot{E}'_q = \dfrac{1}{T'_{d0}}[E_f - E'_q - (X_d - X'_d)I_d] \\ \dot{E}_{fd} = \dfrac{1}{T_E}[-E_{fd} + K_E(V_{ref} - V_t + u_1)] \\ \dot{P}_m = \dfrac{1}{T_g}\Big[- P_m - \tfrac{1}{R}[\omega_{ref} - \omega(t-d)] + u_2 \Big] \end{cases} \tag{13.9}$$

Linearizing (13.9) around an operating point, the following linearized model is obtained:

$$\dot{x} = (A + \Delta A)x + A_d x(t-d) + Bu \tag{13.10}$$

with initial conditions $x(t) = x_0, t \in [-d, 0]$, and

$$A = \begin{bmatrix} 0 & \omega_0 & 0 & 0 & 0 \\ -\dfrac{k_1}{M} & 0 & -\dfrac{k_2}{M} & 0 & \dfrac{1}{M} \\ -\dfrac{k_4}{T'_{d0}} & 0 & -\dfrac{1}{T} & -\dfrac{1}{T'_{d0}} & 0 \\ -\dfrac{k_5 K_E}{T_E} & 0 & -\dfrac{k_6 K_E}{T_E} & -\dfrac{1}{T_E} & 0 \\ 0 & 0 & 0 & 0 & -\dfrac{1}{T_g} \end{bmatrix}, \quad B = \begin{bmatrix} 0 & 0 \\ 0 & 0 \\ 0 & 0 \\ \dfrac{K_E}{T_E} & 0 \\ 0 & \dfrac{1}{T_g} \end{bmatrix} = [B_1, B_2]$$

and with

$$A_d = \begin{bmatrix} 0 & 0 & 0 & 0 & 0 \\ 0 & 0 & 0 & 0 & 0 \\ 0 & 0 & 0 & 0 & 0 \\ 0 & 0 & 0 & 0 & 0 \\ 0 & -\dfrac{1}{T_g R} & 0 & 0 & 0 \end{bmatrix}$$

Matrix ΔA represents system uncertainty due to load variations, which can be represented in a norm-bounded format as $\Delta A = M \times \Delta \times N$. The uncertainty matrices M and N can be easily calculated using the singular value decomposition. The control objective can be stated as follows: design robust reliable stabilization controllers for the system given in Eq. (13.10) for different load conditions. The reliability in this context means that one control signal, either u_1 or u_2, should be able to robustly stabilize the system. Therefore, the following cases are considered:

1. Only controller u_1 is active (u_2 fails).
2. Only controller u_2 is active (u_1 fails).
3. Both controllers are active (no failure).

The above cases are, respectively, represented by the following input matrices:

$$B_1 = \begin{bmatrix} 0 & 0 \\ 0 & 0 \\ 0 & 0 \\ \frac{K_E}{T_E} & 0 \\ 0 & 0 \end{bmatrix}, \quad B_2 = \begin{bmatrix} 0 & 0 \\ 0 & 0 \\ 0 & 0 \\ 0 & 0 \\ 0 & \frac{1}{T_g} \end{bmatrix}, \quad B_3 = \begin{bmatrix} 0 & 0 \\ 0 & 0 \\ 0 & 0 \\ \frac{K_E}{T_E} & 0 \\ 0 & \frac{1}{T_g} \end{bmatrix}$$

In addition to the required reliable stabilization for the above three cases, the oscillations have to be damped out within 10–15 s over different load conditions as required by power engineers. In other words, the uncertain state-delay system of equation (13.10) is said to be robustly stabilized with the degree of stability $|\alpha| > 0.3$ if there exists a state feedback control $u(t) = Kx(t)$, where K is a feedback gain matrix to be determined later, such that the closed-loop system response satisfies the following:

$$||x(t)|| < ||x(0)||e^{-\alpha t}, \quad t > 0$$

13.3.4 Problem 4: PSS/FACTS Reliable Control with Guaranteed Cost and Regional Pole Placement Constraints (Soliman et al. 2011)

The problem is to design a reliable PSS/FACTS control satisfying a desired degree of stability and additional constraints: guaranteed cost and regional pole placement. The latter constraint represents a desired damping ratio. The study system considers a single machine infinite bus power system (SMIB) equipped with thyristor-controlled series capacitors (TCSC), Fig. 13.6.

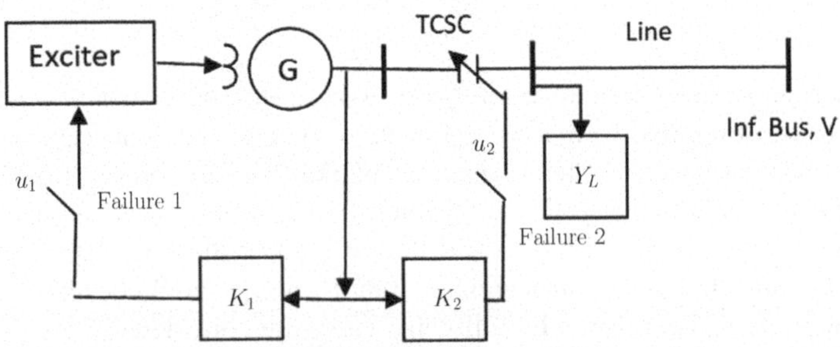

Fig. 13.6 Single machine infinite bus system with TCSC

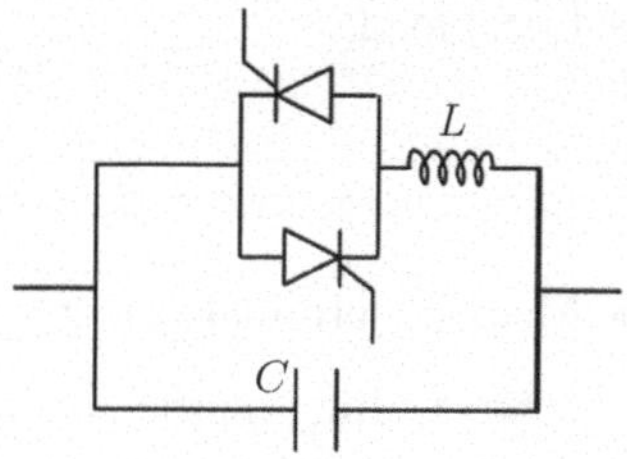

Fig. 13.7 Thyristor-controlled series capacitor (TCSC) topology

The system includes a PSS, K_1, acting on the generator exciter and another controller, K_2, setting the thyristor firing angle which determines the reactance of the TCSC. The TCSC is selected among other FACTS controller because it is economical and effective in damping system oscillations, Fig. 13.7.

The fourth-order model of the system represents the machine dynamics around a certain operating point as given below:

$$\dot{x} = Ax + Bu \tag{13.11}$$

where $x(t) \in \mathbb{R}^n$ is the state vector and $u(t) \in \mathbb{R}^m$ is the control vector given as

$$x = [\Delta\delta,\ \delta\omega,\ \Delta E'_q,\ \Delta E_f]' \tag{13.12}$$

$$u = [u_1,\ u_2]' \tag{13.13}$$

In our case, $n = 4, m = 2$, and $u_2 = \Delta X_c$. It is important to notice that the dynamic characteristic of the TCSC is very fast; its time constant is 0.02 s, as compared with that of the system under study, hence neglected.

The system data are in p.u. as follows:

- Synchronous machine:

$$X_d = 0.973,\ X'_d = 0.19,\ X_q = 0.55,\ \omega_0 = 377\,\text{rad/s},\ T'_{d0} = 7.76\,\text{s},\ M = 9.26\,\text{s}$$

- Exciter:

$$K_E = 50,\ T_E = 0.05\,\text{s}$$

- Transmission line:

$$X_E = 0.997$$

- Machine Loading(nominal):

$$S = P + jQ = 1.0 + j0.015\,\text{pu}$$

- Local load:

$$Y_L = 0.249 + j0.262$$

- Machine terminal voltage: $1.05 + j0$

The resulting matrices of the system model are given below:

$$A = \begin{bmatrix} 0 & 377 & 0 & 0 \\ -0.0588 & 0 & -0.1303 & 0 \\ -0.09 & 0 & -0.1957 & 0.1289 \\ 95.532 & 0 & -815.93 & -20 \end{bmatrix}, \quad B = \begin{bmatrix} 0 & 0 \\ 0 & 0.0704 \\ 0 & 0.0177 \\ 1000 & 93.846 \end{bmatrix}$$

Consider the problem of determining the state feedback, given as

$$u = Kx \tag{13.14}$$

for excitation and TCSC controllers $u = [u_1,\ u_2]'$ that ensure stability when either one fails or when both are active. Therefore, the following decomposition is used:

$$B = [b_1,\ b_2],\ K = \begin{bmatrix} K_1 \\ K_2 \end{bmatrix} \tag{13.15}$$

To cope with the possible fault scenarios: (1) only controller 1 is active (fault in controller 2), (2) only controller 2 is active (fault in controller 1), and (3) no faults; the following matrices are, respectively, defined:

$$B_1 = [b_1,\ 0],\ B_2 = [0,\ b_2],\ B_3 = [b_1,\ b_2] \tag{13.16}$$

It is worth mentioning that the possibility of simultaneous failure of both controllers is very remote, so it is excluded. Thus, the system (13.11) under all possible faults becomes (for $i = 1, 2, 3$)

$$\dot{x} = Ax + B_i u \tag{13.17}$$

It is assumed that (A, B_i) are controllable, $\forall i$.

The objectives of the present work could be briefly introduced as follows. For system (13.17), find a state feedback controller K, such that the faulty closed-loop system

$$\dot{x} = (A + B_i K)x \tag{13.18}$$

will meet either of the following constraints (selected by the designer):

(i) Controlled overshoot and settling time:
To avoid generator shaft fatigue and possible breakdown, the overshoot or equivalently the damping ratio must not be less than a minimum value. If the closed-loop poles lie within the circular region $D(q, r)$ with center at $-q$ and radius $r < q$ (Fig. 13.8b), both damping ratio and settling time can be achieved. Selecting ($\zeta_{min} = 0.25$) and ($\alpha = 0.36$), we find the desired circular region $D(q, r)$ as

$$q = 11, \; r = 10.64 \tag{13.19}$$

(ii) Guaranteed cost control and (i):
Although pole placement in a circular region puts interesting practical constraints on the transient response of power systems, in practice, it might be desirable that the controller be chosen to minimize a cost function as well.
The cost function associated with the faulty system (17) is

$$J = \int_0^{\infty} (x^T Q x + u^T R u) dt \tag{13.20}$$

where $Q = Q' > 0$ and $R = R' > 0$ are given weighting matrices. With the state feedback (13.14), the cost function of the closed loop is

$$J = \int_0^{\infty} x^T (Q + K' R K) x dt \tag{13.21}$$

The guaranteed cost control problem is to find K such that cost function J exists and to have an upper bound $J^{\star}$, i.e., satisfying $J \leq J^{\star}$.

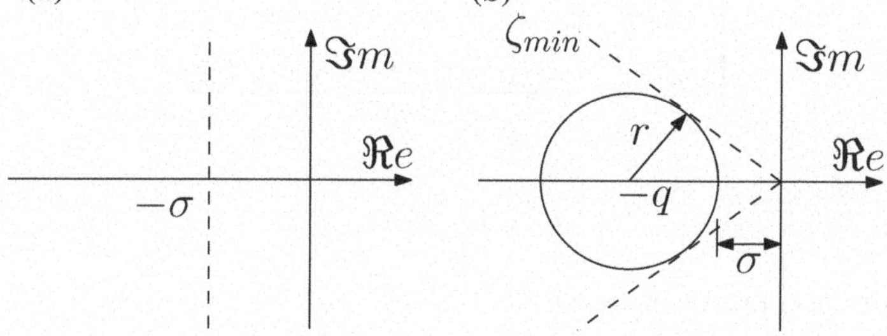

(a) Degree of stability (b) Circular region D

Fig. 13.8 Stability regions: **a** Shifted region, **b** Circular region $D(q, r)$

13.4 Problems' Solution

13.4.1 *Solution of Problem 1 (Soliman et al. 2008a)*

13.4.1.1 Design of K_1, When K_2 Is Off

The following lemma is used.

Lemma *When K_2 is off, K_1 stabilizes the overall system (13.5) if and only if it stabilizes the subsystem G_{11}.*

Proof When K_2 is off, we have

$$\begin{bmatrix} Y_1 \\ Y_2 \end{bmatrix} = \begin{bmatrix} G_{11} & G_{12} \\ G_{21} & G_{22} \end{bmatrix} \begin{bmatrix} R_1 - K_1 Y_1 \\ R_2 \end{bmatrix} \tag{13.22}$$

Designing K_1 to stabilize the subsystem G_{11} via the feedback $U_1 = R_1 - K_1 Y_1$ results in

$$\begin{bmatrix} 1 + K_1 G_{11} & 0 \\ K_1 G_{11} & 1 \end{bmatrix} \begin{bmatrix} Y_1 \\ Y_2 \end{bmatrix} = \begin{bmatrix} G_{11} & G_{12} \\ G_{21} & G_{22} \end{bmatrix} \begin{bmatrix} R_1 \\ R_2 \end{bmatrix} \tag{13.23}$$

Solving (13.23) for Y, we obtain

$$Y = \begin{bmatrix} Y_1 \\ Y_2 \end{bmatrix} = \begin{bmatrix} G^{\star}_{11} & G^{\star}_{12} \\ G^{\star}_{21} & G^{\star}_{22} \end{bmatrix} \begin{bmatrix} R_1 \\ R_2 \end{bmatrix} = G^{\star} R \tag{13.24}$$

where

$$\begin{cases} G^{\star}_{11} = \dfrac{G_{11}}{d^{\star}} \\ G^{\star}_{12} = \dfrac{G_{12}}{d^{\star}} \\ G^{\star}_{21} = \dfrac{G_{21}}{d^{\star}} \\ G^{\star}_{22} = \dfrac{K_1(G_{11}G_{22} - G_{12}G_{21}) + G_{22}}{d^{\star}} \end{cases} \tag{13.25}$$

where

$$d^{\star} = 1 + K_1 G_{11} \tag{13.26}$$

The form of $d^{\star}(s)$ proves the result. □

13.4.1.2 Design of K_2, When K_1 Is Off or On

The controller K_2 stabilizes the overall system G if and only if it stabilizes G_{22} and $G^{\star}_{22}$ simultaneously (13.25). As seen, the reliable stabilization problem is reduced to simultaneous stabilization of G_{22} and $G^{\star}_{22}$.

For the form of the study system (13.3), it can be shown that

$$G_{11}G_{22} - G_{12}G_{21} = 0$$

Thus, $G_{22}^{\star}$ simplifies to

$$G_{22}^{\star} = \frac{G_{22}}{d^{\star}} \tag{13.27}$$

The Kharitonov's theorem, fact 4, gives the necessary and sufficient conditions for robust stability of uncertain polynomials, when the coefficients a_i are independent. If a_i are dependent, the above theorem can be easily modified to be only a sufficient condition for robust stability (resulting in conservative results).

Although the a_i depends on the machine loading vector $[P, Q]$, assume them independent and define their bounds as

$$a_i^- = \min_{P,Q} a_i, \quad a_i^+ = \max_{P,Q} a_i \tag{13.28}$$

and simply construct the polynomial

$$p^{\star} = \sum_{i=0}^{n} [a_i^-, \ a_i^+] s^i \tag{13.29}$$

Then, the robust stability of polynomial (13.29) with dependent coefficients implies the robust stability of polynomial with independent coefficients. However, the converse is not true. Applying the above technique to the study system with loading $0.4 \leq P \leq 1.1$ and $0.2 \leq Q \leq 0.4$, we get the interval plant:

$$G_{11} = \frac{31.4(s + 0.06)}{d}, \quad G_{22} = \frac{[-0.0194, -0.0112]s}{d} \tag{13.30}$$

with

$$d = s^3 + 0.06s^2 + [31.45, 44.664]s + [-6.554, -0.9464]$$

Note that the transfer functions G_{12}, G_{21} are not used in calculations.

The above uncertain system can be reliably stabilized via the following design steps:

- Step 1: Design K_2 which stabilizes G_{22} (when K_1 fails).
- Step 2: Design K_1 that stabilizes simultaneously both G_{11} (when K_2 fails) and $G_{11}^{\star}$ (when K_2 is on).

Since we started the design by K_2, we replace every 1 by 2 and vice versa in equations (13.26) and (13.27).

13.4.1.3 Design of K_2 (When K_1 Fails) to Stabilize the Uncertain Plant G_{22}

Picking an intermediate point in the interval plant G_{22}, a PID stabilizer is selected as

$$K_2 = -\frac{k(s+2)^2}{s} \tag{13.31}$$

where $k > 0$. Forming the closed-loop interval polynomial

$$1 + K_2 G_{22} = 0 \tag{13.32}$$

results in

$$\begin{aligned} &s^3 + [0.06 + 0.0112k, 0.06 + 0.0194k]s^2 \\ &+[31.45 + 4 \times 0.0112k, 44.664 + 4 \times 0.0194k]s \\ &+[-6.554 + 4 \times 0.0112k, -0.9464 + 4 \times 0.0194k] = 0 \end{aligned} \tag{13.33}$$

Note that the polynomial (13.33) is of 3rd order after pole-zero cancellation at origin between G_{22} (13.30) and K_2 (13.31). From fact 4, the above third-order interval polynomial is robustly stable if and only if the following Kharitonov polynomial p_3 is stable:

$$s^3 + (0.06 + 0.0112k)s^2 + (31.45 + 4 \times 0.012k)s + (-0.9464 + 4 \times 0.0194k) = 0 \tag{13.34}$$

In the last equation, the worst conditions are selected in forming the intervals. For good performance, power engineers require that the oscillation should be damped within 10–15 s. Or equivalently, the settling time $t_s = \frac{4}{|\alpha|}$ should be 10–15 s, or the degree of stability $|\alpha|$ should be around 0.3. In other words, the closed-loop poles must lie to the left of the vertical line –0.3. For this, the root locus method can be utilized to stabilize polynomial (13.34) with the desired degree of stability to get

$$K_2 = -\frac{928(s+2)^2}{s} \tag{13.35}$$

13.4.1.4 Design of K_1 to Stabilize Simultaneously G_{11} (K_2 fails) and $G_{11}^{\star}$ (K_2 on)

In this case, $G_{11}^{\star}$ [from Eq. (13.27); replacing 1 by 2 and vice versa]. Selecting K_1 in the form:

$$K_1 = k\frac{(s+2)^2}{s+0.06}, \quad k > 0$$

Table 13.1 Degree of stability for different loads and control failures

Load: P, Q	0.4, 0.2	0.7, 0.3	1.1, 0.4
u_1 on, u_2 fails	0.54	0.44	0.37
u_2 on, u_1 fails	0.62	0.62	0.6
u_1, u_2 on	0.82	0.78	0.74

Proceeding as before, the Kharitonov polynomial p_3 is formed, after pole-zero cancellation at $s = -.06$, and cast it in the root locus format. It is found that

$$K_1 = 0.26\,\frac{(s+2)^2}{s+0.06} \tag{13.36}$$

stabilizes the polynomial with the desired degree of stability 0.3.

13.4.1.5 Design Validation

The designed controllers K_1 and K_2 are tested to stabilize the power system under study over a wide range of operating conditions. It is indicated that the controllers can successfully stabilize the system with good degree of stability (Soliman et al. 2008a, b). Table 13.1 shows the results for some selected loads (P, Q): light, nominal, and heavy loads are, respectively, (0.4, 0.2), (0.7, 0.3), and (1.1, 0.4).

13.4.2 Solution of Problem 2 (Soliman et al. 2008b)

13.4.2.1 Design of K_1 When K_2 Is Off

The main controller (K_1) can be designed to achieve a closed-loop system with a prescribed degree of stability (α) in addition to optimality. The optimal state feedback is a constant matrix K_1 that achieves closed-loop eigenvalues with real parts to the left of $-\alpha$ (Anderson and Moore 1990). The performance index of optimality is given by

$$J_1 = \int_0^\infty e^{2\alpha t}(x^T Qx + u^T Ru)dt \tag{13.37}$$

where the constant symmetrical matrices Q and R are positive semi-definite and positive definite, respectively. Assuming (A, D) is observable and D is any real matrix satisfying $DD' = Q$, the optimal controller K_1 can be obtained by solving the algebraic Riccati equation (Anderson and Moore 1990):

$$X(A+\alpha I) + (A+\alpha I)'X - XB_1R^{-1}B_1'X + Q = 0 \tag{13.38}$$

The solution matrix X is symmetric, positive definite, and K_1 is given by

$$K_1 = -R^{-1}B_1'X \tag{13.39}$$

13.4.2.2 Design of K_2 When K_1 Is on or Off

The redundant controller K_2 must stabilize the system when K_1 is either on or off. If K_1 is off, the controller K_2 must stabilize the system:

$$\dot{x} = A_1x + B_2u_2, \quad A_1 = A \tag{13.40}$$

If K_1 is on ($u1 = -K_1x$), the controller K_2 must stabilize the system

$$\dot{x} = A_2x + B_2u_2, \quad A_2 = A - B_1K_1 \tag{13.41}$$

Since K_2 must stabilize both systems of equations (13.40) and (13.41), the reliable stabilization problem is reduced to simultaneous stabilization of two plants. This can be written compactly as

$$\dot{x} = A_kx + B_2u_2, \quad k = 1, 2 \tag{13.42}$$

The analytical solution of such a problem is of considerable difficulty. However, it can be solved using powerful iterative optimization algorithms. Select K_2 to minimize the following objective function:

$$J_2 = \max \Re e(\lambda_{k,i}), \quad k = 1, 2, \ i = 1, 2 \ldots n \tag{13.43}$$

where $\Re e(\cdot)$ design the real part of $(\cdot)$, and where $\lambda_{k,i}$ is the ith closed-loop eigenvalue of the kth plant. The objective function J_2 is minimized iteratively by the controller K_2 until the desired degree of stability is achieved. This nonlinear min-max optimization problem is best solved using the particle swarm optimization (PSO).

13.4.2.3 PSO (Bonyadi and Michalewicz 2017)

The PSO is a probabilistic optimization approach inspired by the birds' (or particle) motion seeking for food. It is simple, gradient-free, and rarely stuck into a local minimum. At every iteration, every particle moves in a certain way in search of better local minima. Each individual particle remembers the position in the parameter space where this particle achieved the best value of the objective function. This is called the individual best position. In addition, the whole swarm keeps track of the position where the best value of the whole swarm was achieved. Each member of the swarm moves according to a relationship that is influenced by its individual best value and the swarm best value.

Mathematically, we assume that we have N particles (birds) within the swarm moving together in the parameter space. The position and velocity of the ith particle in the kth iteration are denoted by $x_i^{(k)}$ and $v_i^{(k)}$, respectively. At every iteration, these parameters are updated based on the individual and collective knowledge of the swarm. A possible update formula is given by (Bakr 2013)

$$\begin{cases} v_i^{(k+1)} = wv_i^{(k)} + C_1 r_1 (x_i^{\star} - x_i^{(k)}) + C_2 r_2 (x_G - x_i^{(k)}) \\ x_i^{(k+1)} = x_i^{(k)} + v_i^{(k+1)} \end{cases} \tag{13.44}$$

Notice that the velocity vector represents the change in the position of the particle at every iteration. The number of birds N is usually taken as 20–40. The parameter w is called the inertia parameter and is usually given values 0.95–0.99. The parameter C_1 is called the cognitive parameter. Its value is chosen heuristically to be around 2.0. The parameter C_2 is called the social parameter. Its value is usually chosen equal to C_1. The parameters r_1 and r_2 are random numbers satisfying $0 < r_i < 1, i = 1, 2$. The position $x_i^{\star}$ is the position of the point with the best value of the objective function reached so far by the ith particle. x_G is the position of the best point reached by the swarm as a whole. Or instead, the Matlab R2017b command "particle swarm" can be used to carry out the PSO optimization.

For the design of K_1 with the desired degree of stability, the Q matrix is $I_{5\times5}$, and $R = 1$. Solving equations (13.38) and (13.39) resulted in the following optimal PSS:

$$K_1 = [2.1513, \ -43.6825, \ 7.7605, \ 0.9641, \ -3.9962]$$

K_2 is designed to simultaneously stabilize the two systems (13.40) and (13.41) when K_1 fails or active. The design is optimized through PSO, and the resulting governor controller is given as

$$K_2 = [-0.0673, \ 87.518, \ -3.8269, \ -0.2829, \ 1.1294]$$

The above controllers K_1 and K_2 achieve the desired degree of stability. For a cleared three-phase fault on the machine terminal causing angle disturbance ($\Delta\delta = 0.1$), the time response is traced in Fig. 13.9 for different actuator failures. The system could successfully tolerate the outage of any of the two controllers without seriously deteriorating the degree of stability.

The performance of the proposed stabilizers is tested for different loads. The eigenvalue analysis shows the effectiveness and robustness of the proposed reliable stabilizers and their ability to provide good damping of low-frequency oscillations under different loading conditions. It should be pointed out that the difficulty of PSS/governor stabilization problems stems from the notable difference in the speed of response of the two controllers. Nevertheless, the proposed stabilizer provides satisfactory damping of low-frequency oscillations either when both controllers are sound or in the case of failure of one controller. Extension to multi-machine is given in Soliman et al. (2008a, b).

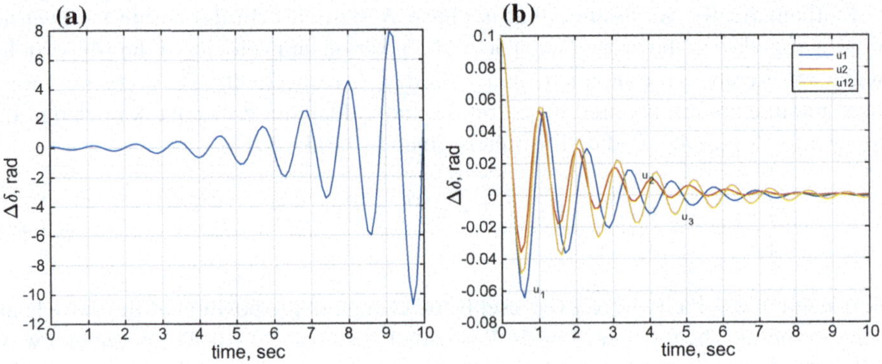

Fig. 13.9 Time response without and with control. **a** No control, **b** with control + different failures

13.4.3 Solution of Problem 3 (Soliman et al. 2016)

The goal is to design a robust and reliable control. The controller has to ensure asymptotic stability of the closed-loop system with degree α for all admissible uncertainties, due to load changes, as well as actuator faults. The design is given in the following theorem.

Theorem 1 *The state-time delay system given in Eq. (13.10) is robustly stable with degree α for different controller failures by a state feedback control $u = Kx$ if there exists a feasible solution to the following LMIs $Y = Y' > 0$, $S = S' > 0$, $Z = Z' > 0$, $\varepsilon > 0$, for $i = 1, 2, 3$:*

$$\begin{bmatrix} [(A+\alpha I)Y + B_i L + \star] + Z + \varepsilon M M' & A_d e^{\alpha d} & Y N' \\ \star & -S & 0 \\ \star & 0 & -\varepsilon I \end{bmatrix} < 0 \tag{13.45}$$

where $Y = P^{-1}$, $L = KY$, $Z = YSY$. The controller gain matrix is then calculated as $K = LY^{-1}$.

Proof The closed-loop system takes the following form, for $i = 1, 2, 3$:

$$\dot{x} = (A + \Delta A + B_i K)x + A_d x(t-d) \tag{13.46}$$

Using the transformation $x = e^{-\alpha t} z$, $t > 0$ yields the following equivalent representation of Eq. (13.31):

$$\dot{z} = (A + \Delta A + B_i K + \alpha I)z + e^{\alpha d} A_d x(t-d) \tag{13.47}$$

The uniform asymptotic stability of equation (13.47) guarantees the α uniform asymptotic stability of the closed-loop system of equation (13.46).

Table 13.2 Loading conditions of SMIB power system

Load (p.u.)	P	Q
Heavy	1	0.5
Nominal	0.7	0.3
Light	0.4	0.1

Now consider the following Lyapunov–Krasovski function:

$$V(z) = z'Pz + \int_{t-d}^{t} z'Szd\tau \tag{13.48}$$

where $P = P' > 0$, $S = S' > 0$.

The time derivative of equation (13.48) is obtained as

$$\dot{V} = \dot{z}'Pz + z'P\dot{z} + z'Sz - z(t-d)'Sz(t-d) \tag{13.49}$$

To ensure stability of the system of (13.47), the derivative of $V(t)$ must be negative definite, i.e., $\dot{V} < 0$. Along the trajectory of (13.47), this condition can be written as

$$\begin{bmatrix} z(t) \\ z(t-d) \end{bmatrix}' \begin{bmatrix} [P(A + \Delta A + B_i K \alpha I) + \star] + S & PA_d e^{\alpha d} \\ \star & -S \end{bmatrix} \begin{bmatrix} z(t) \\ z(t-d) \end{bmatrix} < 0 \tag{13.50}$$

Pre- and post-multiplying Eq. (13.50) by diag($[P^{-1}, I]$), letting $P^{-1} = Y$, $KP^{-1} = L$, and using Fact 1, the LMI of equation (13.45) is obtained. This completes the proof.

The numerical parameters of the system are given in the appendix and the loading conditions of interest are shown in Table 13.2.

At the nominal load, the system matrix A is determined as

$$A = \begin{bmatrix} 0 & 314 & 0 & 0 & 0 \\ -0.1186 & 0 & -0.0906 & 0 & 0.1 \\ -0.1933 & 0 & -0.4633 & 0.1667 & 0 \\ -11.864 & 0 & -511.6 & -20 & 0 \\ 0 & 0 & 0 & 0 & -1 \end{bmatrix}$$

Using the singular value decomposition, the uncertainty matrices M and N are calculated as $M = [0, 0, 0, 6.63, 0]'$, $N = [6.63, 0, -2.1, 0, 0]$. Therefore, the LMI of equation (13.45) is solved using the LMI control toolbox. It has a feasible solution and the robust reliable state feedback matrix K is found to be

$$K = \begin{bmatrix} -1.0521 & 35.126 & -2.222 & -0.0571 & 1.1766 \\ 405.19 & -13580 & 866.54 & 22.497 & -452.01 \end{bmatrix}$$

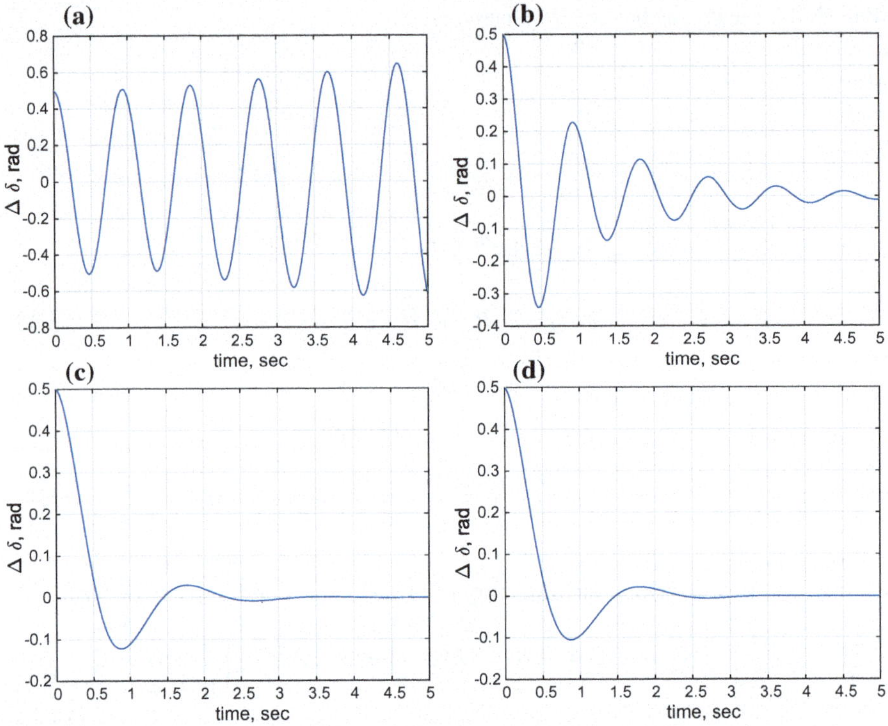

Fig. 13.10 $\Delta\delta$ (rad.), time (s), (Heavy load). **a** No control. **b** Only u_1 is on (u_2 fails). **c** Only u_2 is on (u_1 fails). **d** Both u_1 and u_2 are on (no failure)

The proposed controller is tested at extreme loads when a 0.5 rad change in the torque angle due to a cleared fault at the machine terminal is considered. The following numerical experiments are carried out using Simulink. When the heavy load is applied, simulation results for open-loop, for excitation control (u_1) only, for steam valving control (u_2) only, and for both controllers are shown in Fig. 13.10. Similarly, for the light load, the responses are shown in Fig. 13.11. As shown from Figs. 13.10 and 13.11, the system is poorly damped or even unstable for extreme loads. A remarkable stability improvement is achieved using the proposed controllers, achieving damped oscillations within the desired settling time. Consequently, the proposed controller achieves robust stabilization against load variations and reliability against the failure of either excitation or steam valving control channels.

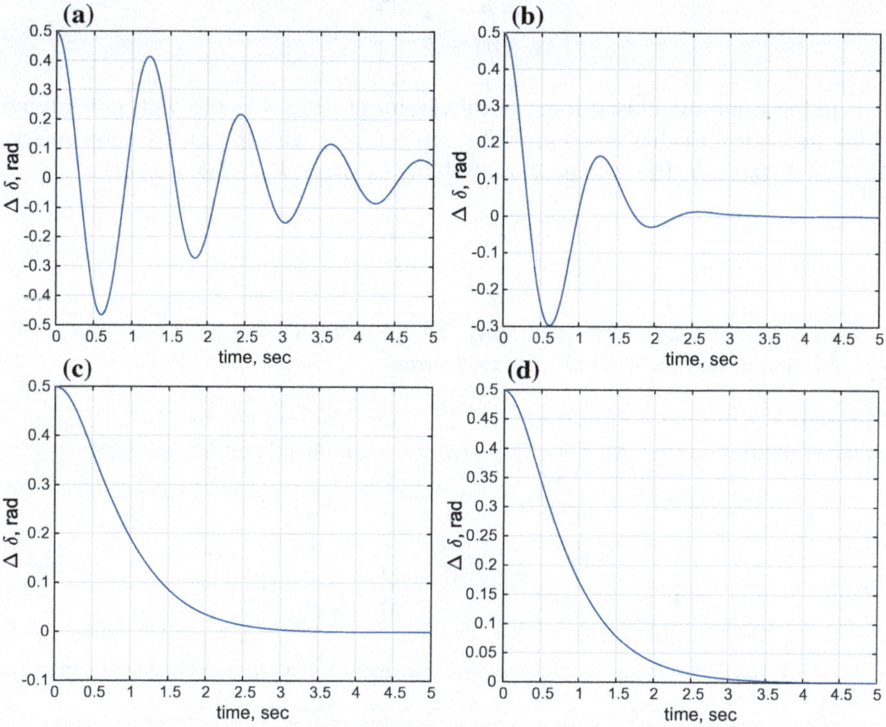

Fig. 13.11 $\Delta\delta$ (rad.), time (s), (Light load). **a** No control. **b** Only u_1 is on (u_2 fails). **c** Only u_2 is on (u_1 fails). **d** Both u_1 and u_2 are on (no failure)

13.4.4 Solution of Problem 4 (Soliman et al. 2011)

Theorem 2 *The closed-loop faulty system (13.18) is reliably stabilized with poles lie in a disk $D(q, r)$, by state feedback $u = Kx$ if there exist matrices $Y = Y' > 0$, S such that the following LMIs, $\forall i$,*

$$\begin{bmatrix} -r^2 Y & \star \\ AY + qY + B_i S & -Y \end{bmatrix} < 0 \qquad (13.51)$$

have a feasible solution. The controller fulfilling reliable stabilization + regional pole constraint $D(q, r)$ is given by

$$K = SY^{-1}$$

Proof The poles of a matrix A lie in the disk $D(q, r)$ if and only if there exists a matrix $P = P^T > 0$, such that

$$\begin{bmatrix} -r^2 P & \star \\ A + qI & -P^{-1} \end{bmatrix} < 0 \tag{13.52}$$

It follows that the closed-loop faulty system (13.18) is stable with poles lying in the disk $D(q, r)$ that is achieved by replacing A with the closed-loop matrix $A_{ci} = A + B_i F, \forall i$. Pre- and post-multiplying by diag($[P^{-1}, I]$), we get

$$\begin{bmatrix} -r^2 P^{-1} & \star \\ (A + qI)P^{-1} & -P^{-1} \end{bmatrix} < 0 \tag{13.53}$$

This is a nonlinear matrix inequality. To convexify the result, we let $Y = P^{-1}$, $S = FY$ and obtain the LMI (13.51) as desired.

Theorem 3 *Consider the faulty system (13.18) and the cost function (13.20), if there exists a feedback matrix K, $Y = Y' > 0$ and S such that the following LMIs*

$$\begin{bmatrix} -r^2 Y & \star & \star & \star \\ AY + qY + B_i S & -Y & \star & \star \\ Y' & 0 & -Q^{-1} & \star \\ S & 0 & 0 & -R^{-1} \end{bmatrix} < 0 \tag{13.54}$$

hold for all possible faults, and then the controller providing reliable stabilization with regional pole constraint $D(q, r)$ and guaranteed cost is given by $K = SY^{-1}$. Moreover, the cost function has an upper bound

$$J^\star = q x_0^T P x_0\,, \quad x_0 = x(0) \tag{13.55}$$

Proof The proof consists of two parts. In the first part, we show reliable stabilization with regional pole constraint $D(q, r)$. In the second part, we provide the guaranteed cost.

For the first part, we build on Theorem 2 and conclude that the controller achieving reliable stabilization with regional pole constraint is given by LMI (13.51). By Fact 3, this LMI is equivalent to the nonlinear matrix inequality, $\forall i$:

$$-r^2 P + (A_{ci} + qI)^T P (A_{ci} + qI) < 0 \tag{13.56}$$

For the second part, we apply the results of Liu and Yao (2016) to reach that the faulty system (13.18) with cost function (13.20) is stable provided that, $\forall i$,

$$-r^2 P + (A_{ci} + qI)^T P (A_{ci} + qI) + K^T R K + Q < 0 \tag{13.57}$$

hold, where ($Q > 0$, $R > 0$). It is clearly evident that if (13.57) is satisfied, it implies that (13.56) is fulfilled as well. Pre- and post-multiplying (13.57) by P^{-1} and using Schur complements operation, the nonlinear matrix inequalities (13.57) can be linearized and expanded to get LMI (13.54). This establishes that controller

(13.54) achieves reliable stabilization with regional pole placement and guaranteed cost.

To show that controller (13.54) provides an upper bound of the cost function, consider a Lyapunov function:

$$V(x) = x^T Px, \quad P = P' > 0$$

Notice that (13.57) is equivalent to, $\forall i$

$$(A_{ci}^T P + \star) < -\frac{1}{q} A_{ci}^T P A_{ci} - \frac{q^2 - r^2}{q} P - \frac{1}{q}(Q + F^T RF) < -\frac{1}{q}(Q + F^T RF) < 0 \tag{13.58}$$

Differentiating $V(x)$ with respect to time and using (13.58), we obtain

$$\dot{V} = x^T (A_{ci}^T P + \star)x \leq -x^T \frac{1}{q}(Q + F^T RF)x$$

Therefore, integrating both sides of the above inequality from 0 to ∞ gives

$$\int_0^\infty x^T \frac{1}{q}(Q + F^T RF)xdt \leq V(x_0) - V(x(\infty))$$

Since the stability of the system has already been established, $x(t) \longrightarrow 0$ as $t \longrightarrow \infty$, it can be concluded that $V(x(t)) \longrightarrow 0$ as $t \longrightarrow \infty$. This completes the proof.

13.5 Simulation Results

The linear matrix inequalities (13.50) and (13.54) are solved using the Matlab LMI control toolbox to get the feedback matrix for the design cases mentioned above. The results are summarized in Table 13.3.

Table 13.3 Proposed controllers

Regional pole placement	$F = \begin{bmatrix} 0.0293 & 110.7953 & -1.8771 & -0.0098 \\ 0.4942 & -103.6476 & 1.9155 & 0.0101 \end{bmatrix}$
Regional pole placement + guaranteed cost	For $Q = I$, $R = I$ $F = \begin{bmatrix} -0.0074 & 120.8538 & -2.1108 & -0.0113 \\ 0.4988 & -103.2079 & 2.0051 & 0.0111 \end{bmatrix}$

Table 13.4 Closed-loop eigenvalues for faulty and sound controllers (two designs)

Design 1 Regional pole placement	Only u_1 is active(u_2 fails)	$-4.235 \pm j7.08$, $-10.77 \pm j2.94$
	Only u_2 is active(u_1 fails)	$-1.6971, -3.3342$, $-10.7392 \pm j3.51$
	u_1, u_2 active: no failure	$-1.518, -14.5624$, $-10.1212 \pm j5.2864$
Design 2 Regional pole placement + guaranteed cost	Only u_1 is active(u_2 fails)	$-4.64 \pm j6.77$, $-11.086 \pm j3.07$
	Only u_2 is active(u_1 fails)	$-1.422, -3.71$, $-10.63 \pm j3.496$
	u_1, u_2 active: no failure	$-1.56, -15.81$, $-10.14 \pm j4.67$

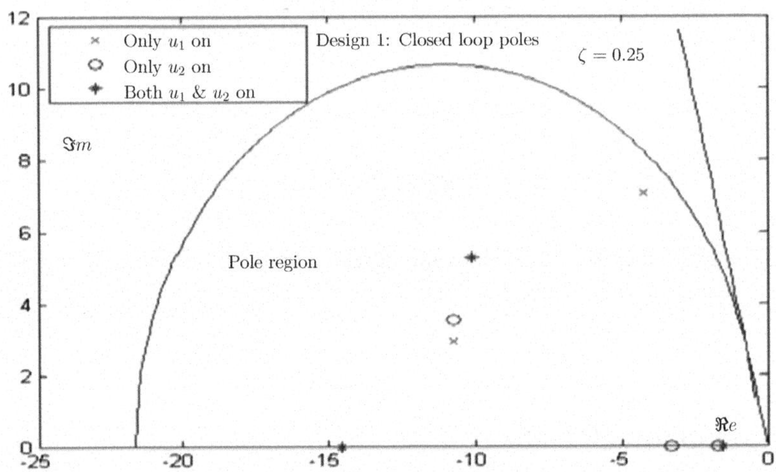

Fig. 13.12 Closed-loop pole locations for all possible faults, design 1: reliable stab. + pole region

The closed-loop poles for every design case representing all possible actuator failure are summarized in Table 13.4.

Figures 13.12 and 13.13 show, respectively, that using designs 1 and 2, the closed-loop pole locations satisfy the required performance, reliable stabilization + regional pole placement, for all possible actuator failures.

For cleared three-phase fault on the machine terminal causing disturbance of $\Delta\delta = 0.1$ rad, the initial condition of the SMIB system is taken as $x_0 = [0, 1, 0, 0, 0]'$. The simulation results when using the proposed controllers are presented in Fig. 13.14, design 2. It is evident that reliable stabilization + regional pole placement + guaranteed cost are satisfied, the angle deviation decays to 0 within 3 s, for any possible failure.

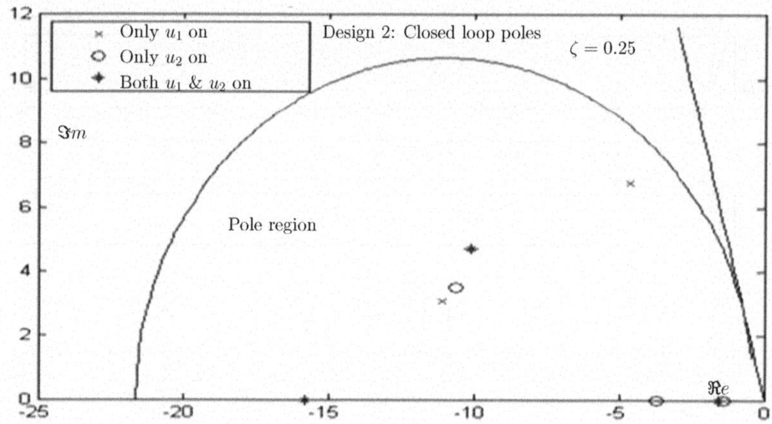

Fig. 13.13 Closed-loop pole locations for all possible faults, design 2: reliable stab. + pole region + guaranteed cost

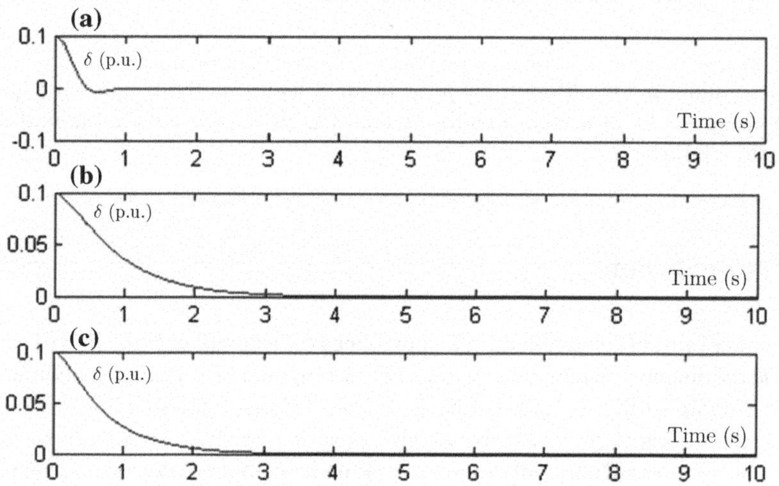

Fig. 13.14 Angle deviation response. Design case 2

To show the effectiveness of the design 2, the conventional linear quadratic regulator (LQR) is used. For the same values used before, the matrices Q and R are taken as I; the conventional optimal feedback regulator is obtained as follows:

$$F = \begin{bmatrix} -0.0620 & 8.9587 & -0.5248 & -0.9761 \\ -0.4697 & -69.8057 & 1.2027 & -0.0910 \end{bmatrix}$$

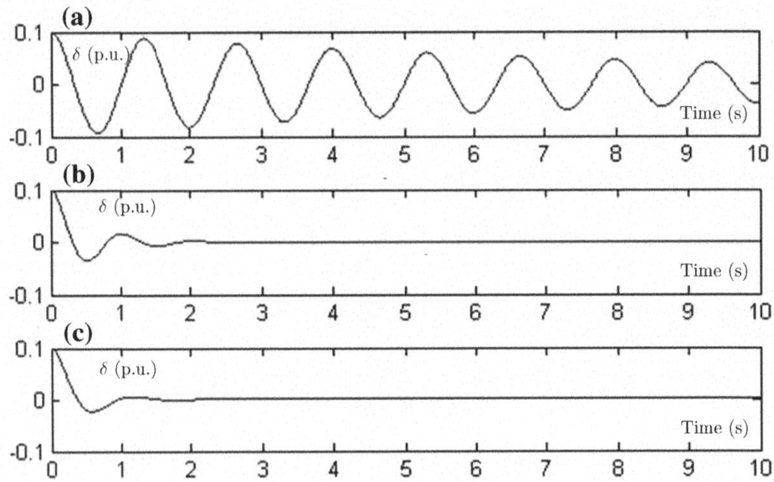

Fig. 13.15 Angle deviation response, conventional LQR

The simulation for different actuator failures using the conventional LQR are shown in Fig. 13.15. It is clear that the standard LQR is very vulnerable in the face of actuators failure.

13.6 Conclusion

This chapter presents different design approaches for reliable stabilization of a single machine infinite bus system. The designs are carried out in the frequency domain and in time domain to find output feedback or state feedback reliable controllers. The two- channel control encloses two stabilizers acting on the excitation/governor or PSS/TCSC of a synchronous alternator operating in an interconnected power system. The governor delay is also considered. Four different control schemes to satisfy desired dynamic performance constraints are derived. In the PSS/TCSC case, taking the guaranteed cost constraint into account, the problem of reliable controller design with regional pole constraint is tackled by LMI approach for a power system subject to actuator failures. Simulation results show that using the proposed controllers, the system performs fairly well. It provides satisfactory stability, transient property, and quadratic cost performance despite possible actuator faults. The proposed controller outperforms the standard LQR.

Appendix

Data for Problem 1

The system data are in p.u., unless otherwise stated, as follows:

- Synchronous machine:

$$X_d = 1.6,\ X'_d = 0.32,\ X_q = 1.55,\ \omega_0 = 314\text{rad/s},\ T'_{do} = 6\,\text{s},\ M = 10\,\text{s}$$

- Transmission line:

$$R_e = 0,\ X_e = 0.4$$

- Machine loading (wide-range):

$$0.4 \le P \le 1.1, \textit{and}\ 0.2 \le Q \le 0.4$$

.

Data for Problem 2

In addition to the data given for problem 1, the synchronous machine under study is provided with a thyristor-based excitation system of transfer function $T_1(s)$, while the governor is represented by the transfer function $T_2(s)$.

- Exciter–AVR:

$$T_1(s) = \frac{K_E}{1 + T_E s},\quad K_E = 25,\quad T_E = 0.05\,\text{s}$$

- Governor:

$$T_2(s) = \frac{1}{1 + T_g s},\quad T_g = 1\,\text{s}$$

- Speed regulation: $R = 0.05$.

Data for Problem 3

The previous data is used with $K_E = 25$, and governor time delay "$d = 1$" s.

References

Anderson, B. O., & Moore, J. B. (1990). *Optimal control: Linear quadratic methods*. Englewood Cliffs, NJ: Prentice Hall.

Bakr, M. (2013). *Nonlinear optimization in electrical engineering with applications in MATLAB*. London, United Kingdom: IET.

Bonyadi, M. R., & Michalewicz, Z. (2017). Particle swarm optimization for single objective continuous space problems: A review. *Evolutionary Computation, 25*(1), 1–54.

Datta, A., Keel, L. H., & Bhattacharyya, S. (2009). *Linear control theory: Structure, robustness, and optimization*. Boca Raton: CRC Press.

Hussein T., Saad M. S., Elshafei A. L., & Bahgat A. (2010). Damping inter-area modes of oscillation using an adaptive fuzzy power system stabilizer. *Electric Power Systems Research, 80*, 1428–1436.

Jin, Y., Fu, J., Zhang, Y., & Jing, Y. (2014). Reliable control of a class of switched cascade nonlinear systems with its application to flight control. *Nonlinear Analysis: Hybrid Systems, 11*, 11–21.

Kaviarasan, B., Sakthivel, R., & Kwonc, O. M. (2016). Robust fault-tolerant control for power systems against mixed actuator failures. *Nonlinear Analysis: Hybrid Systems., 22*, 249–261.

Liu, K. Z., & Yao, Y. (2009). Robust control: Theory and applications. New Jersey: Wiley (Asia) Pte Ltd.; Malik, O. P. Adaptive and intelligent control applications to power system stabilizer. *International Journal of Modelling, Identification and Control, 6*, 51–61 (2016)

Mondal, D., Chakrabarti, A., & Sengupta, A. (2014). *Power system small signal stability analysis and control*. Elsevier: Academic.

Soliman, H. M., & El-Metwally, K. (2017). Robust constrained pole placement for power systems using two dimensional memberships fuzzy. *IET Generation, Transmission & Distribution, 11*, 3966–3973.

Soliman, H. M., El-shafei, A. L., El-Metwally, K. A., & Makkawy, E. M. K. (2008a) Fault-tolerant wide-range stabilisation of a power system. *International Journal of Modelling, Identification and Control, 3*, 173–179.

Soliman, H. M., Morsi, M. F., Hassan, M. F., & Awadallah, M. A. (2008b). Power system reliable stabilization with actuator failure. *Electric Power Components and Systems, 37*, 61–77.

Soliman H. M., Yousef H. A., Al-Abri R., & El-Metwally, K. (2018). Decentralized robust saturated control of power systems using reachable sets. In Complexity 2018 (Vol. 2018, Article ID 2563834). New Jersey: Wiley/Hindawi.

Soliman, H. M., Benzaouia, A., Yousef, H. (2016). Wide-range reliable stabilization of time-delayed power systems. *Turkish Journal of Electrical Engineering & Computer Sciences, 24*, 2853–2864.

Soliman, H. M., & Shafiq, M. (2015). Robust stabilization of power systems with random abrupt changes. *IET, Generation, Transmission & Distribution, 9*, 2159–2166.

Soliman, H. M., Dabroum, A., Mahmoud, M. S., & Soliman, M. (2011). Guaranteed-cost reliable control with regional pole placement of a power system. *Journal of the Franklin Institute, 348*, 884–898.

Chapter 14
Active FTC of LPV System by Adding Virtual Components

Houda Chouiref, Boumedyen Boussaid, Christphe Aubrun and Vicenc Puig

Abstract This chapter has been the subject of a detailed study of linear and nonlinear physical systems with emphasis on a particular class of these nonlinear systems that are LPV systems. These LPV systems are often modeled in affine, polytopic, and fractional forms. In our case study, we chose to model the wind turbine in the LPV affine form. This choice is justified by the nature of the evolution of the wind turbine parameters (the aerodynamic torque, the damping coefficient, and the natural frequency of the blade control system) as a function of the scheduling variables which are the pressure and the wind. A study was also made of some theoretical concepts found in the literature concerning the fault-tolerant diagnosis and control of dynamical systems based on the model. The best known tools are the parity space, the state observation, and the parametric estimation. As we will be dealing with the faults that affect the parameters of the systems, we have detailed the diagnostic methods that are based on the parametric estimation. A special study has been devoted to the robustness of diagnostic algorithms to solve a common problem that affects the system which is uncertainty about physical parameters. Finally, we considered a part for the study of the different techniques of the fault-tolerant control. Our choice was focused on the use of virtual components which consists in masking the defects without resorting to synthesize the nominal controller.

Keywords LPV System · Active FTC · Virtual component · Wind turbine · Actuator fault · Sensor fault

---

H. Chouiref · B. Boussaid (✉)
National School of Engeneers of Gabes, University of Gabes, Gabes, Tunisia
e-mail: boumedyen.boussaid@gmail.com

H. Chouiref
e-mail: houda.chouiref@gmail.com

C. Aubrun
Research Center of Automatic Nancy (CRAN), University of Lorraine, Nancy, France
e-mail: christphe.aubrun@univ-lorraine.fr

V. Puig
Advanced Control Systems Group, Technical University of Catalonia, Barcelona, Spain
e-mail: vicenc.puig@upc.edu

N. Derbel et al. (eds.), *Diagnosis, Fault Detection & Tolerant Control*,
Studies in Systems, Decision and Control 269,
https://doi.org/10.1007/978-981-15-1746-4_14

14.1 Introduction

Fault-tolerant control has been widely developed for linear systems (Chouiref et al. 2013), while systems actually have a nonlinear model. The application of fault-tolerant control on the wind turbine is a recently studied concept such as, for example, in Rothenhagen and Fuchs (2009), Odgaard and Stoustrup (2012), the fault tolerance procedure in these papers is to replace the defective sensor with an estimated quantity.

In Sloth et al. (2011), a study is done on the blade subsystem control in the case of a defect caused by the pressure drop. In this work, four methods have been studied among them: the command that tolerates this fault by adapting the LPV controller to the estimated parameter. This strategy is done based on the optimization task using linear matrix inequality (LMI). In Jain et al. (2013), a tolerance is made using the prior information as well as the actual wind turbine information. The advantage of this method is that it does not require any use of an explicit diagnostic module: in Kamal et al. (2012), Sami and Patton (2012), a fault-tolerant command using a fuzzy model of a wind turbine; in Simani and Castaldi (2013), an adaptive command that modifies the controller parameters following the system identification; and in Sami and Patton (2012), a robust command using a sliding-mode adaptive command.

In Yang and Maciejowski (2012), a command tolerance is made by a hierarchical controller with the pre-predicator predictive model. This compensation is performed using an estimate by the Kalman filter method. In this work, the predictive control method is applied to tolerate a defect in the blade subsystem and in the transmission subsystem. In the first component, the defect is modeled as a change in natural frequency and damping coefficient. In the second subsystem, the defect is considered as variation of the coefficient of friction. In Poure et al. (2007), the fault tolerance in the converter was realized using the method of adding virtual components.

This method is developed for linear systems. But for the LPV class, little research has been done and most of them use the principle of adding virtual components considering a sensor fault while the case of actuator faults is little treated. In the case of a sensor fault, the method consists in using the estimate of the variable instead of its measurement during the command reconfiguration. This approach has recently been applied to the wind system where most of the work in this area is aimed at designing a fault-tolerant control law applied to the wind turbine electrical subsystem. It is within this framework that this work, which is aimed at reconfiguration using the virtual component method in case of sensor and actuator defect in the blade subsystem, will be processed.

For this reason, this chapter aims to develop such an FTC control using real modeling or close to the reality of the process. In this chapter, a control method uses virtual components to tolerate a defect that appears on the blade control system of a wind turbine. This command is based on an LPV model of a wind turbine benchmark.

14.2 Active FTC by Adding Virtual Components

This tolerance can be done in two ways depending on the type of actuator or sensor fault.

14.2.1 *Virtual Sensors*

These virtual sensors are used in the case of sensor fault; this problem is simulated to an observation problem that can be in one of these two forms:

1. Fixed value: In this case, the virtual sensor principle consists in replacing the measurement by its estimate.
2. Gain factor: It consists in estimating this gain.

For the design of this virtual sensor, consider the state representation of a system with the following sensor fault:

$$\begin{cases} \dot{x}_f = Ax_f + Bu_c \\ y_f = C_f x_f \end{cases} \tag{14.1}$$

In this case, the new dynamic is such that

$$\dot{x}_f = Ax_f + Bu_c + L(y_f - C_f x_f) \tag{14.2}$$

where L is obtained to stabilize the couple $(A^T, {C_f}^T)$.

The principle of this approach is shown in Fig. 14.1:

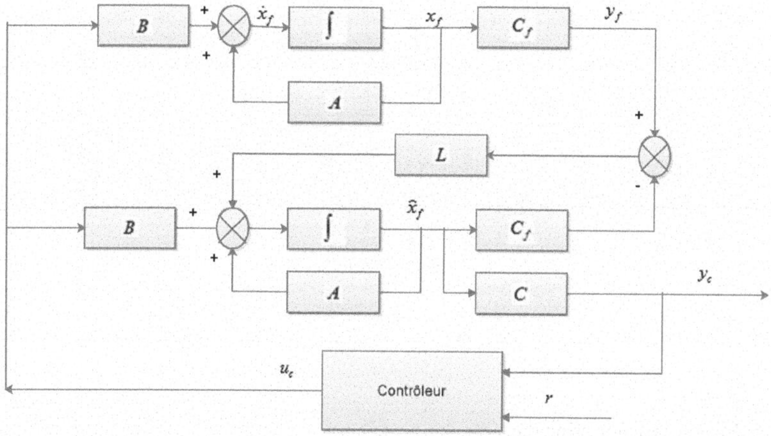

Fig. 14.1 Principle of a virtual sensor

14.2.2 Virtual Actuators

This method is used in case of an actuator fault, and it consists of using the nominal system input signal to transform it into a suitable signal.

To develop this virtual actuator, consider the following nominal system state equation:

$$\begin{cases} \dot{x} = Ax + Bu_c \\ y_c = Cx \end{cases} \tag{14.3}$$

The model in the defective case is given by the following system:

$$\begin{cases} \dot{x} = Ax_f + B_f u_f \\ y_f = Cx_f \end{cases} \tag{14.4}$$

with u_c is the nominal controller output.

In this case, the virtual actuator dynamics is expressed by the following state representation:

$$\begin{cases} \dot{\Delta x} = (A - B_f M)\Delta x + Bu_c \\ y_c = y_f + C\Delta x \\ u_f = M\Delta x \end{cases} \tag{14.5}$$

with M is chosen to stabilize the par (A, B_f).

The principle of this method is shown schematically in Fig. 14.2:

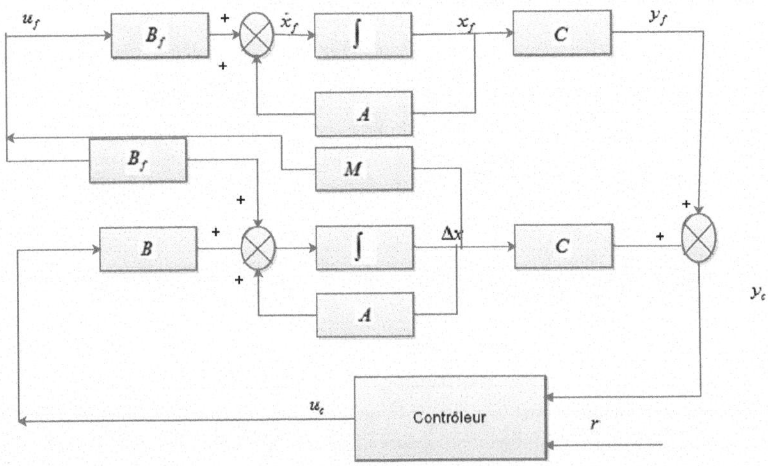

Fig. 14.2 Virtual actuator principle

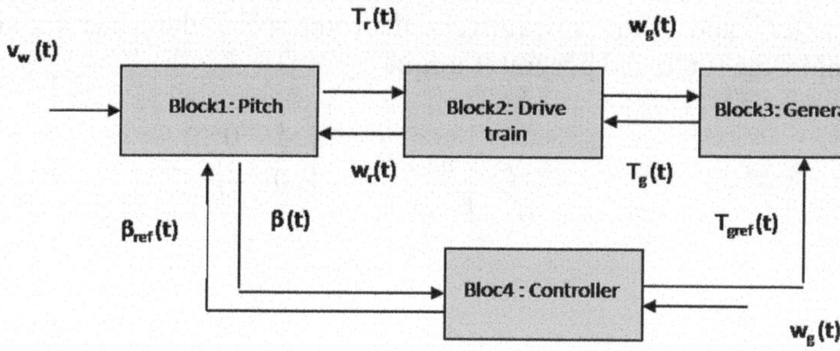

Fig. 14.3 Wind turbine submodels

14.3 Wind Turbine Description

The wind turbine is composed of four subsystems as illustrated in Fig. 14.3: Aerodynamics, pitch, drive train, and generator which are described in the following subsections.

14.3.1 Aerodynamics

The wind turbine aerodynamics is modeled as a torque acting on the blades. This torque, T_r, is given by (Odgaard 2013)

$$T_r(t) = \frac{\rho \pi R^3 C_q(\lambda(t), \beta_j(t)) v_w(t)^2}{6} \tag{14.6}$$

where λ is the tip speed ratio, ρ is the density of the air, R is the radius of blades, v_w is the wind speed, β_j is the angle of the pitch of the blade j, and C_q is the coefficient of the torque.

14.3.2 Drive Train

The drive train is modeled with a flexible two mass system and is used to increase the speed from rotor to generator. The model of the drive train is given as (Sloth et al. 2011)

$$\begin{bmatrix} \dot{w}_r(t) \\ \dot{w}_g(t) \\ \dot{\theta}_\Delta(t) \end{bmatrix} = A_d \begin{bmatrix} w_r(t) \\ w_g(t) \\ \theta_\Delta(t) \end{bmatrix} + B_d \begin{bmatrix} T_r(t) \\ T_g(t) \end{bmatrix} \tag{14.7}$$

where w_r,w_g,θ_Δ,T_r, and T_g are, respectively, the rotor speed, the generator speed, the torsion angle, the rotor, and generator torque

with $A_d = \begin{bmatrix} -\frac{B_{dt}+B_r}{J_r} & \frac{B_{dt}}{N_g J_r} & -\frac{K_{dt}}{J_r} \\ \frac{\eta_{dt} B_{dt}}{N_g J_g} & \frac{-\frac{\eta_{dt}+B_{dt}}{N_g^2} - B_g}{J_g} & \frac{\eta_{dt} K_{dt}}{N_g J_g} \\ 1 & \frac{-1}{N_g} & 0 \end{bmatrix}$ and $B_d = \begin{bmatrix} \frac{1}{J_r} & 0 \\ 0 & -\frac{1}{J_g} \end{bmatrix}$.

14.3.3 Generator

The dynamics of the converter is modeled by a first-order transfer function (Odgaard 2013)

$$\frac{T_g(s)}{T_{rg}(s)} = \frac{\alpha_{gc}}{s + \alpha_{gc}} \tag{14.8}$$

The generator produces a power which is given by

$$P_g(t) = \eta_{gc} w_g(t) T_g(t) \tag{14.9}$$

14.3.4 Pitch

The hydraulic pitch in this benchmark wind turbine is a piston servo mechanism which can be modeled by a second-order transfer function as follows (Odgaard 2013):

$$\frac{\beta(s)}{\beta_r(s)} = \frac{\omega_n^2}{s^2 + 2\zeta\omega_n s + \omega_n^2} \tag{14.10}$$

where β_r corresponds to the reference values of pitch angles, w_n and ξ are, respectively, the natural frequency and the damping ratio.

14.3.5 Wind Turbine Parameters

The parameters of wind turbine case study in nominal case are given in Table 14.1 (Odgaard et al. 2009).

Table 14.1 Wind turbine parameters

Parameter	Notation	Value	Unit
Viscous friction	B_g	47.6	$\frac{\text{MN.m.s}}{\text{rad}}$
Torsion damping	B_{dt}	775.49	$\frac{\text{MN.m.s}}{\text{rad}}$
Friction coefficient	B_r	7.11	$\frac{\text{MN.m.s}}{\text{rad}}$
Gear ratio	N_g	95	
Torsion stiffness	K_{dt}	2.7 e9	$\frac{\text{GN.m}}{\text{rad}}$
Efficiency of the drive train	η_{dt}	0.97	
Generator inertia	J_g	390	kg.m^2
Rotor inertia	J_r	55 e6	kg.m^2
Time constant	α_{gc}	0.05 e-3	
Efficiency of the generator	η_{gc}	0.98	
Radius of blades	R	57.5	m^2
Air density	ρ	1.225	$\frac{\text{kg}}{\text{m}^3}$
Damping ratio	ζ	0.6	$\frac{\text{rad}}{\text{s}}$
Natural frequency	ω_n	11.11	$\frac{\text{rad}}{\text{s}}$

14.4 FTC Control of a Wind Turbine Modeled as LPV

14.4.1 Principle of the Nominal Control on the Blade Control System

In this work, the control in the high-speed zone will be processed in order to maintain the rotor and generator speed at a constant value. This type of control can be done by power regulation or the speed that will be the subject of this part. In this case, the speed regulation is implemented on an integral proportional controller (PI) which is given by the block diagram shown in Fig. 14.4.

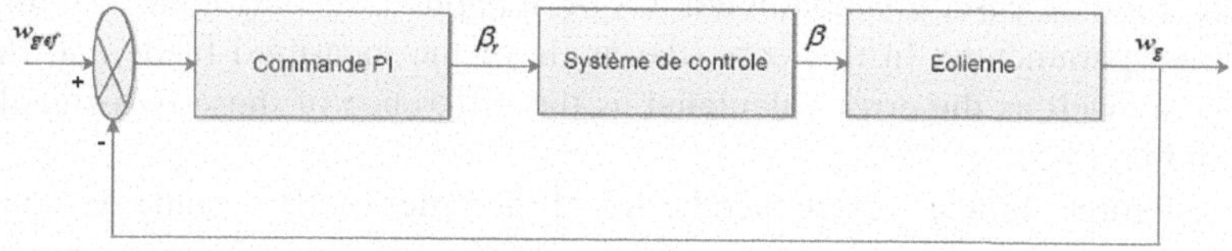

Fig. 14.4 Control strategy of the blade angle

Fig. 14.5 Tracking of the blade angle in the nominal case

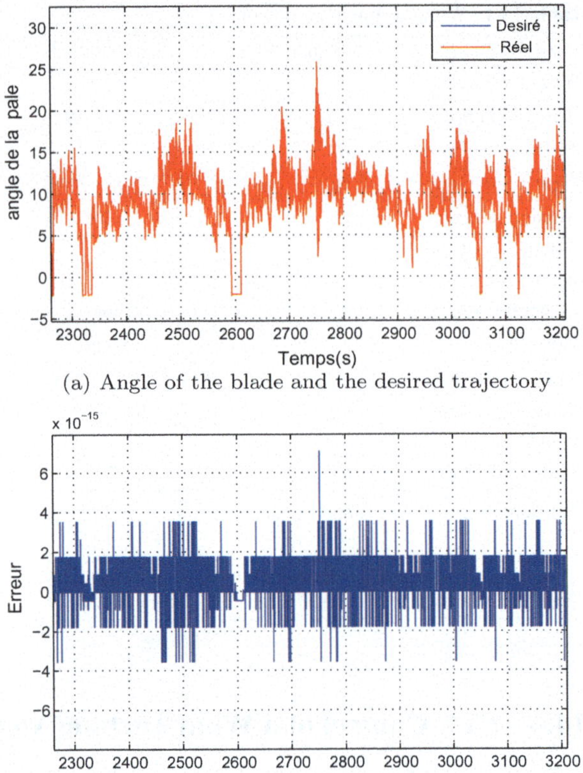

(a) Angle of the blade and the desired trajectory

(b) Tracking error

In this case, the control angle is obtained by a proportional–integral regulator (PI) using the speed error equation calculated as follows:

$$\beta_{ref} = k_p e + k_i \int e \tag{14.11}$$

with $e = w_g - w_{gref}$, k_p is the proportional gain, and k_i is the integrating gain.

In the nominal case, this command has been applied to control the system with its nominal parameters. In this case, the angle of the measured blade and that of reference as well as the error calculated as the difference of these two signals are shown in Fig. 14.5.

These figures show a very reduced error of the order of 10^{-15} and consequently the efficiency of this control law in the regulation of the blade angle of a wind system in the healthy case.

Table 14.2 Defects affecting the benchmark

N	Defect	Symbol	Type
1	Sensor	$\Delta\beta_1$	Fixed value of blade angle 1
2	Sensor	$\Delta\beta_2$	Gain factor in the blade angle 2
3	Sensor	Δw_r	Fixed value of rotor speed
4	Sensor	Δw_r	Gain factor in rotor speed
5	Sensor	Δw_g	Fixed value of generator speed
6	Sensor	Δw_g	Gain factor in the speed of the generator
7	Actuator	ΔT_g	Offset in the torque of the converter control
8	Actuator	$\Delta\beta$(hydraulic)	Changing the response of the blade system
9	Actuator	$\Delta\beta$(Air in the oil)	Changing the response of the blade system
10	System	Δw_g, Δw_r	Dynamic change of the transmission system
11	Sensor	$\Delta\beta_3$	Fixed value of blade angle 3

14.4.2 Study of the Control of a Wind Turbine in the Case of the Defect

In this blade control system, three defects among the defects listed in Table 14.2 will be considered:

(a) First defect scenario (Defect $n°8$): It is an actuator defect that is modeled as a parametric change in the damping coefficient ξ and the natural frequency w_n between 2650 and 2700 s by exceeding their values nominal as shown in Fig. 14.6.

Using the nominal command, the system response and the reference trajectory as well as the tracking error are shown in Fig. 14.7.

By evaluating the performance of this command for this operating situation, Fig. 14.7 shows that the system does not manage to continue its desired trajectory.

(b) Second defect scenario (Defect $n°1$): It is a sensor defect which consists of a fixing of the value given by the sensor of the angle of the blade to a value equal to 5° between the instant 2850 and 2900 s. In this case of default, using the conventional control, the measured and desired blade angle and the error between these two trajectories are shown in Fig. 14.8.

Figure 14.8 shows that between the instant 2850 s instant 2900 s, a considerable tracking error is detected.

(c) Third defect scenario (Defect $n°2$): It is a sensor fault where the measured value of the angle is multiplied by a gain equal to 1.2 during the period of time 2810 and 2815 s. A regulation is made with this type of defect using the nominal law of the control, and the two angles measured and desired as well as the error signal between these angles are shown in Fig. 14.9.

Figure 14.9 shows that the angle of the blade does not reach to reach the set point during the interval of the appearance of defect.

Fig. 14.6 Parametric change of ξ et w_n between 2650 and 2700 s

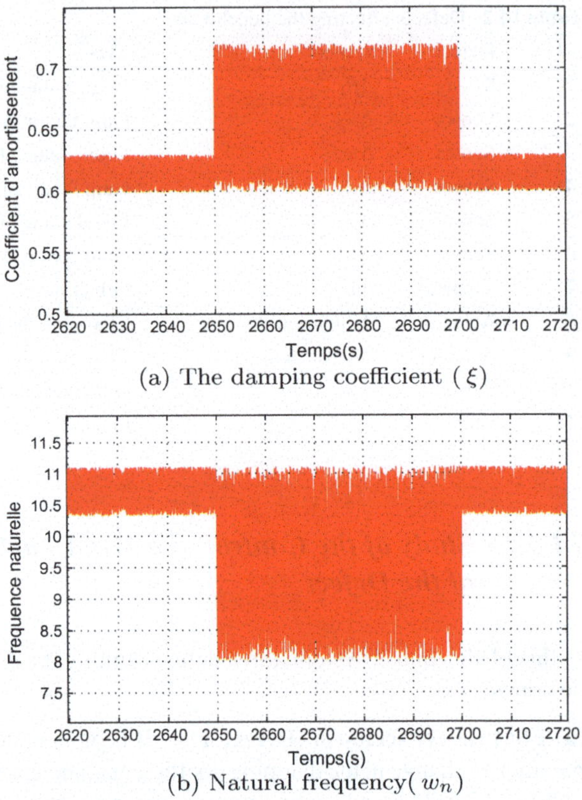

(d) Fourth defect scenario: This is a multiple defect. These are two different types of defects but they affect the residue of the blade angle. The first is an actuator defect between 2735 and 2780 s (Default $n°$ 8), and the second is a defect sensor between 2850 and 2855 s (Defect $n°$ 1). For fault detection in this case, the angle of the blade will be estimated as shown in Fig. 14.10.

Then, this estimated angle will be compared with that measured and a signal of the residue is obtained. This residue signal and the fault indicator are shown in Fig. 14.11.

By applying the nominal command on this system in the case of this fault scenario, the angles of the current blade and that desired and the error signal used for the evaluation of this controller are shown in Fig. 14.12.

Figure 14.12 shows a significant error during the moments between 2735 and 2780 s and between 2850 and 2900 s which justifies the effect of appearance of the defect at these moments.

Fig. 14.7 Pursuit study in case of parametric change

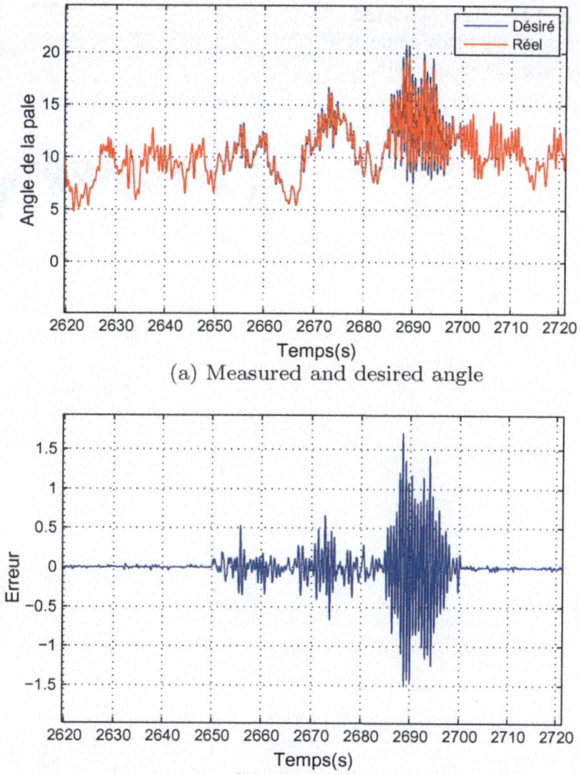

(a) Measured and desired angle

(b) Tracking error

14.4.3 Proposal of an FTC with the Technique of Adding Virtual Components

14.4.3.1 Introduction

This fault-tolerant control technique is based on the virtual sensors and actuators (Seron and Dona 2009; Khosrowjerdi and Barzegary 2013) which allow the tolerance of the fault without affecting the nominal regulator. In this case, the control signal sent to the defective system is cleared of the fault. Indeed, the reconfigured system consists of a reconfiguration block and the nominal controller. This reconfiguration block, an actuator or virtual sensor according to the type of fault, is placed between the nominal controller and the defective system in order to hide the fault and allow the faulty component to behave as in the case before the appearance of this defect. In the case of an actuator fault, the new control signal is obtained from the nominal controller output with fault consideration. In the case of a sensor fault, the reconfiguration block takes the faulty output as input and calculates the faultless one.

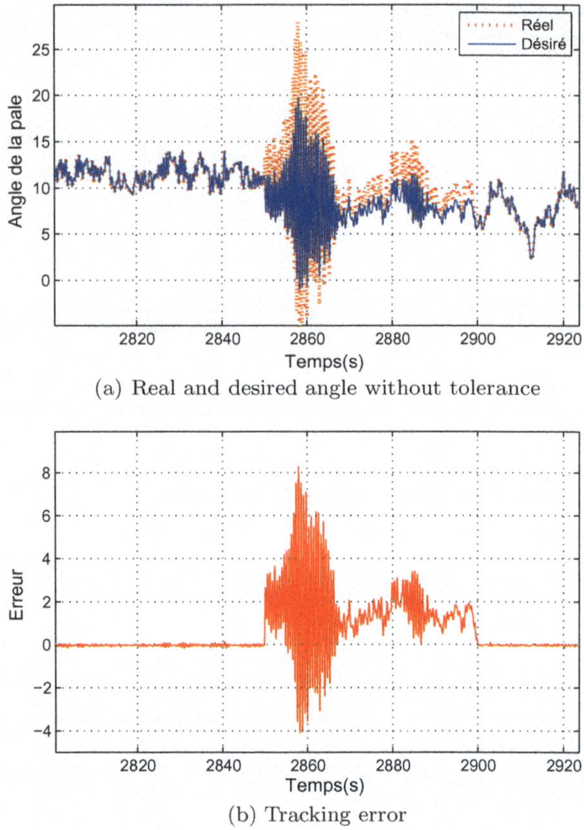

Fig. 14.8 Case of a fixed value of the blade angle at 5° between 2850 and 2900 s

14.4.3.2 Principle of Adding Virtual Sensors/Actuators in LPV Form

– In the case of an actuator fault

Consider the state equation (14.12)

$$\dot{x} = A(\mu)x + B_f(\mu, \psi)u \tag{14.12}$$

with $B_f(\mu, \psi) = B(\mu).diag(\psi_1 ... \psi_{nu})$, ψ_i is the effectiveness of the ith actuator, $\psi_i \in [0, 1]$.

If $rank(B_f) = rank(B)$, the new command signal is obtained by (14.13).

$$u_{FTC} = N(\mu, \psi)u_n \tag{14.13}$$

with u_n is the nominal controller output, $N(\mu) = B_f(\mu, \psi)^+ B$, $B_f(\mu, \psi)^+$ is the pseudo-inverse of B_f.

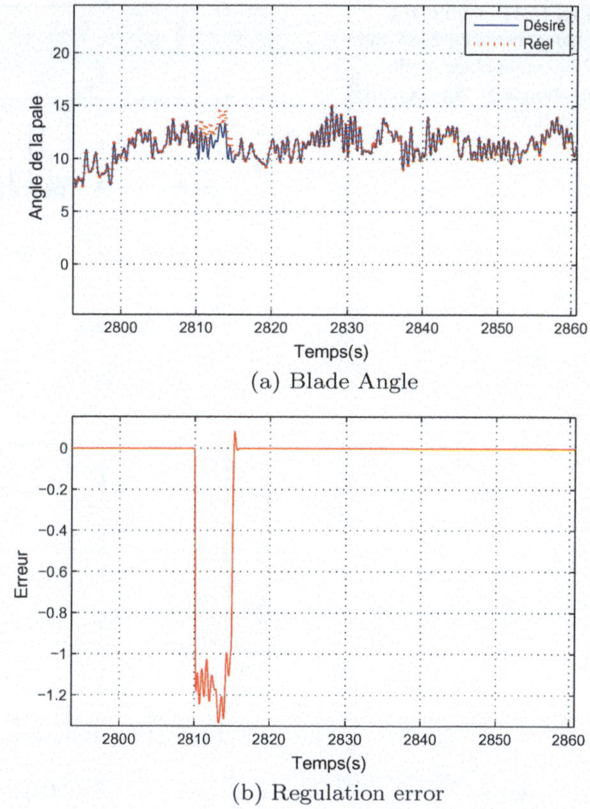

(a) Blade Angle

(b) Regulation error

Fig. 14.9 Case of the angle multiplied by a gain 1.2 during between 2810 and 2815 s

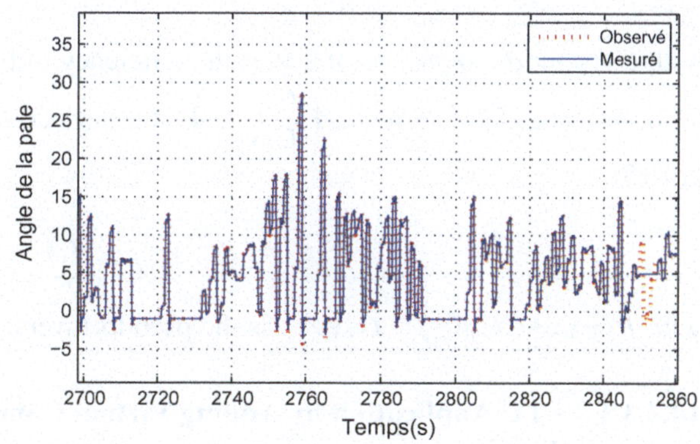

Fig. 14.10 Measured and estimated blade angle

– In the case of a sensor fault

Consider the output equation in the case of sensor fault

$$\begin{cases} \dot{x}_f = A(\mu)x_f + Bu \\ y_f = C_f(\gamma)x_f + f_y \end{cases} \tag{14.14}$$

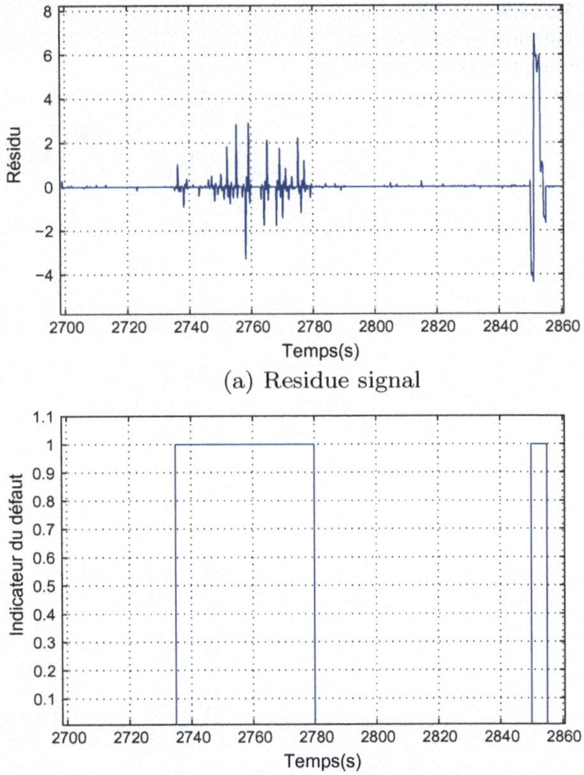

(a) Residue signal

(b) Fault indicator

Fig. 14.11 Parametric change between 2735 and 2780 s and blade angle fixation at 5° between 2850 and 2855 s

with $C_f(\gamma) = diag(\gamma_1 ... \gamma_{ny})C$, γ_i is the efficiency of the ith sensor, $\gamma_i \in [0, 1]$.

If $Rank(C_f(\gamma)) = Rank\begin{pmatrix} C \\ C_f(\gamma) \end{pmatrix}$, the new command signal is obtained by (14.15).

$$y_c = P(\gamma)(y_f - f_y) \tag{14.15}$$

with $P(\gamma) = CC_f(\gamma)^+$, $C_f(\gamma)^+$ is the pseudo-inverse of $C_f(\gamma)$.

14.4.3.3 FTC Application by Adding Virtual Components on a Wind Turbine

In this part of the work, two defects in the control subsystem of the blades will be treated:

(a) Case of an actuator fault: In this case, the transfer function of the actuator is calculated to check the following equality:

Fig. 14.12 Pursuit study without FTC control

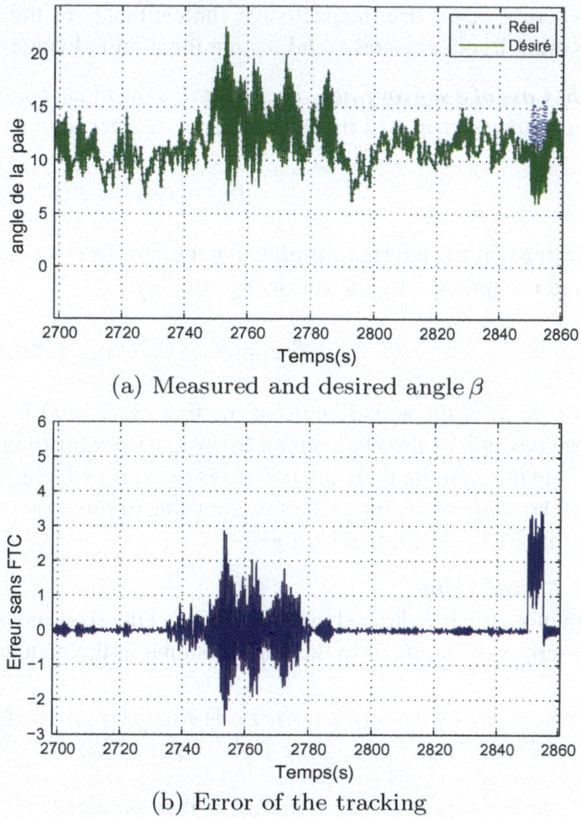

$$FT_a(s)FT_f(s) = FT_n(s) \tag{14.16}$$

with FT_a, $FT_f(s)$, and $FT_n(s)$ are, respectively, the virtual actuator transfer functions, the defective system, and the nominal system.

For the control system, this virtual actuator transfer function is obtained by

$$FT_a(s) = FT_f^{-1}(s)FT_n(s) = \frac{w_n{}^2(s^2 + 2\hat{\xi}\,\hat{w}_n s + \hat{w}_n^2)}{\hat{w}_n^2(s^2 + 2\xi\,w_n s + w_n{}^2)} \tag{14.17}$$

The signal of the FTC command is then given by

$$\beta_{ref\,FTC} = \frac{w_n{}^2(s^2 + 2\hat{\xi}\,\hat{w}_n s + \hat{w}_n^2)}{\hat{w}_n^2(s^2 + 2\xi\,w_n s + w_n{}^2)}\beta_{ref} \tag{14.18}$$

where $\hat{\xi}$ et $\hat{w}_n$ are, respectively, the estimates of the damping coefficient and the natural frequency and ξ and w_n are the nominal values of these same parameters.

(b) Case of a sensor fault: The control of the blade control system in the wind turbine benchmark proposed by Odgaard et al. (2009) is given by

$$\beta_{1c} = \beta_{1ref} + \beta_{1f} \tag{14.19}$$

with β_{1f} is the internal variable that models the error caused by the fault in the sensor and is expressed by the following equality:

$$\beta_{1f} = \beta_1 - 0.5(\beta_{1m1} + \beta_{1m2}) \tag{14.20}$$

where β_1 is the actual angle of the first blade, and β_{1m1} and β_{1m2} are the values of the position of the blade given by two sensors for redundancy.

In this system, there are two ways to express the sensor defect in the measurement of the angle of the blade, that is, the value of this angle remains fixed on a fixed value or it will be multiplied by a gain.

*** A fixed value**

In this case, it is desired to eliminate the defective sensor from the control loop so that a term will be added to the control signal in the manner of the following equation:

$$\begin{aligned} \beta_{1FTC} &= \beta_{1ref} + \beta_{1f} + \Delta\beta_1 \\ &= \beta_{1ref} + \beta_1 - \beta_{1m2} \end{aligned} \tag{14.21}$$

In this case, the variation $\Delta\beta_1$ is then obtained:

$$\Delta\beta_1 = 0.5(\beta_{1m1} - \beta_{1m2}) \tag{14.22}$$

*** A multiplication by a gain**

This type of defect consists in obtaining, with a defective sensor, a false value which is a multiplication of the value measured by a gain κ. To study the case of this defect in the position sensor of the first blade, this defect is modeled with $\beta_{1m1} = \kappa\beta_{1m2}$ where (14.18) becomes equivalent to

$$\beta_{1c} = \beta_{1ref} + \beta_1 - 0.5(\kappa\beta_{1m2} + \beta_{1m2}) \tag{14.23}$$

To correct this defect, the new command is calculated by adding a variation $\Delta\beta_1$ to the nominal command, which will be given by

$$\begin{aligned} \beta_{1FTC} &= \beta_{1c} + \Delta\beta_1 \\ &= \beta_{1ref} + \beta_1 - \beta_{1m2} \end{aligned} \tag{14.24}$$

then the variation in this case is given by the equation:

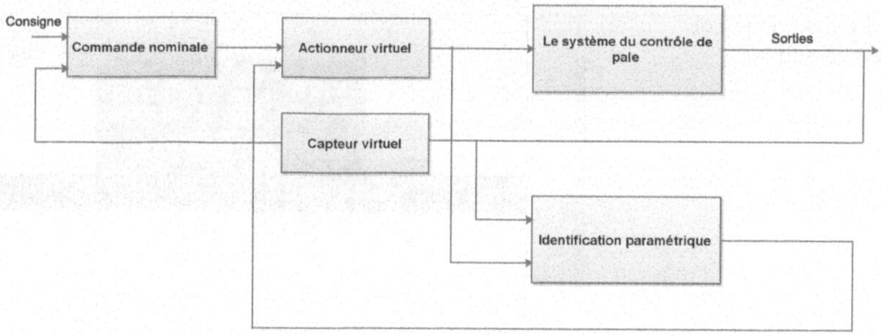

Fig. 14.13 Strategy of the tolerant control to the defect of a wind turbine benchmark

$$\Delta\beta_1 = 0.5(\hat{\kappa} - 1)\beta_{1m2} \tag{14.25}$$

The principle of fault tolerance that can affect the blade control system of a wind turbine is shown in Fig. 14.13 and is summarized by Algorithm 14.1.

14.4.3.4 FTC Simulation of a Wind Turbine

To tolerate defects that may affect the blade system already treated, the commands calculated in the previous paragraph will be implemented in the four cases.

Algorithm 14.1 Tolerance algorithm with virtual components

- Order the system with a nominal control law.
- Using the residue signal that is obtained by comparing the estimated and measured variables and using a detection threshold, determine the time of occurrence of a fault.
- Using the observed fault signature and the theoretical signature matrix table, isolate the fault (actuator fault f_8 or sensor fault f_1).
- If an actuator fault is isolated, update the virtual actuator using Eq. (14.18).
- If a sensor fault is isolated, update the virtual sensor with Eq. (14.22) or (14.25).
- Operate the reconfigured system.

(a) First scenario of defect: This reconfiguration requires a fault indicator as it is already discussed previously. This fault scenario is based on a comparison of the estimated angle and measured at each moment to detect the moment at which the new command will be applied. For this type of fault, an estimate of these two defective parameters is also needed to calculate the fault-tolerant command. These two estimated parameters are shown in Fig. 14.14.

Using (14.15), the new FTC command in the case of parametric change as well as the conventional command are shown in Fig. 14.15.

Fig. 14.14 Estimation of parameters ξ and w_n

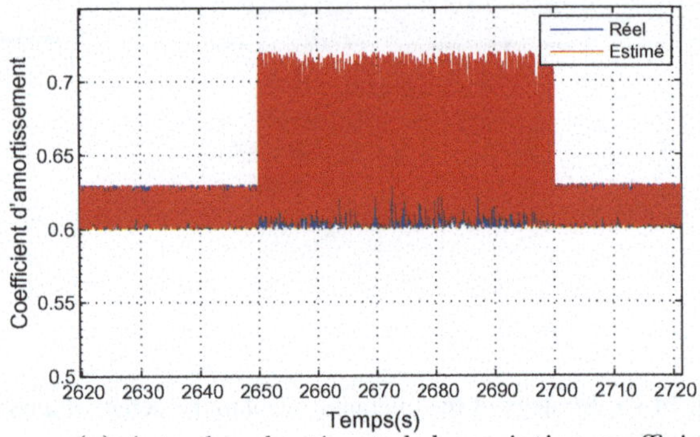

(a) Actual and estimated depreciation coefficient

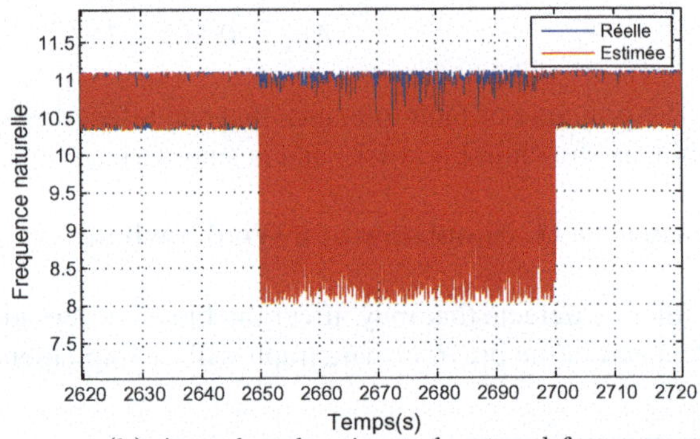

(b) Actual and estimated natural frequency

Fig. 14.15 Nominal and FTC control in the case of parametric change

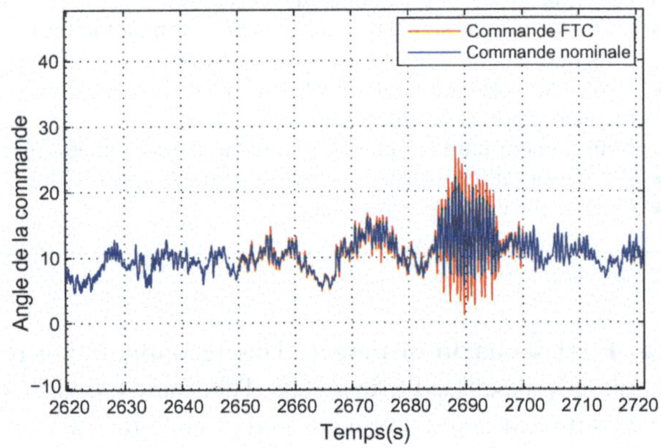

By using this new command which tolerates this type of defect, the two desired and measured blade angles as well as the error signal between them are shown in Fig. 14.16.

Figure 14.16 which corresponds to the case of this defect with reconfiguration of the command shows a maximum value error 0.15.

Fig. 14.16 FTC in the case of parametric change

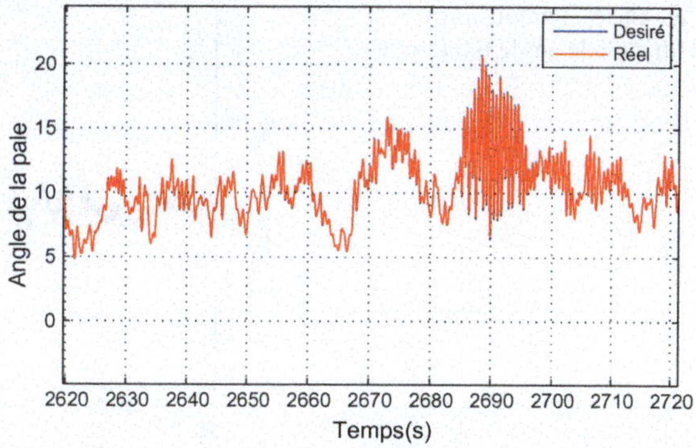

(a) Desired and measured angle, with FTC

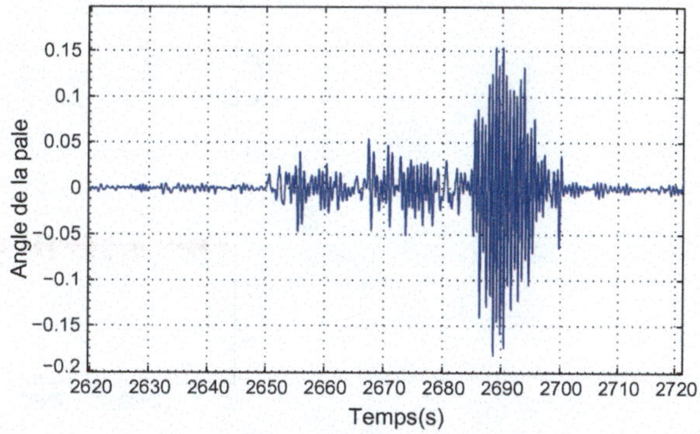

(b) Error of the tracking

Fig. 14.17 Nominal and FTC control in the case of a fixed angle at 5°

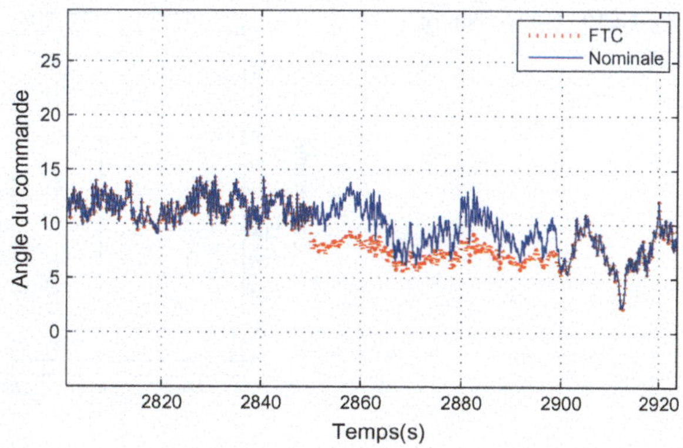

(b) Second scenario of defect: To correct this type of defect, the new control law is obtained by using the equations (14.17) and (14.18). This command tolerant to this type of fault as well as the nominal command is shown in Fig. 14.17.

With this FTC control strategy, the angle of the actual and desired blade and the result of the comparison between the two angles are shown in Fig. 14.18.

Fig. 14.18 Pursuit study with a blade angle fixation at 5°

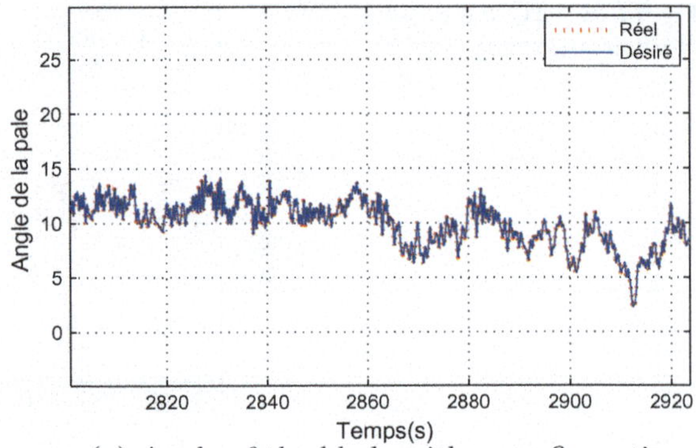

(a) Angle of the blade with reconfiguration

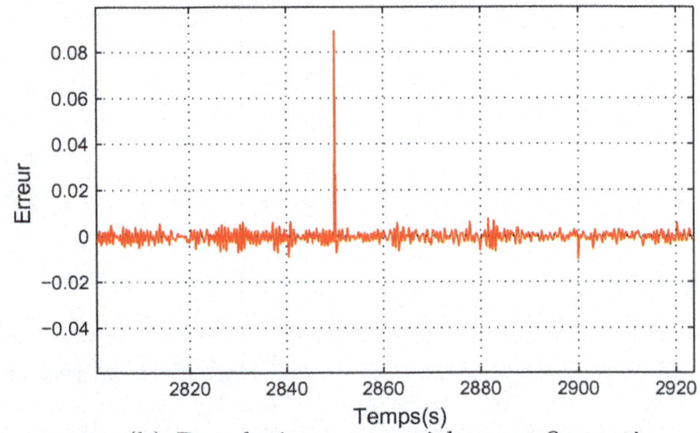

(b) Regulation error with reconfiguration

Fig. 14.19 Estimation of gain

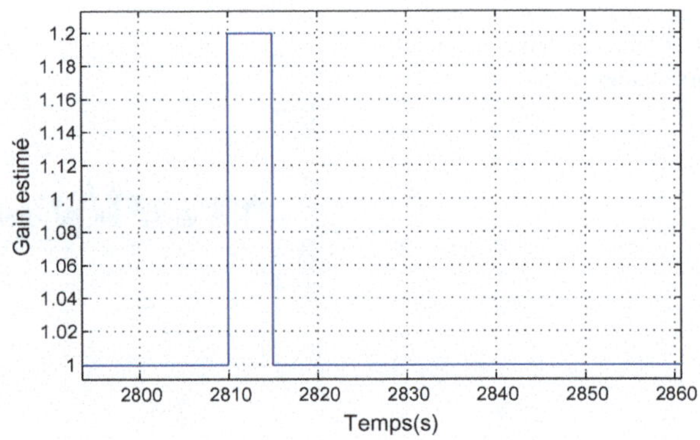

Figure 14.18 which corresponds to the response of the system with reconfiguration of the command shows a maximum value error of less than 0.01.

(c) Third **scenario of defect**: To solve this problem, an FTC command is useful according to Eqs. (14.21) and (14.22). For this reason, an estimation of the multiplication gain is necessary which is, in this case, shown in Fig. 14.19.

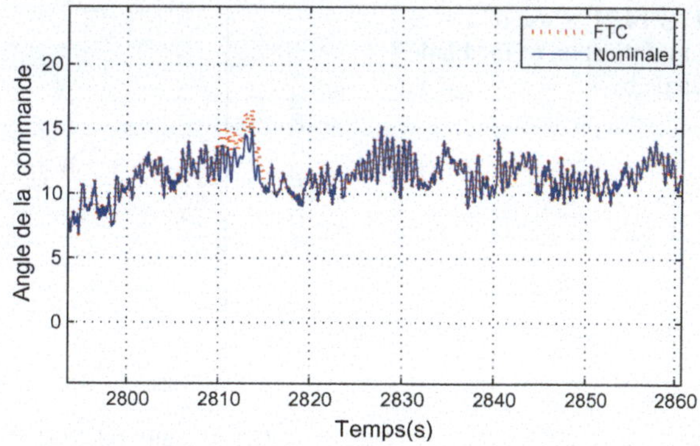

Fig. 14.20 Signal of command

Using this estimated gain and Eqs. (14.21) and (14.22), the new command is computed.

The command that tolerates this type of fault as well as the nominal control in this system is represented, respectively, in red and blue in Fig. 14.20.

By simulating the system with this new FTC command signal, the two signals representing the reference angle, the angle obtained by applying this fault-tolerant command, and the error signal between these two angles are shown in Fig. 14.21.

Figure 14.21 which corresponds to the reconfiguration of the default 3 led to a tracking error that has a maximum value of 0.25.

(d) Fourth **scenario of defect**: To solve the problem of pursuit, a task of reconfiguration of the law of the command is useful. This reconfiguration task requires a fault-type isolation phase to separate the actuator fault from that of the sensor. To complete this isolation phase, the residuals given by

$$\begin{aligned}
r_1(k) &= w_{rm1}(k) - w_{rm2}(k)\\
r_2(k) &= w_{rm2}(k) - \hat{w}_{rm2}(k)\\
r_3(k) &= w_{gm1}(k) - w_{gm2}(k)\\
r_4(k) &= w_{gm2}(k) - \hat{w}_{gm2}(k)\\
r_5(k) &= \beta_{1m1}(k) - \beta_{1m2}(k)\\
r_6(k) &= \beta_{1m2}(k) - \hat{\beta}_{1m2}(k)\\
r_7(k) &= \beta_{2m1}(k) - \beta_{2m2}(k)\\
r_8(k) &= \beta_{2m2}(k) - \hat{\beta}_{2m2}(k)\\
r_9(k) &= \beta_{3m1}(k) - \beta_{3m2}(k)\\
r_{10}(k) &= \beta_{3m2}(k) - \hat{\beta}_{3m2}(k)\\
r_{11}(k) &= \tau_{gm}(k) - \hat{\tau}_g(k)\\
r_{12}(k) &= P_{gm}(k) - \eta_g w_{gm2}(k)\tau_{gm}(k)
\end{aligned} \tag{14.26}$$

These residual signals in case of this fault scenario are shown in Fig. 14.22. These residues will be analyzed against appropriate thresholds, and they will be binarized in the following way: 1 when a failure is detected and 0 in the faultless scenario and they will be grouped in Fig. 14.23.

Fig. 14.21 Case of a multiplication of the blade angle by 1.2

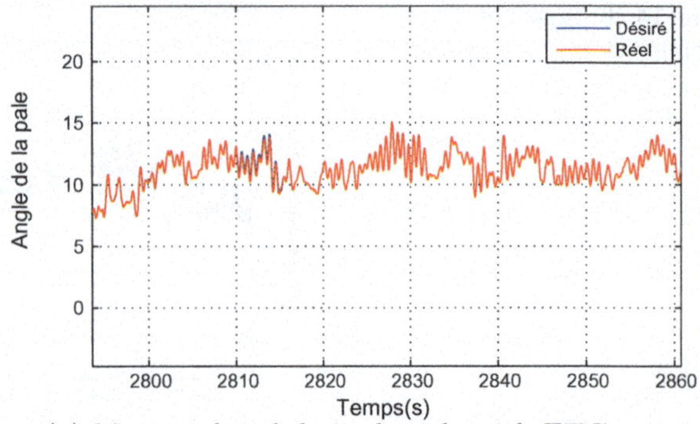

(a) Measured and desired angle with FTC control

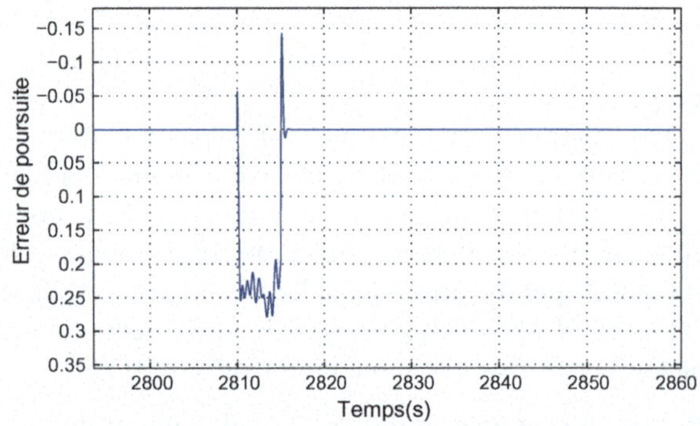

(b) FTC tracking error

To determine the cause of the defect at each moment, these defects will be isolated by comparing the obtained signatures with the theoretical ones by using Table 14.3 which summarizes the effect of each defect on the residues.

To isolate a fault, the first phase to realize is the physical redundancy which consists of measuring the same variable by two different sensors and it is necessary that the signals given by these two redundant sensors are equal in the healthy case of the system. If this difference exceeds a well-defined threshold, then there will be a sensor fault in one between them and in the opposite case there will be the possibility of having another type of fault other than the sensor type. This procedure assumes that only one type of defect can appear at a given moment.

The insulation result in case of this fault is represented by two indicators: the first is for the actuator fault and the second indicates the case of the sensor fault. These two indicators of the defect are, respectively, shown in Fig. 14.24.

After applying the isolation stain in the default case 4 that is detected between the 2735 and 2780 s instants and between 2850 and 2855 s, Fig. 14.24 shows that between 2735 and 2780 s an actuator defect is due to the change in damping coefficient and natural frequency and between 2850 and 2855 s it is a defect sensor. To estimate this

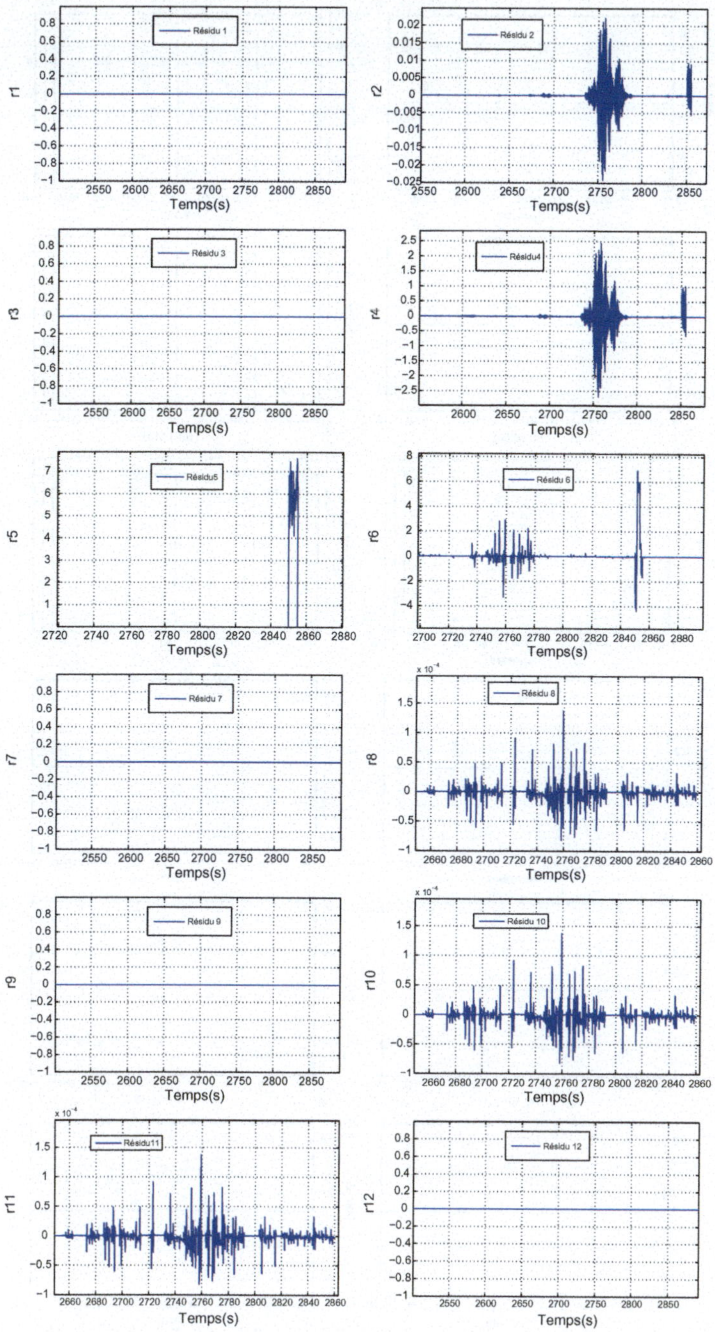

Fig. 14.22 Residue signals in the case of a multiple defect

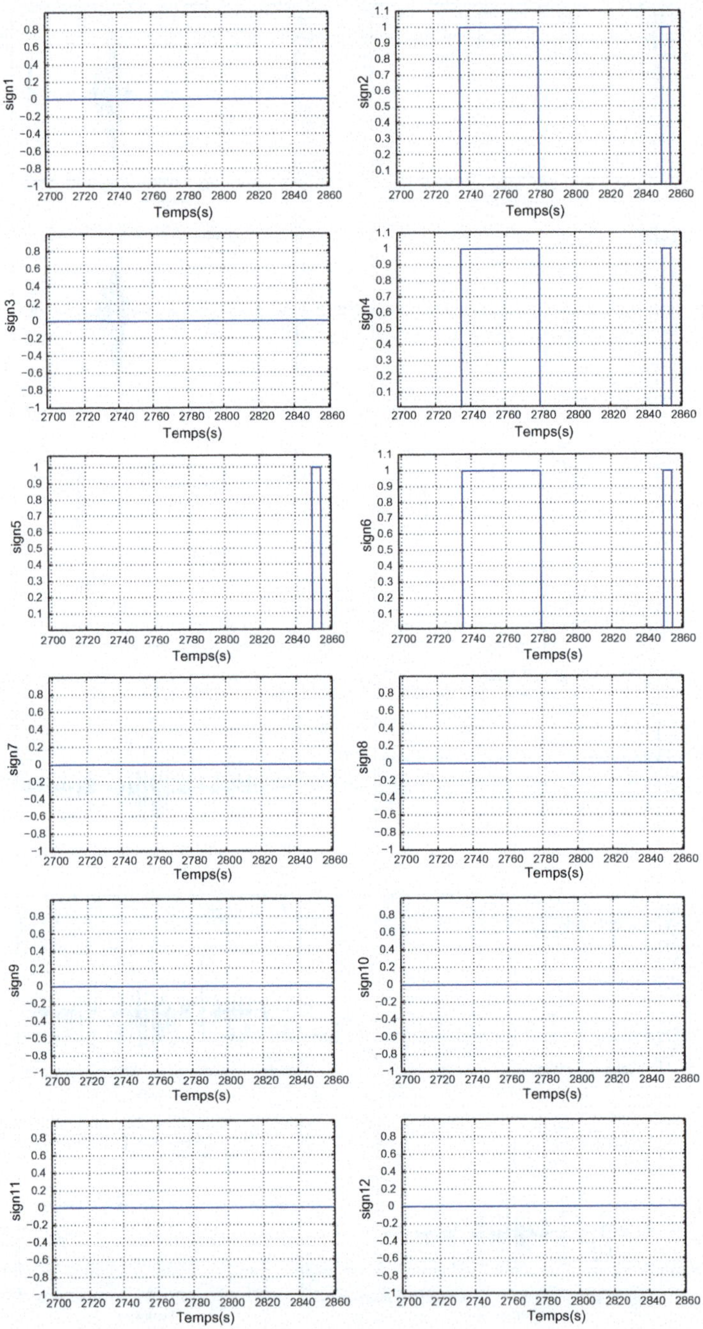

Fig. 14.23 Signatures obtained in case of a multiple defect

Table 14.3 Default signature matrix

	f_1	f_2	f_3	f_4	f_5	f_6	f_7	f_8	f_9	f_{10}	f_{11}
r_1			×	×							
r_2	×	×					×	×	×	×	
r_3					×	×					
r_4	×	×					×	×	×	×	
r_5	×										
r_6	×							×			
r_7		×									
r_8		×							×		
r_9											×
r_{10}											×
r_{11}							×				
r_{12}							×			×	

Fig. 14.24 Fault indicator

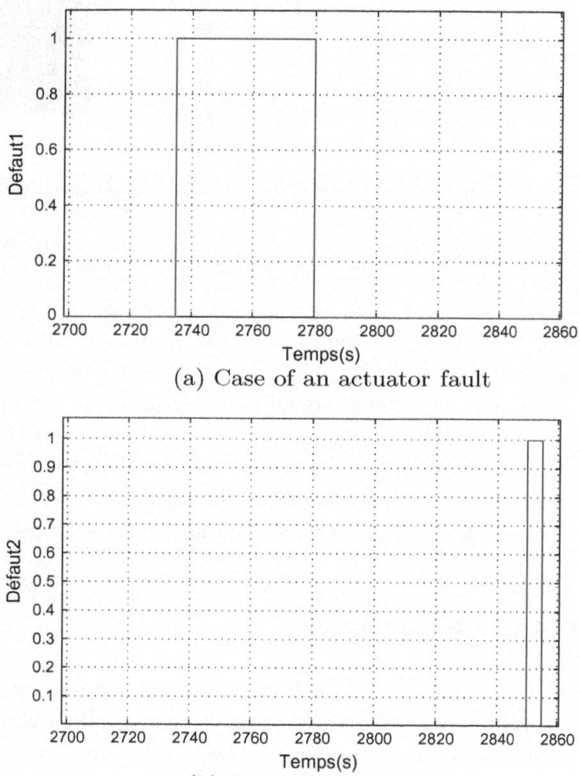

(a) Case of an actuator fault

(b) Case of a sensor fault

Fig. 14.25 Identification of parameters

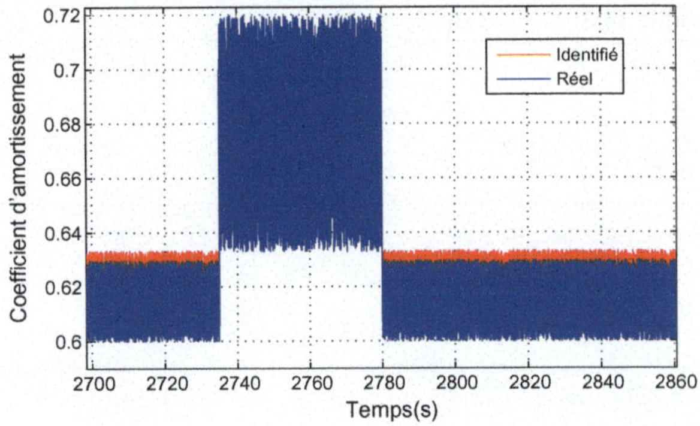

(a) The damping factor

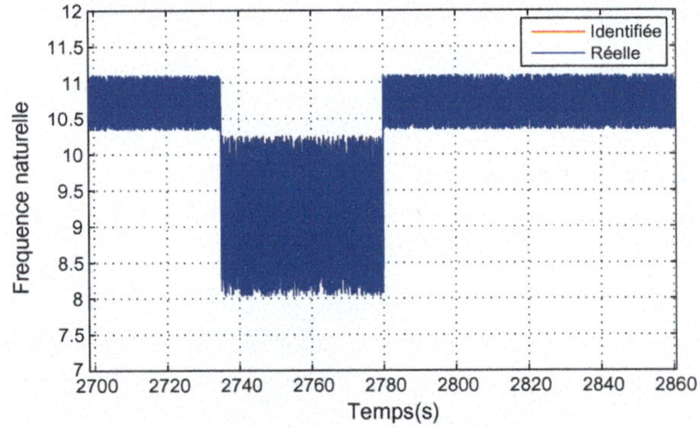

(b) Natural frequency

defect, an identification of damping coefficient and natural frequency is necessary as shown in Fig. 14.25.

The new command signal that tolerates this defect is represented in Fig. 14.26. Using this FTC command, the measured and desired angle as well as the tracking error obtained by the difference between these two signals are shown in Fig. 14.27 which shows a good tracking of the signal.

Figure 14.27 shows that the regulation error is less than 0.3.

14.4.3.5 Interpretation

- ✓ In the case of the first default scenario and applying the nominal command on the system, Fig. 14.6 shows a tracking error that exceeds the value of 1.5, whereas, in Fig. 14.16 which corresponds to the case of this defect with reconfiguration of the command, this error has for maximum value of 0.15.
- ✓ In the case of the second default scenario and applying the nominal command on the system, Fig. 14.7 shows a tracking error that can reach 8, whereas, in

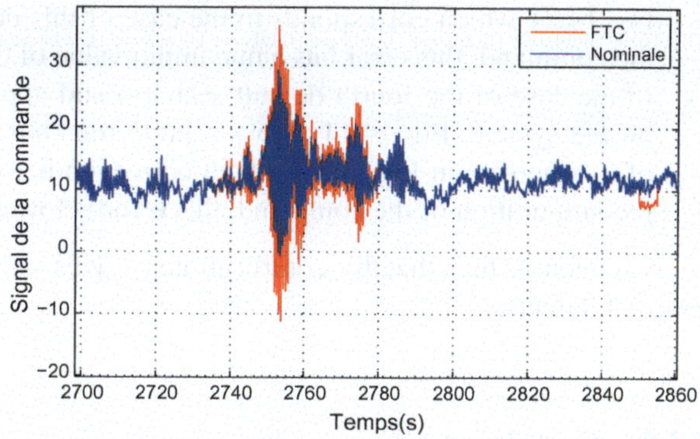

Fig. 14.26 Angle of command

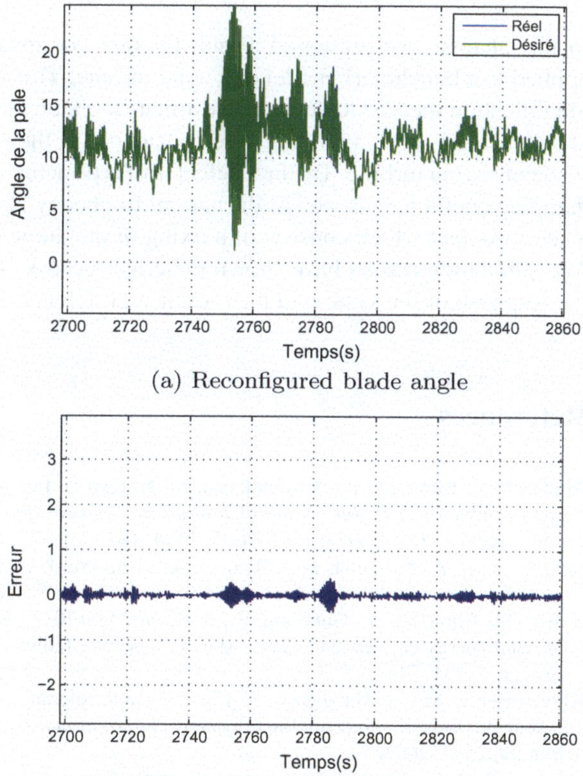

Fig. 14.27 Pursuit study in the case of a multiple defect

(a) Reconfigured blade angle

(b) Error of tracking in the case of the FTC

Fig. 14.18 which corresponds to the case of this defect with reconfiguration of the command, this error is lower than 0.01.

✓ In the case of the third default scenario and applying the nominal command on the system, Fig. 14.8 shows a tracking error that can reach 1.2, whereas, in

Fig. 14.21 which corresponds to the case of this defect with reconfiguration of the command, this error has a maximum value of 0.25.

✓ In the case of the fourth default scenario and applying the nominal command on the system, Fig. 14.11 shows a large tracking error that exceeds the value of 3, whereas, in Fig. 14.27 which corresponds to the case of this defect with reconfiguration of the command, this error is lower than 0.3.

We conclude then that this control strategy gives satisfactory results in the various cases of defects.

14.5 Conclusion

In this chapter, we presented a fault-tolerant command using virtual components applied to a benchmark model of a wind turbine. This command integrates the LPV identification method to detect and estimate the fault already detailed in the previous chapter. In this case, two faults are considered in the subsystem of the control of a blade of a wind turbine. The first defect is of type actuator modeled as a change of the damping coefficient as well as the natural frequency of this system and the second is a sensor defect which consists of a fixing of the angle of a blade β at a fixed value. The simulation results have shown the effectiveness of this technique in correcting the response of the system in the presence of defects.

References

Chouiref, H., Boussaid, B., & Abdelkrim, M. N. (2013). Integrated active fault -tolerant control approach based LMI. *International Journal on Sciences and Techniques of Automatic Control and Computer Engineering (IJ-STA)*, *7*, 1834–1843.

Jain, T., & J, Y., & Sauter, D. (2013). A novel approach to real-time fault accommodation in nrel's5-mw wind turbine systems. *IEEE Transactionson Sustainable Energy*, *4*, 1082–1090.

Kamal, E., Aitouche, A., Ghorbani, R., & Bayart, M. (2012). Robust fuzzy fault-tolerant control of wind energy conversion systems subject to sensor faults. *IEEE Transactions on Sustainable Energy*, *3*, 231–241.

Khosrowjerdi, M., & Barzegary, S. (2013). Fault tolerant control using virtual actuator for continuous-time lipschitz nonlinear systems. *International Journal of Robust and Nonlinear Control*, *24*, 2597–2607.

Odgaard, P. (2013). Fault tolerant control of wind turbines: A benchmark model. *IEEE Transactions on control systems technology*, *21*, 1168–1182.

Odgaard, P., Stoustrup, J., & Kinnaert, M. (2009). Fault tolerant control of wind turbines: A benchmark model. In *The 7th IFAC Symposium on Fault Detection, Supervision and Safety of Technical Processes Barcelona, Spain*.

Odgaard, P., & Stoustrup., J. (2012). Results of a wind turbine FDI competition. In *8th IFAC Symposium on Fault Detection, Supervision and Safety of Technical Processes*.

Poure, P., Weber, P., Theilliol, D., & Saadate, S. (2007). Fault-tolerant power electronic converters: Reliability analysis of active power filter. *IEEE International Symposium on Industrial Electronics* (pp. 3174–3179).

Rothenhagen, K., & Fuchs, F. (2009). Doubly fed induction generator model-based sensor fault detection and control loop reconfiguration. *IEEE Transactions on Industrial Electronics, 56*, 4229–4238.

Sami, M., & Patton, R. (2012). Wind turbine power maximization based on adaptive sensor fault tolerant sliding mode control. In *Proceedings of the 20th Mediterranean Conference on Control and Automation (MED12)*.

Sami. M., & Patton, R. (2012). An FTC approach to wind turbine power maximisation via t-s fuzzy modelling and control. In *Proceedings of the 8th IFAC Symposium Fault Detection, Supervisions, Safety Technical Processes (Safeprocess)*.

Seron, M., & Dona, J. (2009). Fault tolerant control using virtual actuators and invariant set based fault detection and identification. In *48th IEEE Conference on Decision and Control and 28th Chineese Control Conference*.

Simani, S., & Castaldi, P. (2013). Data-driven and adaptive control applications to a wind turbine benchmark model. *Control Engineering Practice, 21*, 678–1693.

Sloth, C., Esbensen, T., & Stoustrup, J. (2011). Robust and fault-tolerant linear parameter-varying control of wind turbines. *Mechatronics, 21*, 645–659.

Yang, X., & Maciejowski, J. (2012). Fault-tolerant model predictive control of a wind turbine benchmark. In *Proceedings of the 8th IFAC Symposium on Fault Detection, Supervision and Safety of Technical Processes*.

The manufacturer's authorised representative in the EU is Springer Nature Customer Service Centre GmbH, Europaplatz 3, 69115 Heidelberg, Germany. If you have any concerns regarding our products, please contact ProductSafety@springernature.com

Printed and bound by CPI Group (UK) Ltd, Croydon, CR0 4YY
25/09/2026
02222901-0012